CHEMISTRY

SECONDARY SCIENCE SERIES

The books in this series adopt the latest approach to science teaching for secondary schools and the new middle schools. The students are encouraged to discover as much as possible for themselves rather than simply to verify what they are told to be true. Scientific knowledge is acquired through a largely experimental approach, related always to everyday life and experience.

Most of the experiments are simple and require the minimum of equipment. To help readers who do not have access to full laboratory facilities, the conclusions to be drawn from investigations are incorporated later in the text. At the end of most chapters there is a *Test your understanding* section to reinforce knowledge and understanding of important points.

SI units have been used throughout the series.

This volume, **Chemistry**, is designed to follow on from the first volume, **Foundation Science**, which provides a two-year course in basic science. It is intended to take pupils up to the standard required by the C.S.E. Chemistry examination, and can be used over a two- or three-year period. It will also be found suitable for G.C.E. 'O' level.

The parallel volumes are:

Biology Book I: **General Plant and Animal Biology**;
Biology Book II: **Human Biology and Hygiene**;
General Science Book I: **Matter and Energy**;
General Science Book II: **Man and His Environment**;
Physics Book I: **Force and Energy**;
Physics Book II: **Atoms and Waves**.

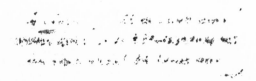

SECONDARY SCIENCE SERIES

CHEMISTRY

R. H. Stone and D. W. H. Tripp

Illustrated by David and Maureen Embry

London, Henley and Boston
Routledge & Kegan Paul

First published in 1971
This revised edition published in 1978
by Routledge & Kegan Paul Ltd,
39 Store Street, London WC1E 7DD,
Broadway House, Newtown Road,
Henley-on-Thames, Oxon RG9 1EN
and 9 Park Street, Boston, Mass. 02108, USA
Set in Times Roman
and printed in Great Britain by
Butler & Tanner Ltd, Frome and London

ISBN 0 7100 8755 1

SECONDARY SCIENCE SERIES

Contents

Introduction to the Second Edition

This book, while complete in itself, follows up the work in Chemistry covered in the first book of the series, *Foundation Science*.

It is intended in particular for those students taking the examinations for the Certificate of Secondary Education, but the authors have throughout kept in mind the needs of those taking the Ordinary Level of the General Certificate Examination. Indeed, this has been essential: a Grade One pass in the C.S.E. is equivalent to a Grade C (or better) in the G.C.E., and the depth of treatment given here is such that this level of treatment is fully catered for. Two new chapters, on Chemical Arithmetic and on the Periodic Table, have been added to this Second Edition to extend its usefulness, and readers will find that the great majority of topics required by the various G.C.E. Boards, particularly with regard to the latest syllabuses, have been covered.

SI units have been used almost exclusively. The recommendations of the Association for Science Education have been adopted with regard to nomenclature. Temperatures are given in kelvin, though the Celsius values are normally given after, in parentheses. In accordance with current practice, the negative index is used. The authors believe that it is no more difficult to comprehend, for example, $g\,dm^{-3}$, than it is to comprehend g/dm^3, particularly as the negative index is now commonly used at all levels to express small numbers. If these ideas are taught from the first, they soon become routine.

All scientific experiments must be carried out with care. In order to help teachers and students, in many cases hazard signs and warnings have been used in the margin where appropriate. These signs and warnings give guidance to some possible dangers, but safe working in a laboratory must, of necessity, rest with the teacher in charge. It is emphasized that the use of these signs and warnings in no way relieves teachers (and students) from their responsibilities in this respect. The fact that no sign or warning is given in a particular case does not imply the absence of risk.

It is assumed that safety glasses will be used at all times and no symbol for their use has been incorporated.

Teachers are advised always to try out an experiment before the class is allowed to undertake it.

Information about hazardous chemicals may be obtained from reference books. *Hazards in the Chemical Laboratory*, published by the Chemical Society, is one of the most complete.

D. W. H. T.
R. H. S.

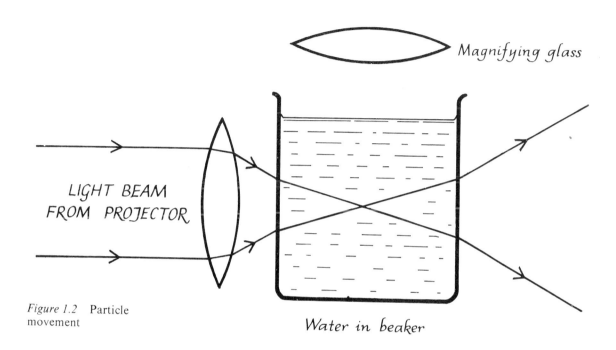

Magnifying glass

LIGHT BEAM
FROM PROJECTOR

Figure 1.2 Particle movement

Water in beaker

1.3. States of matter

We realize that water is converted into a solid, ice, by cooling and into a gas, steam, by heating. The water, ice and steam are all the same type of substance although they do not look alike.

Let us now see how much heat we need to change the solid state, ice, into the liquid state, water, and then to turn the water into the gas state, steam. In Chapter 4 we shall see that heat is a form of energy.

Investigation 1e. Heat needed to change state

Take a lump of ice and break it into small pieces. Now place a gauze on a tripod, place a bunsen burner under the gauze, light

This experiment gives us some idea of the size of the particles of which matter is made. They are so small that it would take at least 1 000 000 particles 'end to end' to measure ten millimetres.

1.2. Movement of particles

We have seen that the particles are very small, so small that we cannot see them. Since they are so small, we shall not be able to see if they are moving, but we can observe some experiments which suggest that these particles are probably moving.

Investigation 1c. Particle motion

Dissolve 1 g of sodium carbonate in 100 cm^3 of distilled water and 1 g of lead(II) ethanoate in a separate 100 cm^3 of distilled water. Take 1 cm^3 of each of these solutions and add to each a further 200 cm^3 of distilled water. Mix these diluted solutions together in a 1 dm^3 beaker. (If the solution turns cloudy, repeat the process but this time add 1 cm^3 of each solution to a larger volume of water before mixing the two solutions.) Arrange the beaker so that it is illuminated from the side by light from a projector and look through the magnifying glass into the solution from above (see Figure 1.2 on page 4). What do you see?

Toxic

The movement that you can see is best explained as being due to the rather larger particles being struck by the very much smaller particles that are moving, in the same way that tennis balls might make a football move.

Investigation 1d. Other ways in which the movement of particles may be observed

Take a moth-ball and place it on the bench in front of you. How do you think the smell reaches your nose?

Take a glass container (a beaker would do quite well) and place a copper(II) sulphate(VI) crystal in the container. Slowly pour water into the container. Note the colour of the water. Stand the container aside for several days and observe the colour as often as you can. What is happening to the colour?

Harmful

These two experiments have shown us that the particles of which matter is made are able to spread. We meet many examples of this every day, for example the way in which smells reach your nose. Such movement of particles is called **diffusion**.

Is diffusion faster in gases or liquids? Why did you decide on your answer?

Do you think it would occur in solids?

Investigation 1g. To find the melting-point of naphthalene

Take a piece of glass tubing about 2 mm in diameter and cut off a few lengths, about 80 mm long (the tube can be drawn out from a soda glass test-tube, if necessary). Hold one end in a bunsen flame in order to seal the end.

Crush some of the substance to be examined to a fine powder, and then place some in the tube that you have prepared. Tap the tube gently on the bench so that the powder falls down to the sealed end. Cut a small ring off a suitable piece of rubber tubing and push it over the end of a thermometer. Now, by pinching the rubber tubing around the thermometer, make space for the glass tube to be pushed through the rubber tubing beside the thermometer. Push it down so that the sample of the substance is beside the bulb of the thermometer. Then support the thermometer and the tube in a boiling tube of water, as shown in Figure 1.5 on page 8. The water in the boiling tube must come above the substance in the tube, but must not overflow into the tube. Arrange a wire stirrer in the boiling tube as shown.

Slowly heat the boiling tube with a small flame and stir the water in the tube. Watch the sample in the small glass tube and notice the temperature at which it melts.

Repeat this experiment using 1,4-nitromethylbenzene. Then repeat the experiment using *mixtures* of 1,4-nitromethylbenzene and naphthalene. What happens now?

We have seen that a large amount of energy is given to a solid when it changes into a liquid, but that no change in temperature occurs until all the solid has melted; further, we see that the temperature at which this melting occurs is fixed for a pure solid. The melting-point can, therefore, be used to help to identify the solid and to indicate its purity.

Flammable

Harmful

1.4. Boiling

Investigation 1h. The boiling of water

Arrange a distillation flask and Liebig condenser as shown in Figure 1.6 on page 9. Notice that the bulb of the thermometer is arranged to be beside the side arm of the distillation flask. If a distillation flask is not available, a flask fitted as shown in Figure 1.7 (on page 10) may be used.

Pour some water into the flask, being careful that none runs down the side arm. Turn on the tap so that a steady flow of water passes through the jacket of the condenser. Heat the flask. Observe

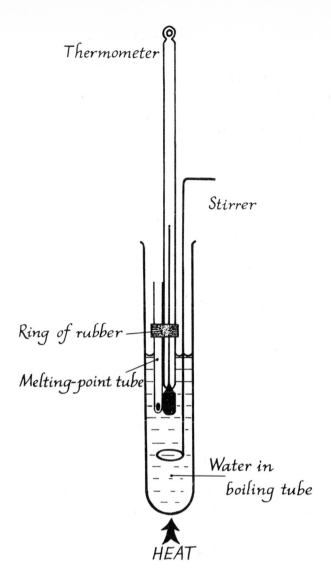

Thermometer

Stirrer

Ring of rubber

Melting-point tube

Water in boiling tube

HEAT

Figure 1.5 Melting-point apparatus

the thermometer and record the temperature every minute. Also watch the liquid in the flask and when it has nearly all disappeared turn off the bunsen burner. Where has the water gone to?

Plot a graph to show how the temperature changed as the time passed. Notice that the liquid boils at a constant temperature, which is known as the **boiling-point** of the liquid. This process is called **distillation** and can be used to separate substances that boil at different temperatures.

Repeat the investigation using a solution of sodium chloride in water. When you have finished, what is in the beaker and what is in the flask?

The substance that has been cooled in the condenser and collected in the beaker is called the **distillate**.

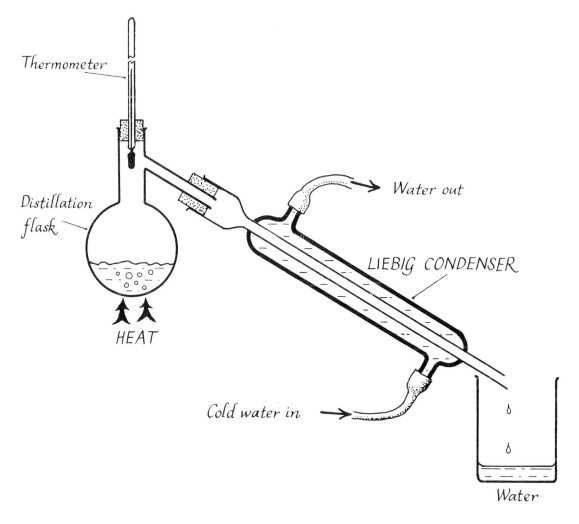

Thermometer

Distillation flask

HEAT

Water out

LIEBIG CONDENSER

Cold water in

Water

Figure 1.6   Distillation

Again, it is often necessary to find the boiling-point of small quantities of liquids. This can be done by using Siwoloboff's method.

Investigation 1i. To find the boiling-point of a liquid by Siwoloboff's method

For this experiment two 80 mm lengths of glass tube are required; these should have external diameters of about 2 mm and 4 mm and the smaller one should slide easily inside the larger tube. Seal one end of each tube by holding it in a bunsen flame. When cold, use a teat pipette to put one or two drops of ethanol into the larger of the two tubes. Make sure that the liquid goes

Flammable

9

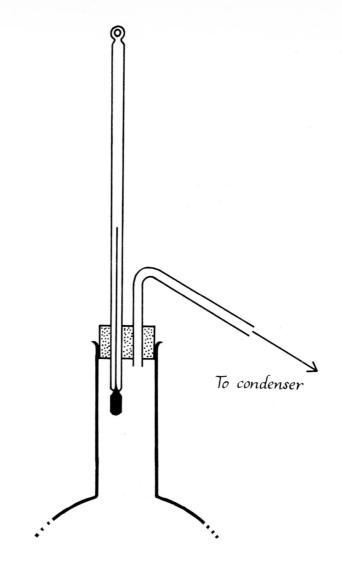

Figure 1.7 Fitting a flask
for distillation

To condenser

to the bottom of the tube. Then invert the smaller tube and place
it, open end downwards, into the larger tube until the open end
is below the surface of the liquid in the larger tube (see Figure
1.8). Arrange these two tubes beside a thermometer as in Investigation 1g.

Slowly heat the water bath. Bubbles of gas will be observed to
pass through the liquid in the outer tube. Continue heating gently
until the bubbling stops. Now allow the water bath to cool slowly
and watch carefully. Suddenly the inner tube will appear to be
full of liquid; this is the temperature at which the vapour in the
tube has condensed and is the boiling-point of the liquid. Why did
the bubbles of gas force their way through the liquid? Why does
the liquid reappear in the inner tube?

We have seen that there is a temperature at which a liquid
changes into gas and which remains constant until all the liquid has

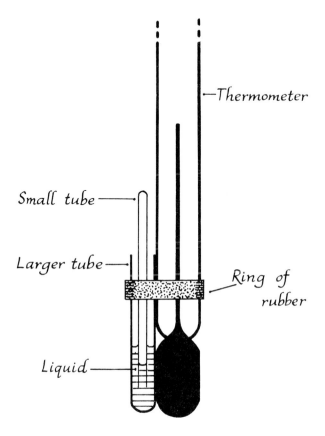

-Thermometer

Small tube

Larger tube

Ring of
rubber

Liquid

Figure 1.8 Siwoloboff's
method

changed state. We must realize, however, that it is possible for
water to be converted to vapour without reaching the boiling-
point. When washing is left out to dry in the winter, water is lost
from it as a vapour; we often see water vapour rising from the
roads after it has rained; and we have noticed how good the wind
is for drying wet surfaces. Hence we can see that water is converted
into vapour below its boiling-point, but this only happens when
the air is able to circulate. You could easily carry out a simple
investigation in the laboratory to see if water can change to vapour
below its boiling-point. When this happens and the amount of
liquid is reduced it is said to **evaporate**.

1.5. Solids to gases

We have seen how it is possible to melt a solid to a liquid and

11

to boil a liquid to a gas with the use of heat. Let us now examine a different solid.

Investigation 1j. The action of heat on ammonium chloride

Take an ignition tube and place a little ammonium chloride in the tube. Hold the tube as shown in Figure 1.9. Heat the end of the tube over a very small flame and watch carefully to see what happens.

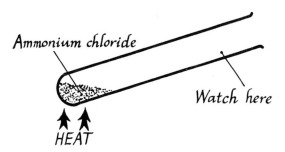

Figure 1.9 Heating ammonium chloride

Substances like ammonium chloride, that change from a solid into a gas without becoming liquids first are said to **sublime**. Notice how the ammonium chloride re-forms on the colder walls of the tube near the open end.

1.6. The use of physical properties

We have seen that changes of state occur at fixed temperatures for a substance. Other than the nature of the substance, the only factors that alter the boiling- and melting-points of a substance are the purity of the substance and the external pressure. We saw, in Investigation 1g, that addition of impurity causes a substance to melt at a lower temperature. The external pressure has an effect on the boiling-point of a liquid because the vapour particles have to escape into the atmosphere; this is more difficult at higher pressure and hence the boiling-point increases with rise in pressure. The melting-point is not affected to such a great extent as the boiling-point by changes in pressure, but you can melt ice by subjecting it to a pressure: this is why one can skate on ice.

Hence, if we know the pressure, it is possible to use melting- and boiling-points to help in the identification of a substance. Table 1.1 gives a list of melting- and boiling-points of some substances at normal atmospheric pressure (760 mm of mercury).

TABLE 1.1. MELTING-POINTS AND BOILING-POINTS OF SOME SUBSTANCES

Substance	Melting-point	Boiling-point
Water	273 K (0 °C)	373 K (100 °C)
Ethanol	156 K (− 117 °C)	351 K (78 °C)
Mercury	234 K (− 39 °C)	630 K (357 °C)
Ethoxyethane (ether)	157 K (− 116 °C)	308 K (35 °C)
Sulphur	389 K (116 °C)	717 K (444 °C)
Lead	600 K (327 °C)	1893 K (1620 °C)
Iron	1808 K (1535 °C)	3273 K (3000 °C)
Oxygen	55 K (− 218 °C)	90 K (− 183 °C)
Helium	1 K (− 272 °C)	4 K (− 269 °C)

Also, finding the boiling-point or melting-point of a substance will enable us to see if it is pure.

Investigation 1k

Break up some ice into small pieces and place them in a 100 cm³ beaker. Stir carefully with a thermometer and notice the steady temperature. Now slowly add some sodium chloride (common salt) whilst continuing to stir. Again observe the temperature. What difference can you see?

We make use of this change in melting-point in winter when we wish to melt ice or snow.

Test your understanding

1. What is matter composed of?
2. What do we know about these pieces of matter?
3. What size are these pieces of matter?
4 What name is given to the process by which gases and liquids spread?
5. What are the three states of matter?
6. What happens at the melting-point?
7. What happens to the melting-point of a substance if an impurity is added?
8. What happens when a substance sublimes?
9. What is a distillate?
10. How is it possible to melt ice without heating it?

Chapter 2

Mixtures and Solutions

In Chapter 1, we examined some of the properties of matter in a very general way. We observed that matter can be divided into very small pieces, which we have called particles, and that these particles, even though very small, have the same properties as larger pieces of the same type of matter. We will now examine in more detail the effect of mixing different types of matter.

2.1. Solutions

We have already used the word 'solution', since it is in common everyday use. Now let us see what a solution is. You will expect this to be a solid mixed with a liquid so that the solid seems to disappear, for example sugar in water. This is an example of a solution, but it is not the only example. A solution is a mixture of two substances, solids, liquids or gases, in which you cannot see the two different substances, but in which you can separate them fairly easily. The substance that is present in the larger amount is called the **solvent**. The substance that is present in the smaller amount is called the **solute**. In other words, the solvent is the substance that dissolves the solute.

Investigation 2a. A solution of a gas in a liquid

Take a bottle of aerated ('fizzy') drink and slowly open the crown cap or bottle top. Whilst you are opening the cap, look at the liquid in the bottle. What do you see? Can you see any spray coming out of the bottle?

When the bottle is opened, bubbles can be seen to rise through the bottle and spray escapes from the top of the bottle. The bubbles are caused by a gas, carbon dioxide, which is dissolved in the drink.

Investigation 2b. Air dissolved in water

Take a large (at least one dm³), round-bottomed flask and fill

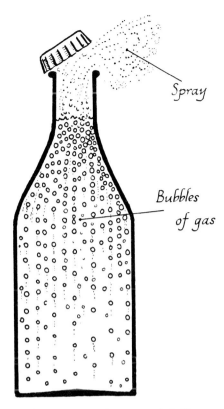

Figure 2.1 A bottle of 'fizzy' drink

Spray

Bubbles
of gas

it to the neck with water. Fit it with a rubber bung and delivery tube as shown in Figure 2.2. The flask should be filled to the top of the neck, so that the delivery tube becomes full of water as the bung is pushed into the neck of the flask. Arrange a small gas jar, full of water, on a beehive shelf over the end of the delivery tube in a bowl of water.

Heat the flask until the water in the flask has boiled for some time and then remove the bunsen burner. What can you see in the gas jar?

If the gas jar is full of gas, carefully slide a cover plate over the bottom end whilst it is still under water, and remove the jar. Remove the cover plate and place a lighted splint in the gas jar. What happens?

(A better yield of air can be obtained if air is bubbled through the water to be used for an hour or two before the apparatus is set up.)

Can you answer this question: 'Do gases dissolve better in hot water or in cold water?'

2.2. Solutions of solids in liquids

We are familiar with solutions of solid substances in liquids, but we must now examine a number of common laboratory chemicals to see which ones dissolve and which do not.

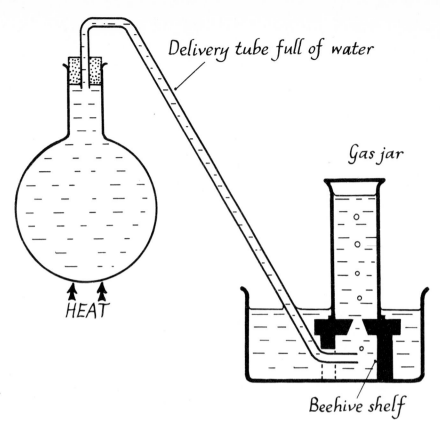

Delivery tube full of water

Gas jar

HEAT

Beehive shelf

Figure 2.2 Collecting air
from water

Investigation 2c. Solubility of chemicals

For this experiment you will need a number of test-tubes. The size is not important so long as they are all the same. Put water to the same depth in each tube (half to two-thirds full would be best). Take some of the chemicals given in the list below and crush them to a fine powder. You will need to be able to measure out the chemicals; a small splint of wood would do quite well. Mark the splint so that you can pick up the same amount of substance each time (approx 5 mm square). Put one measure of the chemical into the tube. Cork the tube and shake it to see if the chemical dissolves. If it does, repeat the process. Make a note of the maximum number of measures that can be dissolved for each chemical.

Use the carbonates, chlorides, nitrates(V) and sulphates(VI) of sodium, potassium, magnesium, calcium, lead(II), copper(II) and silver. (You need not have all the chemicals suggested.)

Draw up a table similar to the one shown and complete it to show the number of measures of solute that can be dissolved in a fixed amount of water.

Try to place the metals in a solubility order starting with the one that has the biggest number of soluble compounds and finishing

Toxic

16

TABLE 2.1. COMPARISON OF SOLUBILITIES OF CHEMICALS

	Carbonate	Chloride	Nitrate(V)	Sulphate(VI)
Sodium				
Potassium				
Magnesium				
Calcium				
Lead(II)				
Copper(II)				
Silver				

with the one that has the least number of soluble compounds. Do the same thing with the carbonates, chlorides, nitrates(V) and sulphates(VI). If there were any chemicals that you could not use, try to decide if they would have dissolved in water or not.

Obviously, if we want to say how much of a substance dissolves in water, we must have some unit in which to measure this. The **solubility** of a substance is the number of grams of solute that can be dissolved in 100 g of solvent, usually water, at a given temperature.

Investigation 2d. The effect of temperature on solubility

Take a test-tube and about one-third fill with water. Add lead(II) chloride to the water, a little at a time, until there is some undissolved solid in the tube. Gently warm the tube and shake. What do you see? Can you dissolve more lead(II) chloride in the hot water?

Toxic

The lead(II) chloride dissolves more readily in hot water than it does in cold water. How does this compare with what you have learnt about the solubility of gases?

Cool the tube containing the lead(II) chloride solution by shaking it in water running from a cold tap. What happens?

The reappearance of crystals of a solid from a solution is known as **crystallization**.

Try this experiment with some other substances and make a list of those that are more soluble in hot water than in cold water. One substance that you should include in your investigation is common salt, sodium chloride.

Investigation 2e. The solubility of potassium chlorate(V) at different temperatures

For this investigation, you will need a − 10 ° to 110 °C thermometer, a small test-tube (about 125 mm × 16 mm), a wire stirrer, a

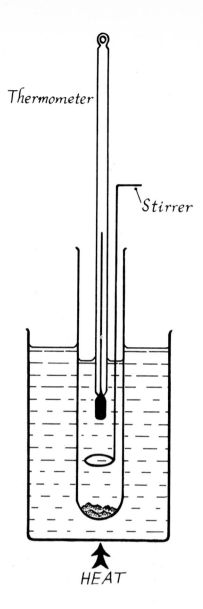

Thermometer

Stirrer

HEAT

Figure 2.3 Apparatus for Investigation 2e

Explosive

Oxidising

beaker of water and a means of heating. Arrange the apparatus as shown in Figure 2.3.

Measure 10 cm³ of water, distilled if possible, into the test-tube and add a weighed amount of potassium chlorate(V) (see Table 2.2). Slowly heat the beaker of water and stir the water in the tube. Note the temperature at which all the potassium chlorate(V) has just dissolved. Allow the solution to cool and note the temperature at which crystallization commences. Repeat the heating and cooling a little more slowly and take the new temperatures for complete solution and the commencement of crystallization; these last two readings should be within 5 kelvin of each other. Make a copy of Table 2.2 and enter the average of these last two readings in the temperature column of your table.

TABLE 2.2.   SOLUBILITY OF POTASSIUM CHLORATE(V)

Mass (g) of Potassium Chlorate(V) in 10 cm³ of Water	Solubility (g in 100 g Water)	Temperature
1.0	10	
1.5	15	
2.0	20	
2.5	25	
3.0	3C	
3.5	35	

Use your results to draw a graph of the change in solubility of potassium chlorate(V) with temperature, of the type shown in Figure 2.4.

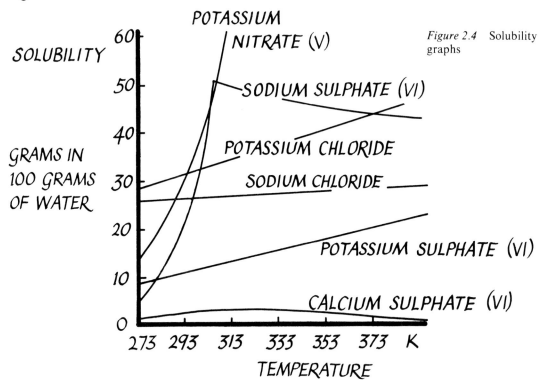

Figure 2.4   Solubility graphs

Notice that generally the solubility rises as the temperature is increased.

2.3. Crystallization

We saw in Investigation 2d and again in Investigation 2e that when a solution is cooled the solid may re-form. We will now examine the process of crystallization a little more carefully.

Investigation 2f. Crystallization of copper(II) sulphate(VI)

Harmful

Take a 100 cm³ beaker and half fill it with distilled water. Stir in copper(II) sulphate(VI) until there is a little undissolved solid. Now warm (do *not* boil) the solution and continue adding copper(II) sulphate(VI) until the hot solution will dissolve no more solute. Such a solution, that is one with some undissolved solute present, is called a **saturated** solution. Now add just enough warm distilled water to dissolve the last of the copper(II) sulphate(VI), followed by a few drops of dilute sulphuric(VI) acid. It is essential to stir continuously during this experiment. Set the hot, nearly saturated solution on one side and cover the top of the beaker with a piece of paper. The paper will keep out dust particles and also slow down the evaporation of the water. After several days some large blue crystals of copper(II) sulphate(VI) will begin to grow at the bottom of the beaker. Remove one crystal and keep the beaker and liquid for the next investigation.

Draw the shape of the crystal. Notice that all the crystals have the same basic shape. Take some small crystals from a bottle of copper(II) sulphate(VI) and examine them under a microscope or a magnifying glass. Try to grow some crystals of other substances. Substances suitable for crystal growth are: alum (aluminium potassium sulphate(VI)), Rochelle salt, nickel(II) sulphate(VI) and sodium chlorate(V). Some common crystal shapes are shown in Fig. 2.5.

Explosive

2.4. Separation of solids from liquids

In the last section we saw that solid crystals can be grown in solution. We now need to separate the crystals from the liquid, which is known as the **mother liquor**.

Investigation 2g. Pouring off liquid

Take the beaker and solution from Investigation 2f and hold a glass rod across the top of the beaker, using the thumb and first two fingers of one hand, so that the rod rests in the lip of the beaker as shown in Figure 2.6. Tilt the beaker and the mother liquor will run down the glass rod, leaving the crystals in the beaker. This process is called **decantation**. Decantation can be used to separate

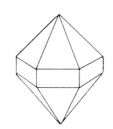
POTASSIUM NITRATE (V)

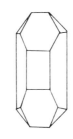
POTASSIUM SULPHATE (VI)

SODIUM NITRATE (V)

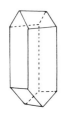
MAGNESIUM SULPHATE (VI)

ZINC SULPHATE (VI)

SODIUM CHLORIDE

AMMONIUM CHLORIDE

ALUMINIUM
POTASSIUM SULPHATE (VI)

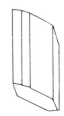
COPPER (II) SULPHATE (VI)

RHOMBIC SULPHUR

MONOCLINIC SULPHUR

SODIUM SULPHATE (IV)

solids from liquids providing that the solid is of a crystalline type or if the solid has separated at the bottom of the container.

If the solid is a fine powder we have to use other methods.

Investigation 2h. Separation of a fine powder from a solution

Take a test-tube and one-third fill it with barium chloride solution. Slowly add dilute sulphuric(VI) acid to the solution. A fine

Figure 2.5 Some crystal shapes

Toxic

Harmful

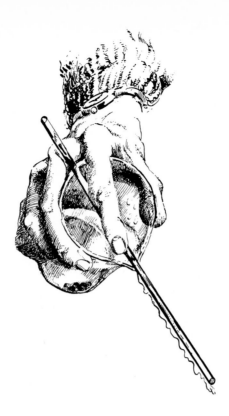

Figure 2.6 Decantation

white powder will be produced as the sulphuric(VI) acid is added –
this powder will slowly settle out at the bottom of the tube. This
process is known as **precipitation**. The powder which settles at the
bottom of the tube is called the **precipitate**. The settling process can
be improved by heating the tube for a few minutes after the precipitate forms.

To separate the solid from the liquid this time, we pour the
liquid through a special piece of porous paper, known as **filter
paper**. Take a round piece of filter paper and fold it in half, and
then in half again, as shown in Figure 2.7. Open out the folds so
that three thicknesses of paper are on one side and one thickness
is on the other. Fit the resulting cone into a filter funnel and
moisten the paper with water. Stand the funnel in another test-tube
as shown. Pour the mixture from the test-tube down the glass rod,
as in Investigation 2g, so that the liquid, which contains some
undissolved solid, runs on to the side of the filter paper in the
funnel. Do *not* pour in so much liquid that the level comes above
the top of the filter paper. Pour in more liquid as needed until all
the mixture from the first tube has been used. The second tube
should now contain a clear liquid, known as the **filtrate**; the solid,
known as the **residue**, should be left on the filter paper. If the solid
comes through, either there is a hole in the filter paper or a finer
filter paper should be used.

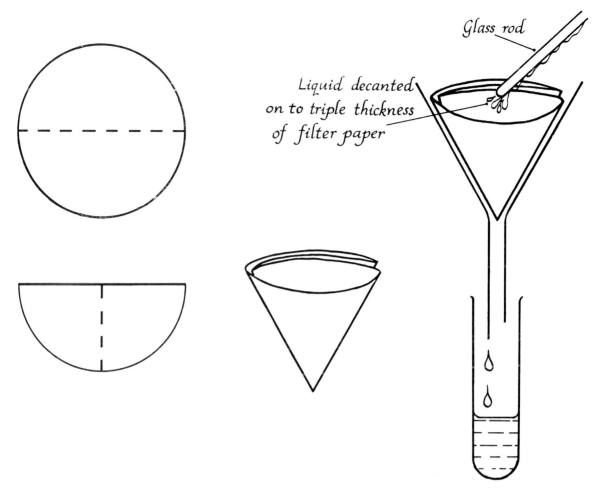

Figure 2.7 Folding and using a filter paper

On the industrial scale, filtration is usually carried out through granular material. For example, sand is used to filter drinking water.

If it is necessary to separate the solute in a solution from the solvent, then a distillation process is used (see Section 1.4). If, however, it is not essential to recover the solvent, a simpler method can be used.

Investigation 2i. Recovering the solute from a solution

Make a solution of common salt, sodium chloride, in a 100 cm³ beaker. About 30 cm³ of solution will be ample. Pour the solution into an evaporating dish or basin and place this on a wire gauze supported by a tripod. Make sure that the bottom of the basin is dry. Light a bunsen burner and gently heat the basin. The flame should be waved under the basin before the burner is stood under the tripod. As the dish warms up the bunsen flame may be increased. Water vapour will be seen to rise from the basin and the

23

solid will slowly begin to reappear, at first around the sides of the dish. As soon as the mixture begins to 'spit', the heating should be reduced by turning the gas down a little, and the solid which has formed should be stirred in order to allow the water vapour to escape.

2.5. Division of solutes between solvents

If we wish to dissolve some paint, we do not use water but some kind of paint thinner, e.g. white spirit. We do this because paint is not very soluble in water but it is fairly soluble in white spirit. We can examine this difference in solubility with a simple investigation.

Investigation 2j. A solute divided between two solvents

Warning – do not handle iodine and do not inhale vapour of methylbenzene.

Flammable   Harmful

Take a test-tube and half fill it with water. Using a spatula or wooden splint, pick up an iodine crystal. Drop the iodine crystal into the water. Cork the tube and shake it well. Some of the iodine will dissolve giving the water a straw colour. Decant the solution into a clean tube which contains about 1 cm³ of methylbenzene. Cork the second tube and shake it well. Set the tube on one side for a little while. You will see that there are two layers in this tube. This is because methylbenzene and water do not mix: we say that they are **immiscible** and the lighter one will float on top of the other one. Which is the lighter, methylbenzene or water? Which is the better solvent for iodine? When a solute is divided between two layers of solvent in this way, the process is known as **partition**.

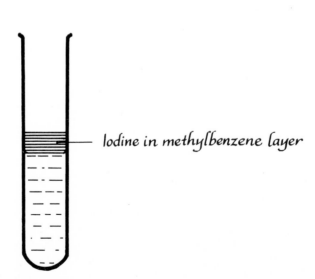

Iodine in methylbenzene layer

Figure 2.8 Partition of iodine between methylbenzene and water

24

We have seen that some solutes are more easily dissolved in one solvent than another. If we find the correct solvent the solute can be 'washed' out quickly. Alternatively, we could separate two solutes if one dissolved more readily in a certain solvent than the other. This idea is used in a means of separation known as **chromatography**.

Investigation 2k. A simple example of chromatography

Take a filter paper circle and cut a wick about 10 mm wide from the edge to the centre as shown in Figure 2.9. Place one drop of an indicator or food colouring at the centre of the filter paper and allow it to dry out. (Universal indicator is very good for this experiment.) Pour a little ethanol into a 100 cm³ beaker (sufficient so that the wick will go just below the surface when bent down). When the spot is dry, bend the wick down and balance the filter paper on the edge of the beaker, so that the bottom 5 mm of the wick dips into the ethanol. Stand this aside and examine it at intervals.

Flammable

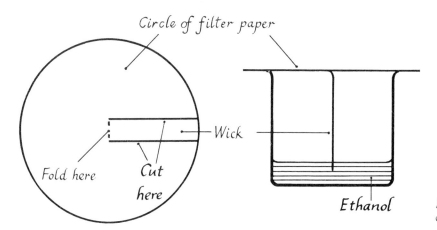

Circle of filter paper

Fold here

Cut here

Wick

Ethanol

Figure 2.9 Ring chromatography

Different rings are seen because the component dyes dissolve in the water in the filter paper and in the ethanol that comes up the wick, that is, partition of the dyes occurs between the water and the ethanol. Hence this form of chromatography is called partition chromatography. The ethanol, which is used to 'wash' the dyes through the paper, is known as the **eluent**. Although this method depends largely upon partition of the solute, it also involves the 'holding' or **adsorbtion** of the solute by the paper itself. We can examine adsorbtion chromatography in another simple experiment.

Investigation 2l. Adsorbtion chromatography

Take a stick of blackboard chalk, preferably *not* of the 'dust free' type, and place the flat end in some indicator solution on a watch glass. Allow the 'chalk' to soak up a little indicator, then remove it and stand it, flat end downwards, in a little water in a 100 cm³ beaker. Examine the chalk at intervals to see the different colours of the indicator dyes separating through adsorbtion on the 'chalk'.

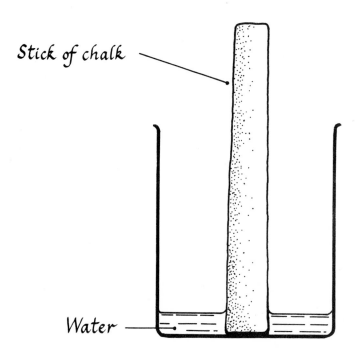

Stick of chalk

Water

Figure 2.10 Adsorbtion chromatography

Considerable use is made today of chromatographic methods for separation and analysis, that is recognition of liquids and gases.

2.6. Some solvents

We saw in Investigation 2j that methylbenzene was a better solvent for iodine than water. We will now examine some common solvents to find out the uses to which they may be put.

★ *Warning – care should be taken not to inhale the vapour of these solvents since some of them are harmful.*

Investigation 2m. Uses of some solvents

Take a number of test-tubes and examine the ability of the solvents listed to dissolve the solutes given in Table 2.3. Make a copy of this table and put a tick in the correct box when a solvent dissolves a particular solute.

TABLE 2.3. THE USES OF SOLVENTS

Solvents	Solutes					
	Oil Paint	Candle Wax	Engine Oil	Salt	Green Colour from Crushed Leaves	Lard
Methylbenzene						
Tetrachloro-methane						
Ethanol (methylated spirit)						
White spirit						
Petrol						
Water						

Toxic

Flammable

Harmful

2.7. Osmosis

Under certain conditions, if a solvent is separated from a solution by a 'skin-like' substance the solvent will dilute the solution. This is known as **osmosis**.

Investigation 2n. Osmosis in nature

Take a dried prune and place it in water for a few days. What happens?

Now remove it, with care, from the beaker of water (the skin must not be broken) and place it in a beaker of strong common salt solution. Leave it again for a few days. What do you observe this time?

We have observed a simple example of osmosis, a process which is very important in controlling the food supply of plants and animals. The solvent, water, is able to pass through the skin of the prune. Since the prune skin allows the solvent but not the solute to pass through it, it is said to be a **semi-permeable membrane**.

Investigation 2o. Osmosis in the laboratory

We can examine the nature of osmosis better if we carry out a simple experiment in the laboratory. Take a 10 cm length of Visking tubing and close one end tightly by tying it round with thread. Make a solution of sugar in water and pour some of the solution into the tubing. Seal the other end of the tubing with thread in the same way, so that you have a 'floppy' sausage about 7–8 cm long. Take a large beaker and nearly fill it with water, and then lower your piece of tubing into the water so that it is completely under the surface of the water.

Stand the beaker aside and watch what happens during the next twenty-four hours.

Water passes in through the Visking tubing to dilute the solution inside. This will continue until the concentration of the solution inside is the same as the concentration outside or until the tubing bursts with the increased pressure from inside. Which happens first?

A careful study of osmosis can provide important information about the size of particles of a solute.

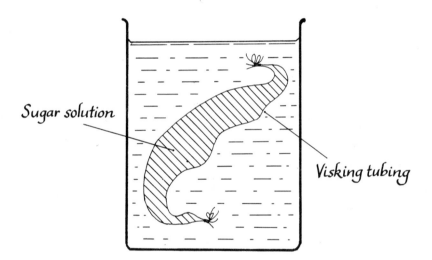

Figure 2.11 Laboratory investigation of osmosis

Test your understanding

1. What is the name given to a substance that is used to dissolve a solute?
2. Do gases dissolve better in hot or in cold liquids?
3. Which salt of a metal is more soluble, carbonate or nitrate(V)?
4. What is meant by the solubility of a substance?
5. How is a fine powder separated from the liquid in which it is suspended?
6. When a solution contains as much solute as possible it is said to be be....
7. When a solute goes into solution in two immiscible solvents the process is called....
8. Which liquid would you select to remove grass stains from white clothes?

9. What is the name given to the process by which a solvent tends to dilute a solution?
10. How would you attempt to discover if a green-coloured dye was a single colour or a mixture of colours?
11. The solubility of sodium nitrate(V) at different temperatures is given in the table.

Temperature/K	280	290	300	310	320	330	340	350
Solubility/g dm^{-3}	750	825	905	990	1 080	1 170	1 275	1 390

Plot a graph of solubility against temperature. From your graph, deduce the solubility of sodium nitrate(V) in water at (a) 295 K and (b) 335 K.

Chapter 3

Atoms

The experiments that have been carried out so far can be explained very well by the idea that all substances are made up of very small particles. The time has now come for us to consider this idea more fully.

3.1 Scientific models

A scientific model is simply a picture that helps us to understand more clearly what we have observed in our experiments and helps to explain our observations. Models can sometimes also be used to suggest further experiments that we can try.

The more observations that a particular model can explain, the more likely it is to be a useful picture. Scientific models are usually very limited in the range of observations that they can explain, and often more than one model may be used to give a picture of the same thing – each model being useful in helping us to understand a different series of observations. So long as we realize that our models can give us only a very imperfect and limited picture of the real thing, then we can make good use of them.

3.2. The idea of atoms

It was a Greek philosopher called Democritus who, about 400 B.C., first used the word **atom** when considering the nature of matter. The word 'atom' means 'indivisible', and Democritus suggested that all matter was made up of small, indivisible pieces. He imagined that if he took a piece of, say, iron and broke it into smaller and smaller pieces, a time would come when he would no longer be able to break it up any further. We must remember that Democritus' idea of an atom was not the same as our idea today, but the suggestion he made that there is a limit to the divisibility of matter is correct.

It was John Dalton who, in 1803, first put ideas about atoms on to a firm basis. By that time it had been found that, while most substances could be broken up into simpler substances with

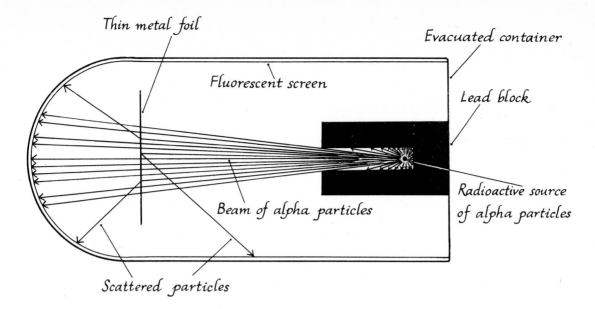

Labels on figure:
Thin metal foil
Evacuated container
Fluorescent screen
Lead block
Beam of alpha particles
Radioactive source of alpha particles
Scattered particles

picture of the atom as consisting largely of empty space with a massive, positively charged central part.

Figure 3.2 Rutherford's scattering experiment

The existence of isotopes was clearly shown by J. J. Thompson. In 1919, Aston invented an instrument called a **mass spectrograph**. In this device, a beam of electrically charged atoms or **ions** (see Section 5.1) is passed through strong magnetic and electric fields. These fields cause the ions to be deflected from their paths by amounts which depend on their mass and on their speed. A modern form of the mass spectrograph is shown in diagram form in Figure 3.3. In this apparatus, only ions travelling at the same speed are allowed to enter the main part of the apparatus.

3.5. The use of symbols

Each element has been given one or two letters which may be used when we wish to refer to that element. It is very important to understand that the symbols are not just abbreviations for the names of the elements, but indicate an actual quantity (see Section 3.7). A table of the elements, with their symbols and atomic numbers, is given on page 36 (Table 3.2).

It is possible to indicate all the different atoms that can exist by writing the symbol for the element concerned with two numbers in front of the symbol (one above and one below), for example, $^{238}_{92}U$. This indicates a uranium atom with an atomic number of 92 and a total number of 238 nucleons. Thus, this particular sort of uranium atom has $(238 - 92)$, which makes 146, neutrons in its nucleus. The total number of nucleons in the nucleus of an atom is called its **mass number**.

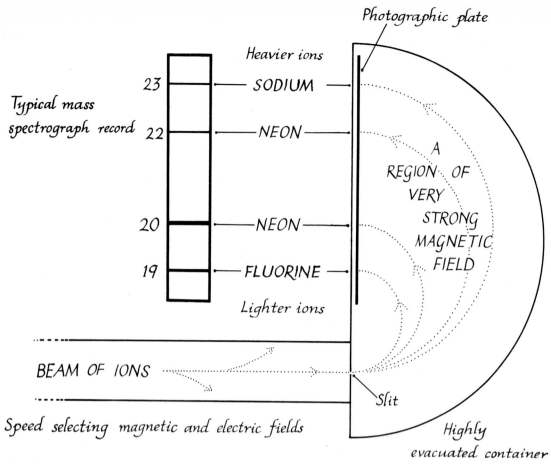

Typical mass spectrograph record

Heavier ions

23 —— SODIUM ——

22 ——NEON——

20 ——NEON——

19 —— FLUORINE ——

Lighter ions

Photographic plate

A
REGION OF
VERY
STRONG
MAGNETIC
FIELD

Slit

BEAM OF IONS

Speed selecting magnetic and electric fields

Highly evacuated container

Figure 3.3 A mass spectrograph

3.6. Atomic mass and atomic weight

To provide an accurate figure to which the masses of atoms, ions and sub-atomic particles may be compared, we need a standard unit to which we can make reference, just as we use the standard metre for comparisons of length and the standard kilogram as the standard of mass on the large scale. It has been agreed that the atom of the element **carbon** which has six protons and six neutrons in its nucleus shall be used as the standard, and *one-twelfth part of the mass of this particular isotope*, $^{12}_{6}C$, is taken as the standard unit of mass on the atomic scale.

The comparative mass of a particular sort of atom on the basis of this unit is called its **relative atomic mass**. The relative atomic mass of an element is the *average* mass of the atoms in the normal, naturally occurring mixture of isotopes, compared with the accurate unit of mass as described in the previous paragraph. Thus chlorine, which has two isotopes, $^{35}_{17}Cl$ and $^{37}_{17}Cl$, has a relative atomic mass of 35.456. This is because the normal natural isotopic mixture of chlorine atoms is made up of about three parts of $^{35}_{17}Cl$ and one part of $^{37}_{17}Cl$.

3.7. The mole

This is the unit of **quantity of matter**. It is a number and is sometimes known as **Avogadro's number** or **Avogadro's constant**. The number is very large and for most purposes may be taken as approximately 6×10^{23}. It is the number of $^{12}_{6}C$ atoms needed to make exactly 12 g of carbon. When we use symbols in our work, then each symbol stands for one mole of the element concerned. For example, S indicates one mole, that is 6×10^{23} atoms, of sulphur, and will have a mass of 32 g because the relative atomic mass of sulphur is 32; H_2O indicates one mole, that is 6×10^{23} molecules, of water, and has a mass of 18 g; $SO_4{}^{2-}$ shows one mole, that is 6×10^{23} ions, of sulphate(VI), and has a mass of 96 g. The use of symbols and formulae is dealt with in Chapter 7.

3.8. Stable and unstable nuclei

Some combinations of protons and neutrons give nuclei which can exist without change for an indefinitely long time; that is, they are **stable**. Other combinations are **unstable**. Sooner or later, the unstable nuclei break up. Nuclei of this sort are said to be **radioactive**. (For a fuller discussion of radioactivity, see Chapter 37.)

3.9. The arrangement of the electrons in atoms

The electrons in atoms are moving at such high speeds (approaching that of light which, in a vacuum, is 3×10^8 m s^{-1}) that for all practical purposes, they may be thought of as being 'spread out' in the space which surrounds the nucleus, thus forming a 'cloud of electricity'.

To get a better understanding of this spreading-out, we can think of how the picture is formed on the screen of a television receiver. The whole picture is formed by one tiny spot of light, caused by the impact of a beam of electrons on the special surface of the picture tube. This spot moves across the screen in a number of lines – usually 625 – and, as it does so, the brightness is continuously varied to give the light and dark parts of the picture. Twenty-five complete pictures are formed in one second. A little simple arithmetic shows that, for a 480 mm (19 inch) screen, the spot of light is moving at about 7.5 km s^{-1} and, to our eyes, seems to be spread out over the whole picture. This speed is much less than the speed of the electrons in atoms and, of course, the size of an atom is far smaller than that of a television screen.

The space that an electron fills as it moves round the nucleus is known as an **electron cloud**.

TABLE 3.2.

PERIODIC TABLE OF THE ELEMENTS

THE ELEMENTS ARE ARRANGED IN VERTICAL GROUPS WHICH SHOW
SIMILAR CHEMICAL PROPERTIES BECAUSE THEY HAVE SIMILAR
ELECTRONIC ARRANGEMENTS.
THE NUMBERS ARE THE ATOMIC NUMBERS

ELECTRON ENERGY LEVELS

Electron energy level	Elements
1	1 Hydrogen H — 2 Helium He
2	3 Lithium Li, 4 Beryllium Be, 5 Boron B, 6 Carbon C, 7 Nitrogen N, 8 Oxygen O, 9 Fluorine F, 10 Neon Ne
3	11 Sodium Na, 12 Magnesium Mg, 13 Aluminium Al, 14 Silicon Si, 15 Phosphorus P, 16 Sulphur S, 17 Chlorine Cl, 18 Argon Ar
4	19 Potassium K, 20 Calcium Ca, 21 Scandium Sc, 22 Titanium Ti, 23 Vanadium V, 24 Chromium Cr, 25 Manganese Mn, 26 Iron Fe, 27 Cobalt Co, 28 Nickel Ni, 29 Copper Cu, 30 Zinc Zn, 31 Gallium Ga, 32 Germanium Ge, 33 Arsenic As, 34 Selenium Se, 35 Bromine Br, 36 Krypton Kr
5	37 Rubidium Rb, 38 Strontium Sr, 39 Yttrium Y, 40 Zirconium Zr, 41 Niobium Nb, 42 Molybdenum Mo, 43 Technetium Tc, 44 Ruthenium Ru, 45 Rhodium Rh, 46 Palladium Pd, 47 Silver Ag, 48 Cadmium Cd, 49 Indium In, 50 Tin Sn, 51 Antimony Sb, 52 Tellurium Te, 53 Iodine I, 54 Xenon Xe
6	55 Caesium Cs, 56 Barium Ba, 57 Lanthanum La, 72 Hafnium Hf, 73 Tantalum Ta, 74 Tungsten W, 75 Rhenium Re, 76 Osmium Os, 77 Iridium Ir, 78 Platinum Pt, 79 Gold Au, 80 Mercury Hg, 81 Thallium Tl, 82 Lead Pb, 83 Bismuth Bi, 84 Polonium Po, 85 Astatine At, 86 Radon Rn
7	87 Francium Fr, 88 Radium Ra, 89 Actinium Ac

| Lanthanides | 58 Cerium Ce, 59 Praseodymium Pr, 60 Neodymium Nd, 61 Promethium Pm, 62 Samarium Sm, 63 Europium Eu, 64 Gadolinium Gd, 65 Terbium Tb, 66 Dysprosium Dy, 67 Holmium Ho, 68 Erbium Er, 69 Thulium Tm, 70 Ytterbium Yb, 71 Lutetium Lu |
| Actinides | 90 Thorium Th, 91 Protactinium Pa, 92 Uranium U, 93 Neptunium Np, 94 Plutonium Pu, 95 Americium Am, 96 Curium Cm, 97 Berkelium Bk, 98 Californium Cf, 99 Einsteinium Es, 100 Fermium Fm, 101 Mendelevium Md, 102 Nobelium No, 103 Lawrencium Lw |

3.10 The shapes of electron clouds

The two most important shapes are the sphere and the dumb-bell. In the first, the electron moves in a spherical space which has the nucleus of the atom as its centre. In the second, the nucleus is at the mid-point of two pear-shaped regions in which the electron moves (see Figure 3.4).

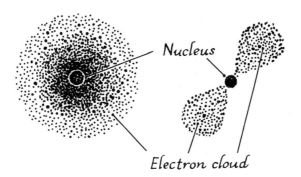

Nucleus

Electron cloud

Figure 3.4 The two main electron-cloud shapes

We must remember that this is only a model of how the electrons in atoms behave, and gives us only a very rough, but nevertheless very useful, picture which helps us to understand how and why atoms join together, as we shall see in Chapters 5 and 6.

The electrons in an atom also have different amounts of energy associated with them. There are seven main 'levels' of energy, and this, coupled with the fact that the shapes are repeated in each level of energy, gives rise to a repetition in the chemical properties of the elements. Thus, the elements can be grouped into 'families', the members of each family having similar electron arrangements (see Table 3.2).

3.11. The occurrence of the elements on earth

There are eighty-nine elements which occur naturally. An estimate of their relative abundance in the air, the seas and other waters, and in the earth's crust to a depth of twenty-four miles has been made by F. W. Clarke. Oxygen accounts for about one-half of the total and silicon for over one-quarter. Third on the list comes aluminium. About 99 per cent of the total is accounted for by some twenty elements. Figures 3.5 and 3.6 illustrate the relative abundances of the elements in two different ways.

Some elements are found widespread over the earth's surface – some in large quantities and others in much smaller quantities.

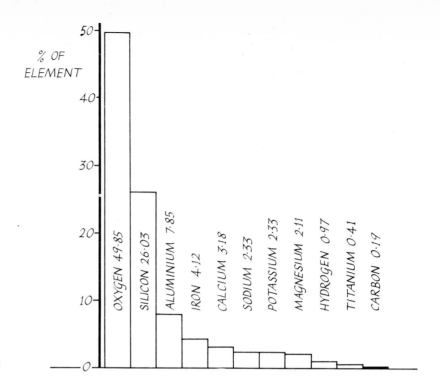

Figure 3.5   The relative
abundance of the elements

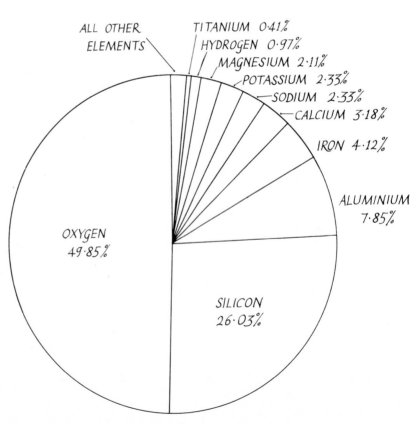

Figure 3.6   The relative
abundance of the elements

Other elements are found only in very restricted areas and occur only in very small amounts.

3.12. Metals and non-metals

Investigation 3a. Conductors and non-conductors of electricity

Take pieces of each of the following elements: sulphur (roll), zinc (foil or granulated), iodine (a flake), copper (foil or wire) and iron (gauze or a nail). Set up the electric circuit as shown in Figure 3.7.

Corrosive

★ *Warning – do not touch iodine.*

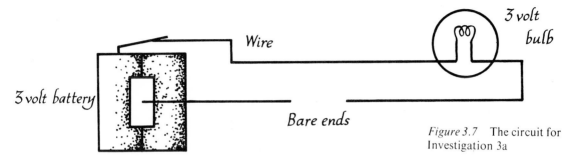

Figure 3.7 The circuit for Investigation 3a

Check that, when the bare ends of the wires are touched together, the bulb lights. Now, take the pieces of the different elements in turn and complete the electric circuit by using them to bridge the gap between the bare ends of the wires. Which pieces allow the bulb to light? Try pieces of other elements.

In the cases of the elements which allow an electric current to pass through them, one or more of the electrons which are in the atoms are relatively easily detached. In fact, in these cases, the 'loose' electrons are already moving from atom to atom in the piece of the solid and, in so doing, are binding the atoms together. Elements of this sort are known as **metals**.

The other elements are called **non-metals**. They do not have 'loose' electrons and act as **insulators**. Non-metal atoms are capable, under the right conditions, of picking up electrons.

Carbon, which is normally counted as a non-metal, does, in the form of **graphite**, have 'loose' electrons, and is a good conductor of electricity.

Test your understanding

1. What are the uses of scientific 'models'?
2. What is an element?
3. Name the three chief sub-atomic particles. What are their relative masses? What electric charges do they carry?
4. What are 'isotopes'?

39

5. Name two elements that have isotopes. For each of the elements you suggest, give (i) the atomic number, (ii) the mass numbers (relative atomic masses) of the isotopes.
6. Name three scientists who helped to give us our modern picture of the structure of the atom.
7. What does a mass spectrograph do?
8. Explain the meanings of the terms (i) relative atomic mass, (ii) mole, (iii) ion.
9. What do you understand by a 'radioactive nucleus'?
10. What is an 'electron cloud'? Describe the shapes of the two most important electron clouds.
11. What is there about the structure of metals that allows them to conduct an electric current?
12. Name, in order of decreasing abundance, the eight most common naturally occurring elements.
13. Name a non-metal that will conduct electricity.
14. If you were given a piece of an unknown element, how would you find out if it could conduct an electric current?
15. Fluorine, chlorine, bromine and iodine are a 'family' of elements (the 'halogens') that are very similar chemically. What is the reason for their chemical similarity?
16. Give the names of the elements forming two other families.

Chapter 4

Energy

Scientists are particularly interested in what they call 'energy' and, to a scientist, the term has a very exact meaning.

4.1. The meaning of the term 'energy'

In everyday life, we say that a person who is always 'on the go' is energetic. Of course, he remains an energetic person even when he is resting or asleep and is not actually rushing about!

Scientists think of energy as the *ability to do work*; that is, the ability to make objects move against forces trying to stop them moving.

There are only two main kinds of energy:

a. **Potential energy**. This is 'stored' energy. It is shown, for example, in the wound-up spring of a clockwork toy. The work done and put into the spring, as it is being wound up, is stored in the spring and may be released later. As the stored energy is released, it makes the toy 'work'.

Another good example of potential energy is to be found in a grandfather clock. Here, the 'winding-up' process consists of raising heavy weights so that they are left hanging some distance above the bottom of the clock case. Energy is stored in them because of their weight and the distance that they are above the bottom of the clock case. As they slowly fall towards the bottom of the case, their stored energy is used up to drive the 'works' of the clock and thus make it go.

Potential energy is rather like that of the energetic person mentioned above, while he is resting; it is held, for the moment, under check.

b. **Kinetic energy.** This is energy of movement. The word 'kinetic' has the same basic meaning as our word 'cinema' (the movies). Kinetic energy is shown, for example, by the 'push-and-go' toys. When such a toy is pushed across a rough surface, the movement of the wheels is conveyed by a series of cogs to a heavy wheel, or **flywheel**. The flywheel is set in rapid motion and, when the toy is put down again on the floor, the energy of its motion is transferred back to the wheels. The toy moves across the floor until all the energy in the flywheel has been used up and it comes to rest. All moving objects have kinetic energy.

Although the cases mentioned above have been used to illustrate the meanings of the terms 'potential' and 'kinetic' energy, there are many other ways in which energy can be stored. How many ways of storing energy in a potential form can you think of?

4.2. The law of conservation of energy

The most important of all the scientific laws is the law of conservation of energy. This law states that the total amount of energy in the universe is constant. This means that energy can neither be created nor destroyed in any change, either physical or chemical; all that can happen is that it may be changed from one form to another.

Consider, for example, a waterfall. At the top of the fall the water has potential energy by reason of its weight and the distance that it is above the bottom of the fall (compare the weights in a grandfather clock when the clock is fully wound). As the water falls over the edge, it starts to move faster and faster. As it moves faster and faster, so it gains more and more kinetic energy, but it is, of course, closer to the ground, and thus has less potential energy. The sum of the potential and kinetic energies is always the same: at the top of the fall the energy is all potential, at the bottom it is all kinetic. In between, it is a mixture of both forms.

At the moment of impact on the ground at the bottom of the fall, the kinetic energy disappears and an exactly equivalent amount of a new form of energy, **heat**, appears, causing the water to rise in temperature. Heat energy is a form of kinetic energy. The average speed of the water molecules becomes greater. The amount of heat energy produced is exactly equal to the amount of potential energy that the water had originally at the top of the fall. (In this example, the relatively small amount of kinetic energy due to the speed of flow of the water in the river has been neglected.)

Changes of energy from one form to another are very common. A few examples of such changes are given in Table 4.1. How many more can you add to the list?

Here energy is POTENTIAL

Here energy is both POTENTIAL and KINETIC

Here energy is all KINETIC at moment of impact

Water is warmer here than at the top of the fall

Figure 4.1   The energy of a waterfall

TABLE 4.1.   CHANGES IN THE FORMS OF ENERGY

Original Form of Energy	New Form of Energy	Method of Change
Kinetic	Electrical	Generator or dynamo
Chemical	Electrical	Cell or battery
Electrical	Chemical	Accumulator charging
Light	Chemical	Photography
Chemical	Light	Flash-bulb
Electrical	Heat	Electric fire
Electrical	Light	Light-bulb
Chemical	Heat	Many chemical changes

4.3. Chemical energy

This is a form of potential energy. All substances have chemical energy stored in them.

Investigation 4a

Take a match from a box of 'safety' matches and examine it closely. Describe what you see as fully as you can. Now, examine the 'special' material on the side of the matchbox. Feel it with your fingertips. Describe its colour and any other points that you may notice about it.

Now, strike the match (away from you) on the side of the box. Describe all that happens. Hold the match with the flaming end slightly higher than the other end until the flame has burned about half-way down the stick and then blow out the flame. Describe carefully the appearance of what is left.

Here we have an excellent example of changes in the forms of energy. The original substances, both in the head of the match and on the side of the box, contain potential energy in chemical form. In the course of the chemical change that takes place when the match is struck, some of this stored energy is changed into the forms of heat and light. Do you think that you could get as much energy from what is left over? How would you set about getting as much energy as you could from what is left?

4.4. Energy is needed to start chemical changes

If you lay a match down on the table next to the special side of the box and just wait for the match to catch fire, you will never stop waiting! It is absolutely necessary for the temperature of the match-head to be made high enough before the chemical changes involved in the burning of the match can start to take place. This rise in temperature is brought about by **friction** when the head of the match is rubbed against the specially prepared side of the box. The heat energy thus given to the head of the match and the side of the box is sufficient to start the reaction. Once the reaction is under way, the energy given out in the chemical change is enough to keep the reaction going. The energy that is needed to start a chemical reaction is called the **activation energy**.

In the case of reactions such as the burning of a match, where energy is given out to the surroundings and the remaining substances contain less energy than the starting substances, the chemical reaction is said to be an **exothermic** change.

We can illustrate the energy changes for this type of reaction by means of a diagram, or graph, as shown in Figure 4.2.

It is necessary for enough energy to be given so that the 'hill' may be got over before the reaction can take place.

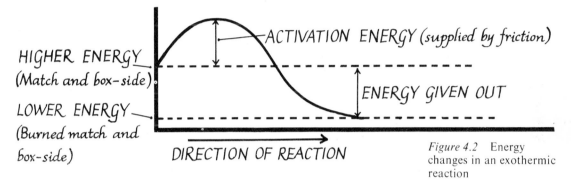

Figure 4.2 Energy changes in an exothermic reaction

4.5. Endothermic chemical changes

In the case of a match being struck, energy in the forms of heat and light is given out to the surroundings, and the remaining substances do not have as much energy stored in them. This is not always the case with chemical changes, although the greater number are of this kind.

There are chemical changes where the energy left in the substances after the change has taken place is *greater* than that which was stored at the start. Such reactions are said to be **endothermic**. It is clear that the extra stored energy has to come from somewhere and, in general, it has to be supplied throughout the course of the change, for example, by heating with a bunsen burner, by using an electric current, or by shining a light on the reacting substances.

We can illustrate such an endothermic change by means of another energy diagram, as shown in Figure 4.3.

4.6. Exothermic and endothermic: an industrial example

An important industrial process is that which is used for the manufacture of **water-gas**. Water-gas is a mixture of two gases,

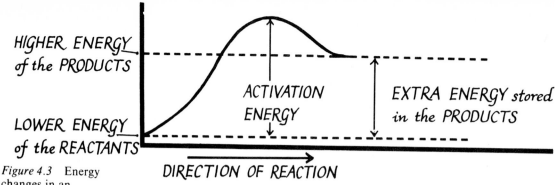

Figure 4.3 Energy changes in an endothermic reaction

hydrogen and carbon monoxide, in equal proportions by volume. It is a useful fuel (see Chapter 21).

Water-gas is made by passing very hot steam over white-hot coke. The chemical change is very endothermic and is only kept going by the energy from the white-hot coke. The coke gradually cools down during the reaction, and, if nothing were done, the temperature would, sooner or later, fall so far that there would not be enough energy available for the activation of the reaction (we must remember that we put fires out by pouring water on them):

steam + coke → hydrogen + carbon monoxide

However, when the temperature of the coke has fallen to that of about red-heat, the supply of steam is switched off and, instead, a blast of air is forced through the coke. This brings about another chemical change and another gas, **producer-gas**, is formed. Producer-gas is a mixture of carbon monoxide and nitrogen, together with carbon dioxide. It is another useful industrial fuel, but the really important point is that this reaction is exothermic. As a result, the coke is raised to white-heat again (we must remember that we use bellows to 'blow-up' a fire that is dying down):

air + coke → nitrogen + carbon monoxide + carbon dioxide

Thus, by alternately passing steam and air through the coke, the reaction can be kept going. The men working the process call the passing of the steam the '**run**', and the passing of the air the '**blow**'. Figure 4.4 shows a simple diagram of a water-gas/producer-gas plant.

46

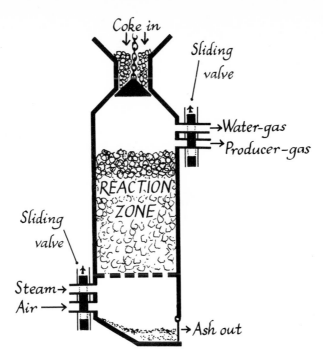

Coke in

Sliding valve

→Water-gas
→Producer-gas

REACTION ZONE

Sliding valve

Steam→
Air →

→Ash out

Figure 4.4 Diagram of a water-gas/producer-gas plant

4.7. Spontaneous chemical changes

Although, in general, chemical changes need the help of energy from the outside to start them going, this is not always the case. For example, if a small piece of sodium metal is put in some water (see Chapter 11) there is an instantaneous and very vigorous reaction. The reaction is very exothermic and the activation energy is easily got from the reaction itself. This is possible because the sodium atoms in a piece of sodium do not all have the same energy; some will have less than, and some will have more than, the average amount. Some of the more energetic atoms have enough energy to take them over the 'activation hill' (see Figure 4.2). When they react with the water, they give out enough energy to activate other atoms, and so the reaction can continue.

Investigation 4b. A spontaneous endothermic change

Weigh out about 20 g of ammonium nitrate(V) crystals. Examine the crystals carefully and note their shape. In a small (100 cm³) beaker put about 20 cm³ of cold water and take the temperature with an ordinary −10° to 110 °C thermometer. Now, while *gently* stirring with the thermometer (take care not to break it) add the ammonium nitrate(V) a little at a time, but fairly quickly, to the water. What is the lowest temperature shown on your thermo-

★ *Warning – this should only be done by an experienced teacher.*

Explosive

Flammable

Harmful

Explosive   Oxidising

meter? If the beaker with the water is placed on a small wood block in a few drops of water before the solid is added, it is usually possible to freeze the beaker to the block of wood. The arrangement is shown in Figure 4.5.

As the water becomes cooled, together, of course, with the beaker, the thermometer, the ammonium nitrate(V) and the water outside (if you are using it), it is clear that energy is being taken in by the substances involved: that is, when the ammonium nitrate(V) dissolves in water, it is taking energy from the water and the surroundings and is doing this spontaneously.

The 'driving-force' for spontaneous endothermic changes such as this is not energy, but an entirely different one which scientists call **entropy**. It is too difficult an idea to discuss in detail in this book, but, putting it as simply as possible, there is a natural tendency for things in nature to become more 'mixed-up' or disorganized. The beautiful, regularly shaped crystals of ammonium

Figure 4.5 The freezing of a beaker to a wood block

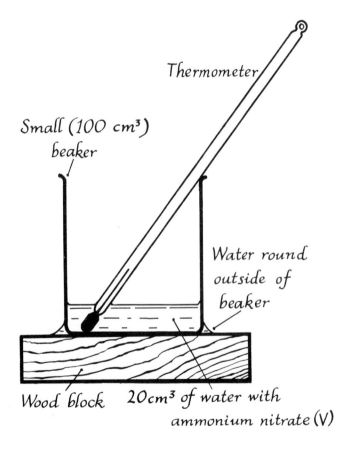

Thermometer

Small (100 cm³) beaker

Water round outside of beaker

Wood block   20 cm³ of water with ammonium nitrate (V)

nitrate(V), with their complicated organization, have become spread out and disorganized within the solution. In doing so, they have taken in energy from the surroundings and the whole has become cooler.

Test your understanding

1. What do you understand by the word 'energy'?
2. Name the two main kinds of energy and explain the difference between them.
3. What happens to the energy of a moving car when the brakes are applied?
4. What does a flywheel do?
5. What is meant by the words 'exothermic' and 'endothermic'?
6. What is 'activation energy'?
7. How might kinetic energy be changed into heat energy?
8. How is water-gas made? What is its composition?
9. The manufacture of water-gas is an endothermic process. What is done in order to keep the temperature of the reaction vessel up?
10. What is 'entropy'?
11. How is it that spontaneous endothermic changes can take place?
12. What are the main natural sources of energy?
13. Describe the energy changes that take place when a match is struck.
14. What energy changes are involved in (i) an electric torch, (ii) charging an accumulator?

Chapter 5

The Result of Atoms Reacting
— Ions

We have already seen that all matter is made up of atoms and that these atoms are small parts of simple substances known as elements. There are 105 different elements but very many more substances. These substances are formed when two or more different atoms become joined together in some way, and are called **compounds**.

5.1. Ion formation

We saw, in Chapter 3, that an atom was made up of three different particles, the type of element being determined by the number of protons in the nucleus of the atom. In the neutral atom, the number of extra-nuclear electrons is the same as the number of protons in the nucleus. If an electron is taken away from the atom or another electron is added to the atom, the number of protons does not change and so the element does not change. Now, however, it no longer has the same number of protons and electrons and so it is not electrically neutral, but carries a charge. A charged atom is called an **ion**.

In Chapter 3, we learned that elements in the same vertical group in the periodic table had similar chemical properties because they had similar electronic arrangements. Elements with similar properties are believed to have the same number of electrons in their highest occupied energy level. For example, lithium is believed to have one electron in the second energy level, sodium to have one in the third and potassium one in the fourth. Similarly, fluorine, chlorine, bromine and iodine are all thought to have seven electrons in their highest energy levels. We can represent the arrangement of electrons in sodium and chlorine diagrammatically, as shown in Figure 5.1.

We must remember that the electrons are moving very rapidly

and that Figure 5.1 only shows the relative number of electrons in each energy level. Also, we must remember that these electrons may occupy different shapes of cloud. Notice that there are eleven protons in the nucleus of the sodium atom and seventeen protons in the nucleus of the chlorine atom. Hence, the attraction between the nucleus of the chlorine atom and its *third* energy level electrons is greater than the attraction between the nucleus of the sodium atom and its *third* energy level electron.

• *Indicates an electron*

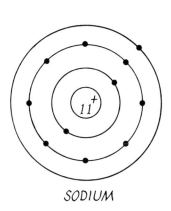

SODIUM

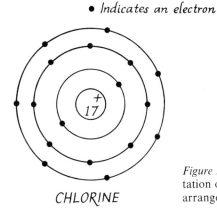

CHLORINE

Figure 5.1 A representation of the electronic arrangement of atoms

Further examination of Table 3.2 shows that helium, neon, argon and krypton are similar. These elements which, with the exception of helium, have eight electrons in their highest energy level are known as the **noble gases**. They only form compounds with difficulty – hence they are said to be unreactive or **noble**. In 1916, Kossel suggested that the arrangement of the electrons in the noble gases was very stable (that is unlikely to change), and this is thought to be responsible for their unreactive nature.

Hence, atoms can obtain a stable structure if electrons can be gained or lost in order to produce the same arrangement as in a noble gas. This electronic rearrangement can occur if the energy needed to cause it to take place is available.

Figure 5.2 shows how a sodium atom and a chlorine atom might form ions. What charge is carried by each of these two ions?

Can you suggest what might happen if a sodium ion and a chlorine ion were close to each other?

Investigation 5a. Heating common salt

Take a small test-tube and place in it enough common salt (sodium chloride) to come about 10 mm up the tube. Support the tube in some kind of holder and heat it in a bunsen flame. When does the salt melt?

Repeat the experiment with calcium chloride and with

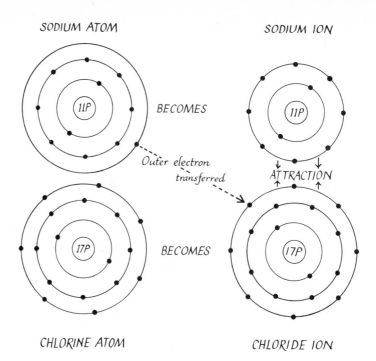

SODIUM ATOM   SODIUM ION

BECOMES

Outer electron transferred

ATTRACTION

BECOMES

CHLORINE ATOM   CHLORIDE ION

Figure 5.2 The formation of ions of sodium and chlorine

magnesium oxide. In each experiment observe what happens and notice when the solid melts.

The sodium ion in Figure 5.2 has a single positive charge and the chlorine, known as chloride ion, has a single negative charge. If a number of sodium ions and a number of chloride ions are near to each other, each sodium ion will attract *all* the chloride ions and each chloride ion will be attracted to *all* the sodium ions.

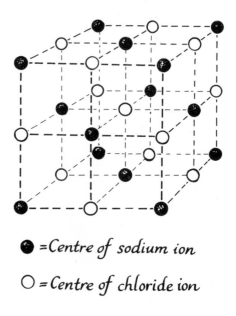

● = *Centre of sodium ion*

○ = *Centre of chloride ion*

Figure 5.3 A sodium chloride lattice

The sodium chloride, which was heated in Investigation 5a, did not melt easily because each sodium ion was attracted to all the chloride ions. Hence a lot of energy is needed to make the ions move apart.

The arrangement shown in Figure 5.3 is known as a **lattice**. Compounds formed in this way are known as **ionic compounds**. They have a definite shape and are not easily melted.

5.2. The electrical properties of ionic compounds

We have suggested that certain compounds are formed by the attraction of positive and negative ions. Let us now see if it is possible to separate these charged ions.

Investigation 5b. Passing electricity through copper(II) bromide

Melt some copper(II) bromide in a crucible, supported over a bunsen flame by a silica triangle and a tripod. When the copper(II) bromide has melted, add some more and stir with a steel knitting needle or a nail. Continue adding copper(II) bromide until the crucible is about three-quarters full of molten copper(II) bromide. Now lower in the **electrodes** and connect them, as shown in Figure 5.4, to a d.c. supply of about 9 volts. The circuit should include either an ammeter capable of reading up to 5 amperes or a suitable bulb to show that electricity is passing through the molten copper(II) bromide. Allow the current to pass for about ten minutes. Take out the negative electrode and allow it to cool. When it is cold, examine the end that was dipped into the molten copper(II) bromide. Can you see anything on the negative electrode? Whilst the electricity was passing through the molten copper(II) bromide, did you observe anything around the positive electrode?

★ Warning – this investigation should be carried out by a teacher. A fume cupboard should be used.

Toxic

In Investigation 5b, we see that electricity can pass through molten copper(II) bromide. The negative electrode becomes coated with reddish-brown copper. While electricity is passed through the molten substance a reddish coloured gas can be seen to escape near the positive electrode; this gas is bromine (see Chapter 31). The breaking up of a compound, in a manner such as this, is known as **decomposition**. A substance that allows the passage of electricity, when molten or when dissolved in water, and which is decomposed while the electricity is passing through it is known as an **electrolyte**. The passage of electricity through an electrolyte is called **electrolysis.**

5.3. Water and ionic compounds

We have seen that when a compound is made by attraction between ions carrying opposite charges, a piece of matter of

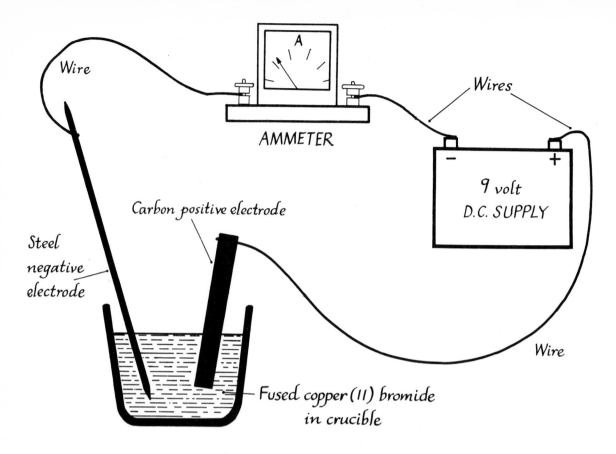

Wire

A

AMMETER

Wires

Steel
negative
electrode

Carbon positive electrode

9 volt
D.C. SUPPLY

−   +

Wire

Fused copper (II) bromide
in crucible

Figure 5.4 Passage of
electricity through
copper(II) bromide

definite shape is produced, because the ions are arranged in a
lattice. This shape is known as a **crystal**.

Investigation 5c. The effect of heat on some crystals

Take some small test-tubes and heat the following substances
in the same way that sodium chloride was heated in Investigation
5a. Use:
  a. copper(II) sulphate(VI) crystals,
  b. sodium carbonate crystals (washing soda),
  c. magnesium sulphate(VI) crystals (Epsom salts).
Observe what happens in each case. Do not discard the residue,
but keep it to compare with the original crystals.
When the residue from the copper(II) sulphate(VI) crystals is
cool, tip some out on to a small watch glass. Make a careful note
of its appearance and colour. Carefully add a few drops of water
to the residue. Does anything happen? Does the substance become
hot or cold?
When the substances in Investigation 5c were heated, water
vapour was seen to escape from the tube, and each substance lost
its crystal shape and became a powder. The washing soda con-

tained so much water that it first dissolved in this water to form a solution; the powder formed as the water was boiled away. In the case of the copper(II) sulphate(VI) the blue colour disappeared and the substance became an almost white powder; when water was put on to this powder the blue colour was restored.

Harmful

Hence it can be seen that water forms a definite part of some crystal lattices. This water, which gives a definite crystal shape to a compound, is called the **water of crystallization**. When a compound has lost its water of crystallization it is said to be **anhydrous**.

5.4. Water in the air

Investigation 5d. Substances exposed to the air

Place one watch glass containing calcium chloride and another containing sodium hydroxide pellets on the bench. Take a third watch glass and place some sodium carbonate crystals on it. Place this third watch glass in a warm dry part of the room. Leave these watch glasses to stand whilst you prepare the apparatus for the next two investigations.

★ *Warning – do not touch the sodium hydroxide with your hands.*

Corrosive

Investigation 5e. Damp substances

Take some copper(II) oxide from a bottle that has been in use for some time and place it in a small test-tube. Gently warm the substance at the bottom of the tube, whilst holding the tube as shown in Figure 5.5. Carefully observe the walls of the test-tube near to its mouth.

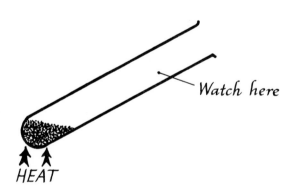

Watch here

HEAT

Figure 5.5 Heating copper(II) oxide

Investigation 5f. Drying substances

Take a test-tube and place some calcium chloride in the bottom of the tube. Make a loose plug of cotton wool and slide it half-way down the tube. On this plug place a few small crystals of copper(II) sulphate(VI). Cork the tube firmly and allow it to stand for a week.

Harmful

55

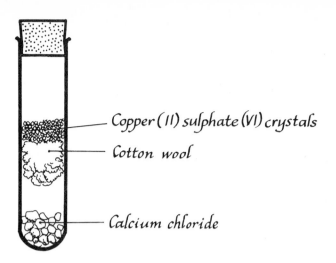

Copper (II) sulphate (VI) crystals

Cotton wool

Calcium chloride

Figure 5.6 Drying copper(II) sulphate(VI)

In Investigation 5d, the calcium chloride and sodium hydroxide are seen to have become moist, and the solids will eventually dissolve in the water that has been picked up from the air. Substances that take up water from the air and dissolve in this water are said to be **deliquescent**. Calcium chloride is so deliquescent that it goes on taking up water from the air even after the solution has been formed.

In Investigation 5e, if the tube was heated carefully, water would have been seen to condense on the walls of the tube. Copper(II) oxide would not, however, dissolve in the water even if left out for a very long time. Substances that become damp when exposed to the air but do *not* dissolve in the water taken up are said to be **hygroscopic**.

The sodium carbonate crystals used in Investigation 5d have not become wet; if the air is dry enough they will become covered with a white powder. This powder is dehydrated sodium carbonate. The crystals have lost some water of crystallization to the air. Substances that lose water of crystallization in this way are said to be **efflorescent**.

In Investigation 5f, we see that calcium chloride has such a great affinity for water that it removes some of the water of crystallization from the crystals of copper(II) sulphate(VI). We can observe this because the copper(II) sulphate (VI) crystals lose their characteristic colour. Other substances that are good **dehydrating agents** are silica gel and concentrated sulphuric(VI) acid.

Test your understanding

1. What is the name given to a combination of different atoms?
2. What is an ion?
3. What is thought to cause atoms of different elements to have similar chemical properties?
4. How many electrons are thought to be in the highest energy level of neon and krypton?
5. What charge do you think a potassium ion would carry?
6. Sketch a sodium chloride lattice.
7. What is the name given to substances that are formed in the same way that sodium chloride is formed?
8. What is an electrolyte?
9. What sort of substances are electrolytes?
10. Some substances owe their distinctive shape to the presence of water. This water is called . . . .
11. What colour is anhydrous copper(II) sulphate(VI)?
12. What is the name given to substances that take water from the air and dissolve in this water?
13. What is a hygroscopic substance?
14. If sodium carbonate is left in dry air it becomes covered with a white powder. This is an example of an . . . substance.

Chapter 6

The Result of Atoms Reacting — Molecules

We saw, in Chapter 5, that some atoms form ions and that these ions are mutually attracted to each other to form a chemical compound. Ions are formed if the number of electrons to be gained or lost is small and if the energy needed for this gain or loss is available. Many more compounds are formed by the sharing of electrons between two or more atoms.

6.1. A simple combination

If an element can gain the electronic structure of a noble gas it has obtained a stable structure (see Section 5.1). Let us now see how two atoms of chlorine can obtain, between them, such a structure. Each chlorine atom has seven electrons in its highest energy level; that is, each atom needs to gain one electron to have the structure of argon, a noble gas. Since both these atoms are of the same element, neither will be able to *capture* electrons from the other; hence the only way by which the argon structure could be obtained is by equal sharing of a pair of electrons, one from

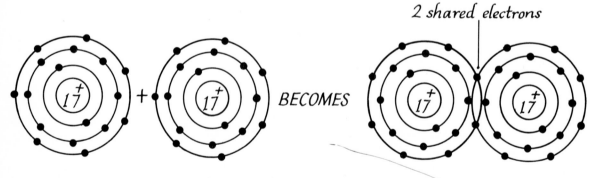

Figure 6.1 Combination of two chlorine atoms

each atom. Figure 6.1 illustrates how such a combination could occur.

Compounds formed in this way are said to be **covalent**, and the combined atoms are known as **molecules**.

Similar electron sharing leads to the formation of molecules in the case of most gaseous elements, the exception being the noble gases. Figure 6.2 (a) illustrates the formation of a hydrogen molecule with the same electronic structure as helium. Figure 6.2 (b) illustrates how an oxygen molecule might be formed. In the case of the oxygen atoms, each atom needs to obtain a share of a further two electrons (the oxygen atom has six electrons in its highest energy level) in order to obtain the same structure as neon.

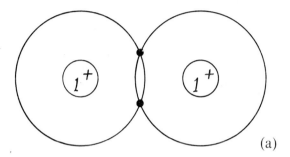

(a)

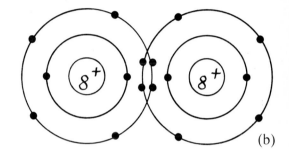

(b)

Figure 6.2 Hydrogen and oxygen molecules

6.2. The directional properties of an atom

In the substances that we have considered in Section 6.1, the molecules have comprised two atoms only. Hence the nuclei of the two atoms are bound to be in a straight line. When there are three or more atoms, however, the nuclei of the atoms need not be in a straight line. They could take up any arrangement, for example, three atoms could lie anywhere in a plane. Three different arrangements for three atoms are shown in Figure 6.3.

$$A-B-C \qquad A-B_{\diagdown C} \qquad \begin{matrix} A-B \\ | \\ C \end{matrix}$$

Figure 6.3 Possible shapes for a molecule with three atoms

The relative positions of the atoms can have a considerable effect on the properties of a chemical compound.

6.3. Formation of a molecule

When an atom bonds covalently to more than one other atom, each bond involves a pair of electrons, usually one from each of the atoms involved in the bond. The electrons that are involved in forming the bond are the electrons in the highest occupied energy level.

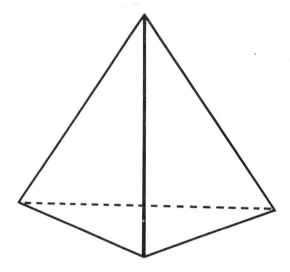

Figure 6.4   A tetrahedron

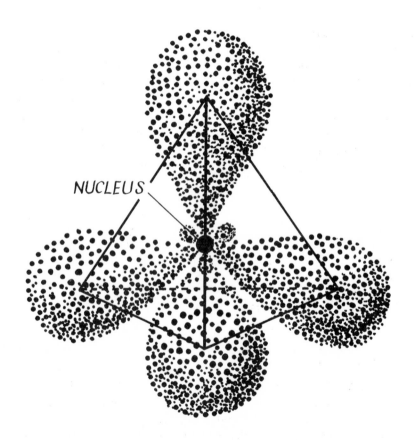

NUCLEUS

Figure 6.5   Charge clouds
around a carbon atom

In the case of carbon, there are four electrons in the highest occupied energy level and hence carbon bonds with up to four other elements in order to have the eight electrons of a noble-gas structure. The maximum number of electrons that can be in any one charge cloud is two. Hence when carbon combines with four other elements, and attains the noble-gas structure with eight electrons, these electrons will occupy four charge clouds. Since the electrons will repel each other, these charge clouds separate as far as possible in space by moving to the corners of a tetrahedron (see Figure 6.4).

The nucleus of the carbon atom is at the centre of the tetrahedron and the four charge clouds are directed outwards from the nucleus to the corners of the tetrahedron, as shown in Figure 6.5.

When carbon combines with hydrogen, to form a molecule of methane, the carbon atom contributes one electron to each bond, and each hydrogen atom contributes one electron. Hence four hydrogen atoms combine with one carbon atom and carbon attains the structure of neon and each hydrogen atom attains the structure of helium. All the charge clouds are identical and so methane is a molecule with perfect three-dimensional symmetry. That is, the hydrogen atoms are arranged evenly all around the central carbon atom. We can represent the arrangement of electrons in the methane molecule in a two-dimensional way as shown in Figure 6.7. Each dot represents a valency electron. Alternatively, the pair of electrons, which form the covalent bond, may be represented by a straight line (see Figure 6.8).

When covalent compounds are formed, a simple molecule results. We can represent this molecule by a formula, which in the case of methane is CH_4. Compare this with the formation of a large ionic lattice (see Section 5.1).

We have seen that only a particular combination of carbon and hydrogen will satisfy the requirements for the formation of a covalent molecule. The number of pairs of electrons used to form the molecule is known as the **valency** (see Section 7.5). These pairs of electrons are known as **bond pairs**.

6.4. The shape of a molecule

When carbon and hydrogen combine there are four identical charge clouds because carbon contributes four electrons, and the four hydrogen atoms contribute one electron each to the formation of the four charge clouds.

However, when nitrogen combines with hydrogen, the nitrogen atom contributes five electrons and so there are only three hydrogen atoms required, each contributing one electron, to make up the eight electrons of the neon structure. Hence in the molecule that

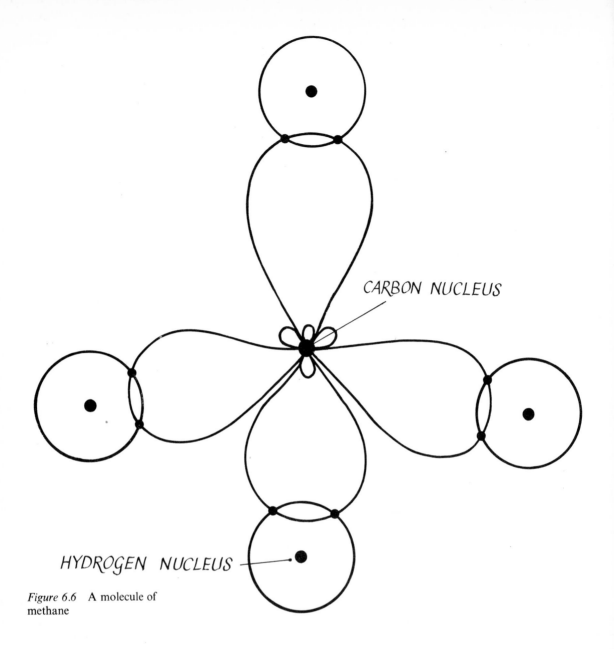

CARBON NUCLEUS

HYDROGEN NUCLEUS

Figure 6.6 A molecule of methane

$$H : \overset{\displaystyle \cdot\cdot}{\underset{\displaystyle \cdot\cdot}{C}} : H$$

with H above and below

Figure 6.7 Electrons in a methane molecule

$$H - \overset{\displaystyle H}{\underset{\displaystyle H}{C}} - H$$

Figure 6.8 A methane molecule

is formed, one of the charge clouds will have two electrons but they will not be used to form a bond between the nitrogen atom and a hydrogen atom. Such a pair of electrons is known as a **lone pair** (see Figure 6.9).

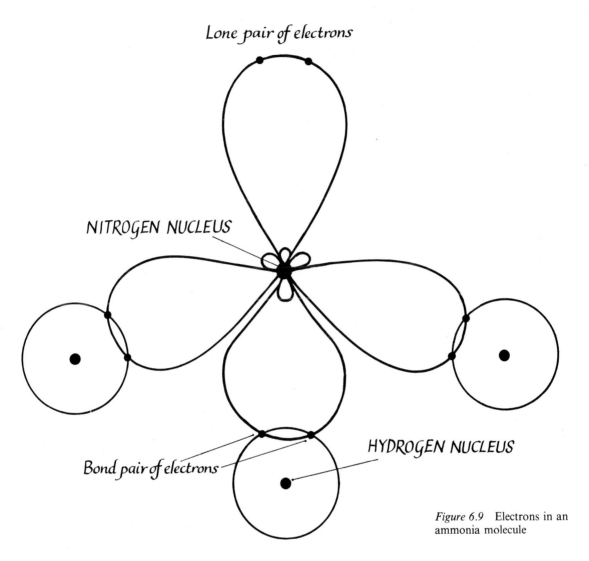

Figure 6.9 Electrons in an ammonia molecule

Hence the combination is between one nitrogen atom and three hydrogen atoms to form a molecule of ammonia, **NH$_3$**. The shape is still tetrahedral, but is no longer regular. The shapes of some molecules and their electronic arrangements are shown in Figure 6.10.

6.5. Distribution of charge

When the molecule of ammonia is formed, there is an unused lone pair of electrons. This causes one side of the molecule, the lone-pair side, to have a negative charge (see Figure 6.11). Since the molecule is electrically neutral, the other side is positive. Such an arrangement is said to produce a **dipole**. The existence of this dipole gives the molecule a number of special properties, especially as a solvent. Water, H_2O, has a dipole in a similar way to ammonia.

The dipoles that we have observed so far have been due to the presence of lone pairs of electrons on one atom of the molecule. Dipoles are also formed when one nucleus exerts a greater attraction for the bond pair of electrons than the other nucleus.

A. All valency electrons used to form two bonds   Linear structure

Carbon dioxide $\ddot{O} :: C :: \ddot{O}$

B. Two bonds formed but one or more "lone pairs"   Angular structure

Water

C. All valency electrons used to form three bonds   Planar structure

Boron trichloride

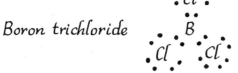

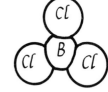

D. Three bonds formed but one "lone pair"   Pyramid structure

Phosphorus trichloride
(like ammonia)

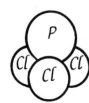

E. Four bonds formed   Regular tetrahedron

Tetrachloromethane
(like methane)

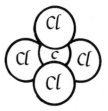

Figure 6.10 Some molecular shapes

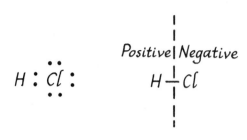

Lone pair

Negative

Positive

Figure 6.11 The dipole in an ammonia molecule

A very important example of this form of dipole is found in the hydrogen chloride molecule.

Positive | Negative

$H : \overset{..}{\underset{..}{Cl}} :$

$H \underset{|}{\overset{|}{-}} Cl$

Figure 6.12 The electronic arrangement and dipole of a hydrogen chloride molecule

Atoms which exert a strong attraction for electrons are fluorine, oxygen, chlorine and, to a lesser extent, bromine, sulphur and nitrogen. A dipole is only formed if the molecule is not symmetrical. Tetrachloromethane, CCl_4, contains chlorine, but since it has the same symmetrical shape as methane, it does not have a dipole. Compounds possessing a dipole are said to be **polar**.

6.6. Properties of covalent compounds

In Investigation 5a, it was seen that ionic compounds could only be melted with difficulty.

Investigation 6a. The boiling-points of covalent compounds

a. Use a list of properties to find out the boiling-points of the following substances that we have seen are covalent compounds: oxygen, chlorine, hydrogen, methane and hydrogen chloride.

b. Use Siwoloboff's method (Investigation 1i) to find the boiling-points of the following liquids: water and tetrachloromethane.

Compare the results of your investigations with the melting- and boiling-points of some ionic compounds such as sodium chloride.

Toxic

We have also seen that ionic compounds are conductors of electricity if they are melted or if they are dissolved in water. Let us now try to find out if covalent compounds will conduct electricity.

Toxic

Take two thick copper wires, about 100 mm long, and connect them to a six-volt battery and a bulb. Dip the copper wires into a small beaker that contains water, as shown in Figure 6.13. Make sure that the wires do *not* touch. Is the water a conductor of electricity? Repeat the investigation with some tetrachloromethane in the beaker.

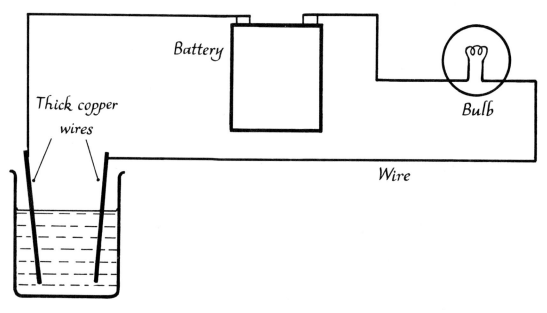

Figure 6.13   Tests on covalent compounds

From these investigations and others like them, we learn that covalent compounds have low melting-points and boiling-points and that they are poor conductors of electricity.

6.7. Larger covalent compounds

So far, we have examined covalent molecules made up of a small number of atoms. These have low melting- and boiling-points. Not all covalent compounds, however, are made up of small molecules. Carbon, as an element, provides two examples of large covalent structures. The purest forms of carbon are diamond and graphite.

In diamond the carbon atoms are situated at the corners of a regular tetrahedron. The atoms in one tetrahedron are covalently bonded to other tetrahedra, so that almost every carbon atom is covalently bonded to four other carbon atoms, producing a large solid mass.

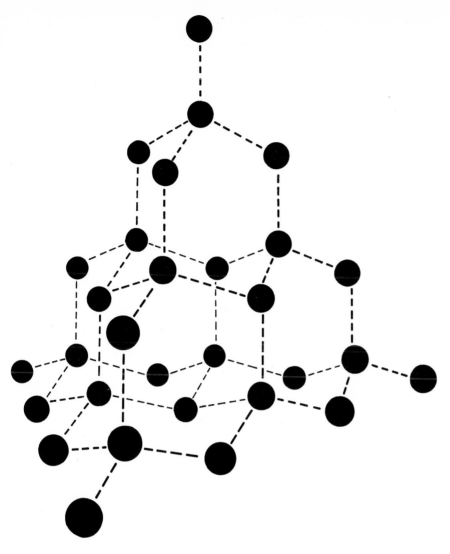

Figure 6.14 The structure of diamond

Graphite is composed of plates in which the atoms are at the corners of regular hexagons. In graphite, the plates are separated by a greater distance than the distance between atoms in the plates. You will notice that each carbon atom is bonded to only three other carbon atoms; this means that there is an unused electron. These unused electrons, one for each carbon atom, serve to hold the layers or plates of graphite together in a similar way to the way in which metals form a lattice (see Section 6.9). These unused electrons are also responsible for the electrical conduction of graphite (diamond does not conduct electricity).

When a chemical element exists in two different forms, like carbon, the element is said to display **polymorphism** or **allotropy**.

Long *chain-like* molecules are also formed by covalent bonds. Examples of these are now in common use: polythene, nylon and

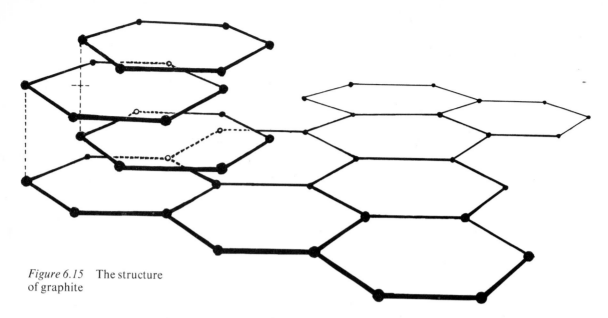

Figure 6.15 The structure of graphite

terylene (see Chapter 36). Other large arrangements are also produced by a number of smaller molecules if they are polar. A good example of this is seen in ice. When water freezes to form ice it expands (this is the cause of burst pipes). When ice is formed, a definite crystalline pattern is produced.

Investigation 6c. Examination of ice

Take a very small flake of snow, or scrape a small flake of ice from the ice-box of a refrigerator. Examine the flake carefully under a microscope or a hand lens.

6.8. Co-ordinate bonds

We have seen that covalent bonds are formed by the sharing of electrons. The examples that we have studied are ones in which the electrons are provided by both the atoms forming the bond, for example in the hydrogen molecule, in which each hydrogen atom contributes one electron, and in the oxygen molecule, in which each oxygen atom contributes two electrons. There are, however, some compounds in which one atom contributes both of the bond pair electrons. Such a bond is known as a **co-ordinate** or '**dative**' bond.

The properties of such a bond are the same as the properties of the normal covalent bond. An example of the electronic arrangement in such a compound, ammonium chloride, is shown in Figure 6.17. The lone pair from the nitrogen atom forms a co-

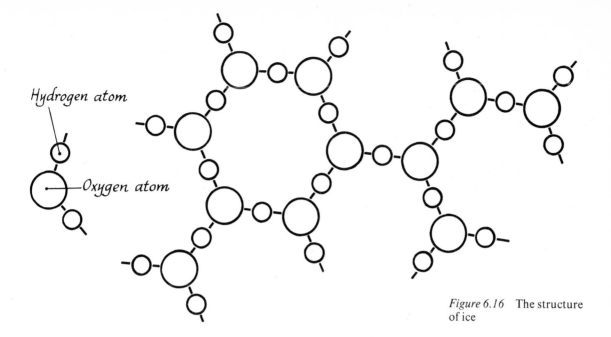

Hydrogen atom

Oxygen atom

Figure 6.16   The structure of ice

ordinate bond with the hydrogen atom of the hydrogen chloride and the chlorine withdraws the bond pair of electrons from the hydrogen–chlorine bond and becomes a chloride ion, Cl^-.

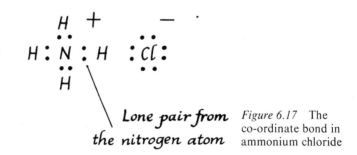

Lone pair from the nitrogen atom

Figure 6.17   The co-ordinate bond in ammonium chloride

6.9. Bonds in metals

In Section 3.12, an investigation was carried out to divide a number of elements into metals and non-metals. This was done by finding which ones permitted the passage of electricity. We have also seen, in Chapter 5, that metals tend to lose electrons when forming ionic compounds. In a solid metal, the atoms tend to lose electrons and form ions with the 'free' electrons moving through the lattice. The lattice is held together by the attraction of large numbers of ions for all the electrons that have been released.

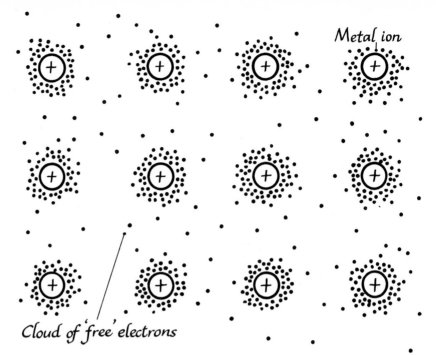

Figure 6.18 Bonds in metals

Metal ion

Cloud of 'free' electrons

Test your understanding

1. What is a 'molecule'?
2. Before atoms can combine they need to gain....
3. What valency is shown by carbon in methane, and what is the shape of the methane molecule?
4. What properties do you normally associate with covalent bonds?
5. A molecule where the electrical charge is not evenly distributed is said to be....
6. What is meant by 'allotropy'?
7. Why does water form a larger structure when it freezes?
8. How does a metal lattice differ from an ionic lattice?

Chapter 7

Compounds and Valency

In the last two chapters, we have discussed the behaviour of the electrons in atoms when chemical compounds are formed. We must remember that we have only discussed a model of an atom and the possible effects on it due to the presence of a model of another atom. We are unable to see the electrons in an atom; all we can do is to estimate the expected properties of a compound from our ideas and then compare these with the observed properties of the substance.

7.1. Covalent and ionic compounds compared

In Chapter 5, it was suggested that an ionic compound was formed by the gain and loss of one or two electrons and, in Chapter 6, that a covalent compound was formed by the sharing of electrons. We must not think that every compound will form completely in one way or the other, but we must realize that these two forms of bonding represent the two extreme cases. Most chemical compounds can be regarded as being in between these two extremes. Thus sodium chloride is an example of a substance that is almost entirely ionic and methane is an example of a substance that is almost entirely covalent, but hydrogen chloride is a substance that shows both covalent and ionic properties. Hence a compound that is polar, like hydrogen chloride, shows properties that are intermediate between ionic and covalent.

Investigation 7a. To examine the type of bonding in different compounds

In order to find out what sort of bonding is present in a compound, we carry out experiments to find out about its properties. In this investigation, we will take sodium chloride as an example of an ionic compound and naphthalene as an example of a covalent compound. We shall use these two substances to discover the properties of ionic and covalent compounds.

Try the following experiments with both substances and make a careful record of your observations.

Flammable

Harmful

Warning – take care not to inhale the vapour of methylbenzene.

1. Place a small amount of the solid in a small soda-glass test-tube (75 mm × 10 mm is suitable). Heat the tube in the same way that you did in Investigation 1j. Does the substance melt easily or with difficulty?
2. Quarter-fill a test-tube (150 mm × 19 mm) with water and sprinkle a little of the substance into the tube. Shake the tube. Does the substance dissolve? If it does, add some more of the solid and keep the solution for use later.
3. Quarter-fill a test-tube with methylbenzene and sprinkle a little of the substance into the tube. Shake the tube. Does the substance dissolve?
4. If the substance dissolved in water, test the solution to see if it conducts electricity (see Investigation 6b).

Compare your results with the information given in Table 7.1.

TABLE 7.1. COMPARISON OF THE PROPERTIES ASSOCIATED WITH THE TWO TYPES OF BONDING

Ionic Compounds	Covalent Compounds
Solid substances with high melting-points	Gases or liquids, or solids with low melting-points
Often soluble in water and other polar solvents	Usually insoluble in water
Insoluble in non-polar solvents (for example, methylbenzene)	Soluble in non-polar solvents
When molten undergo electrolysis	Do not undergo electrolysis when molten

Remember that we do not expect compounds to show all ionic or all covalent properties; some compounds will show properties of both types.

7.2. Water as a solvent

We have seen, in Chapter 6, that water is a covalent compound, but we have found that it is a good solvent for ionic compounds. This is due to the dipole that is possessed by the water molecule. An ionic compound is composed of a number of positively and negatively charged ions in a lattice. The oxygen atom in water is negatively charged, when compared with the hydrogen atoms. Hence, when water molecules surround an ionic lattice, there will be an attraction between the negatively charged oxygen atom of a water molecule and the positive ion in the lattice. This attraction can lead to the removal of the ion and its subsequent surrounding by water molecules. This process is illustrated in Figure 7.1.

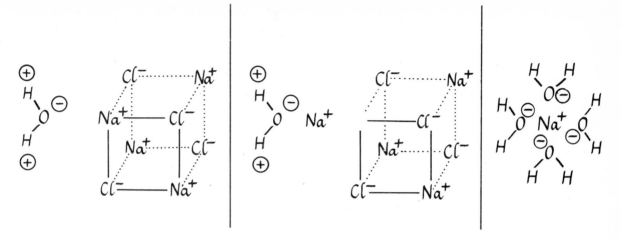

The positively charged hydrogen atoms in the water molecule can have a similar effect on the negative ions in the lattice.

The polar nature of water molecules can have a considerable effect on the behaviour of a compound. For example, hydrogen chloride is a covalently bonded gas, when pure, but it dissolves in water to show ionic behaviour because of the interaction of its own dipole with the dipoles of the water molecules. Figure 7.2 illustrates this interaction, the effects of which will be examined further in Chapter 30.

Figure 7.1 The dissolving of an ionic substance in water

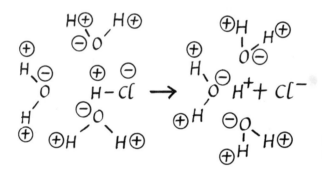

Figure 7.2 The action of water on hydrogen chloride

7.3. Electrons gained or lost?

Investigation 7b. The type of bonding associated with an element

Make a list of all the compounds that you have met that contain chlorine. Do not forget the chlorine molecule itself. Find out and write down the type of bonding that you think is found in the compound, that is ionic, covalent or a combination of both types (see Table 7.1). Does this list suggest that chlorine forms one type of compound only?

In this investigation, we should have seen that the type of bonding that is present in chlorine compounds depends upon the other element that is present. In Section 5.1, we saw that some elements were more likely to gain electrons and that some were more likely to lose electrons. Table 7.2 gives a list of some elements; the ones at the top are more likely to lose electrons and the ones at the bottom are more likely to gain electrons in forming ions. The further apart two elements are in this list, the more ionic the compound that is formed.

Investigation 7c. To classify elements

a. In Table 7.2, we have a list of some elements classified according to the way they behave when forming compounds. What similarities do you notice about the first few elements and what similarities do you notice about the last few? (Hint – refer to Section 3.12.)

b. Look at Table 3.2 and the list of elements given in Table 7.2. Where in Table 3.2 do you find the first and last six elements given in Table 7.2?

We see that metals, which are found in the bottom left-hand corner of the periodic table, tend to lose electrons and that non-metals, which are found in the top right-hand corner of the periodic table, tend to gain electrons.

TABLE 7.2. ELEMENTS IN ORDER OF
ABILITY TO FORM POSITIVE IONS

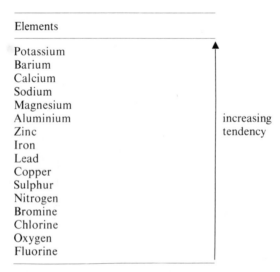

Elements

Potassium
Barium
Calcium
Sodium
Magnesium
Aluminium
Zinc   increasing
Iron   tendency
Lead
Copper
Sulphur
Nitrogen
Bromine
Chlorine
Oxygen
Fluorine

7.4. Complex ions

Many of the common laboratory chemicals, for example copper(II) sulphate(VI), are not made from only two atoms of dif-

74

ferent elements, but from several atoms. None the less, if you were to examine the properties listed in Table 7.1, you would decide that copper(II) sulphate(VI) was an ionic compound. This is because the sulphate group of atoms exerts a strong attraction for electrons, in the same way that the chlorine atom does. This is largely due to the *electron-seeking* nature of the oxygen atoms in the sulphate group of atoms. Hence this complex ion is like a non-metal. The ones most commonly met are listed in Table 7.3.

TABLE 7.3   SOME COMMON COMPLEX IONS

Sulphate (VI)	SO_4^{2-}
Carbonate	CO_3^{2-}
Nitrate(V)	NO_3^-
Sulphate(IV)	SO_3^{2-}
Nitrate(III)	NO_2^-
Ammonium	NH_4^+

The atomic symbols indicate the number and type of atoms in the complex ion; the positive or negative signs indicate the size and type of charge carried by the ion. Hence the sulphate(VI) ion contains one sulphur atom, four oxygen atoms and carries two negative charges. The negative ions, when combined with hydrogen, form acids (see Chapter 13) and are often called **acid radicals**. The ammonium ion behaves like a metal ion and is said to be a **metallic radical**.

7.5. Valency

We have seen that when an ionic compound is formed, it is because a metal atom donates one or two electrons to a non-metal atom or to an acid radical group of atoms. For example, in sodium chloride, the electron from the highest energy level in the sodium atom is donated to the chlorine atom and they both form ions which have the electronic structure of a noble gas. Notice that it is only the highest energy level electrons that can be involved in this way.

Let us now examine the case of sodium sulphate(VI). A sodium atom can only lose one electron to form an ion, but the sulphate(VI) group of atoms needs to gain two electrons to become ionic. This means that each sulphate(VI) group needs to gain one electron from each of *two* sodium atoms. Hence sodium sulphate(VI) will be a lattice which contains twice as many sodium ions as sulphate(VI) ions. The number of electrons gained or lost in the formation of an ion is known as the **valency** of the element or radical. The ratio of the number of ions or atoms present in an ionic compound is represented by an empirical formula. For example, sodium sulphate (VI) can be represented as Na_2SO_4. This does *not* mean that molecules of two sodium, one sulphur and four oxygen atoms exist, *but* that in the lattice the ratio of the total number of atoms present is $2:1:4$ for sodium, sulphur and oxygen (see Section 8.6).

A covalent compound is formed by the sharing of one or more pairs of electrons between two atoms. In the case of an atom forming a covalent compound, the valency is the number of pairs of electrons that are shared to form bonds. For example, in methane,

TABLE 7.4   VALENCIES OF SULPHUR

Compound	Formula	Bond Pairs of Electrons	Valency Shown by Sulphur
Hydrogen sulphide	H_2S	2	2
Sulphur(IV) oxide	SO_2	4	4
Sulphur(VI) oxide	SO_3	6	6

TABLE 7.5   COMMON VALENCIES

Element or Radical	Symbol	Valency Most Usually Found	Type of Bond Most Usually Found
Sodium	Na	1	Ionic
Potassium	K	1	Ionic
Hydrogen	H	1	Covalent
Chlorine	Cl	1	Both types common
Bromine	Br	1	Both types common
Iodine	I	1	Covalent
Mercury	Hg	2	Ionic
Copper	Cu	2	Ionic
Magnesium	Mg	2	Ionic
Oxygen	O	2	Both types common
Calcium	Ca	2	Ionic
Sulphur	S	2, 4 and 6	Covalent
Iron	Fe	2 and 3	Ionic
Zinc	Zn	2	Ionic
Tin	Sn	2 and 4	Both types common
Lead	Pb	2 and 4	Ionic
Nitrate(V)	NO_3	1	Ionic
Nitrate(III)	NO_2	1	Ionic
Hydroxide	OH	1	Ionic
Hydrogencarbonate	HCO_3	1	Ionic
Hydrogensulphate(VI)	HSO_4	1	Ionic
Sulphate(VI)	SO_4	2	Ionic
Sulphate(IV)	SO_3	2	Ionic
Carbonate	CO_3	2	Ionic
Phosphate(V)	PO_4	3	Ionic
Ammonium	NH_4	1	Ionic

N.B. The nitrate(III) ion is often called nitrite and the sulphate(IV) ion is often called sulphite.

carbon shares eight electrons to form four bonds – hence it is showing a valency of four.

Since the valency depends upon the number of chemical bonds formed, the valency of an atom will depend both upon the other atom with which it is joined and also upon the chemical compound. Only a few elements appear to have a fixed valency, although in many cases an element has one valency which is more usual than others. An example of variation of valency is given in Table 7.4.

Table 7.5 gives a list of elements and radicals showing the valencies displayed and noting the type of bonding that is most commonly found.

Test your understanding

1. Why is water a good solvent?
2. What simple tests would enable you to decide if a compound was largely ionic or covalent?
3. Name TWO metals that lose electrons without the need for much energy change.
4. Name TWO elements that are very 'electron-seeking'.
5. What is a 'complex ion'?
6. What charge is carried by the ammonium ion?
7. How do we decide the valency of an element or group of atoms in (a) an ionic compound, (b) a covalent compound?
8. Give an example of an element that has various valencies.
9. Give an example of a substance that consists of (a) a giant structure of ions, (b) a giant structure of atoms.

Chapter 8

Chemical Equations

As you will know, scientific 'laws' are only exact statements of what scientists believe to happen in the normal course of events. The time has now come for us to consider two of the most important laws of chemistry and also a third law that was predicted by John Dalton as a necessary consequence of his atomic theory. This third law was, in due course, investigated and shown to be correct.

8.1. The law of conservation of matter

Investigation 8a

Obtain a flash-bulb, such as is used to provide the 'flash' for a photograph. Wipe the outside with a damp cloth and then dry it thoroughly with another cloth. Then find the mass of the bulb as accurately as your most accurate balance allows. Set up the simple electric circuit as shown in Figure 8.1.

The switch in the circuit should be well away from the bulb and the bulb should be placed behind a safety screen, or enclosed in a tin so that, in the event of the bulb breaking during the firing, particles of glass cannot cause injury. The bulb can, of course, be used in an actual camera or flash-unit (and a photograph taken at the same time – if the camera is loaded with film).

Set off the flash. The bulb will become very hot (do not touch it) and at the same time light is given out. Both these effects are signs of the chemical change that is taking place. Give the flash-bulb plenty of time to cool down to room temperature. Remove the bulb from its holder, wipe it carefully as before with a damp cloth, dry it thoroughly and again find its mass on the same balance. How do the masses of the bulb, before and after the flash, compare with each other?

It was the French chemist, Lavoisier, who, in 1774, heated some tin in a closed container and found that the total mass of the apparatus and contents, both before and after the reaction, was exactly the same.

The law of conservation of matter may be stated formally as

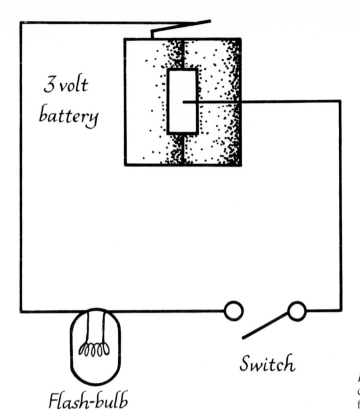

3 volt battery

Flash-bulb

Switch

Figure 8.1 The electric circuit for setting off the flash-bulb

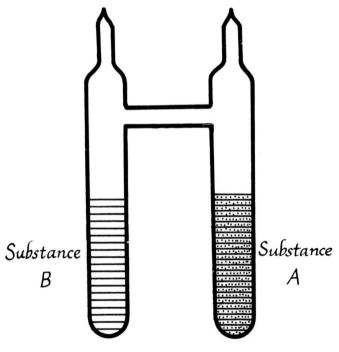

Substance B

Substance A

Figure 8.2 Landolt's apparatus

79

follows: **In a chemical reaction, no change in the total mass of the products as compared with the total mass of the reacting substances has ever been observed.**

A very accurate investigation of this law was carried out between 1893 and 1905 by H. Landolt. Landolt used pieces of glass apparatus as shown in Figure 8.2. Different chemical substances were placed in each of the two limbs of the apparatus. The apparatus was then sealed and its mass found. Next, it was inverted so that the substances could come into contact with each other and could react. After the reaction was over, the apparatus was allowed to stand for some months so that it could recover completely from any changes due to expansion or other heating effects. Its mass was then found again. (Why did you wipe the flash-bulb with a damp cloth and then dry it on each occasion before you found its mass?)

Landolt used such a sensitive balance, and such large masses of substances, that he could have detected a change in mass of one part in ten million (1/10 000 000). He was unable to detect any change.

8.2. The law of constant composition

This law is also sometimes known as the law of definite proportions. It states that **any particular substance, with its own set of properties, has always exactly the same composition.** That is, it always contains exactly the same elements in exactly the same proportions by mass. This law was assumed to be true before experiments were carried out in order to test if it was indeed true.

It was put forward in an exact form in 1802 by Proust, a French chemist, who said that the proportions of the elements which join together to form a compound are fixed by nature and that it is impossible to alter what nature has fixed. A year later, the law was strongly attacked by another chemist, Berthollet, who said that compounds do *not* always have the same composition.

The two opposed views were tested by many experiments carried out very accurately during the nineteenth century, in particular by Stas and Berzelius. Their experiments seemed to show conclusively that pure compounds do always have a fixed composition. Today, in normal practice, we accept the law as being correct for almost every compound, and it does hold with the greatest degree of accuracy. Some figures for common salt (sodium chloride) are given in Table 8.1.

When the proportions by mass of the elements in 100 g of a compound are given, as in Table 8.1, the figures give the *percentage composition* of the compound. Of course, the figure would be exactly the same in any other units: 100 t of salt contain 39.34 t of sodium and 60.66 t of chlorine.

TABLE 8.1. THE COMPOSITION OF SODIUM CHLORIDE

Source of Compound	Mass of Sodium in 100 g	Mass of Chlorine in 100 g
Atlantic Ocean	39.34 g	60.66 g
Siberian salt mines	39.34 g	60.66 g
Cheshire salt deposits	39.34 g	60.66 g
Compound made in the laboratory	39.34 g	60.66 g

Investigation 8b. The percentage composition of black copper oxide

For this investigation you will need a hard-glass (Pyrex) test-tube (150 mm × 19 mm); a length of hard-glass tubing (which will fit very easily inside the test-tube and which is about 250 mm long) fitted with a loose plug of Rocksil wool, soaked in methanol (see Figure 8.3); a length of rubber tubing by which the hard-glass tubing may be attached to the gas tap and other pieces of normal laboratory apparatus.

Flammable

Harmful

The test-tube should be weighed as accurately as possible and the mass noted. It may be found convenient to have a loop of thin copper wire fixed under the rim of the test-tube so that it may be suspended from the hook of the balance. Some black copper oxide (about 3 g) is placed in the tube. The tube is then fixed vertically in a retort stand and is heated gently with a small blue bunsen flame. This is done because copper oxide is rather hygroscopic, that is, it picks up moisture from the air and becomes damp. The moisture is driven off by this heating and may be seen to condense on the cooler parts of the test-tube. If a piece of filter paper is rolled up tightly, it may be used to absorb the moisture from the walls of the test-tube.

When the tube has cooled down thoroughly, it is weighed again and the mass noted. Now, the apparatus is set up as shown in Figure 8.3.

Care should be taken to spread out the copper oxide along the bottom 50 mm of the tube. The tube should be clamped fairly close to the rim end. A slow stream of gas is then allowed to pass through the supply tube and is lit at the end of the tube before the tube is inserted in the test-tube. The gas will continue to burn as it emerges from the mouth of the test-tube (see Figure 8.3).

It will be necessary to heat the copper oxide gently from outside the tube with a small bunsen flame in order to ensure that the reaction is properly completed. It is also necessary to stir the copper oxide with the end of the gas supply tube. If this is done carefully, and care is taken to see that no solid matter sticks to the supply tube at the end of the investigation, it will not affect the result.

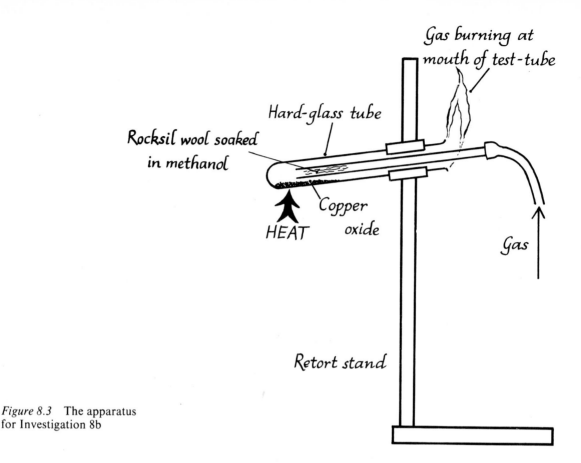

Gas burning at mouth of test-tube

Hard-glass tube

Rocksil wool soaked in methanol

Copper oxide

HEAT

Gas

Retort stand

Figure 8.3 The apparatus for Investigation 8b

The gases will react with the oxygen in the copper oxide to form water. This water may be seen on the cooler parts of the test-tube. The solid left behind is copper metal.

When the reaction appears to be completed, the supply of gas is turned down to a minimum so that only a very small flame keeps burning at the mouth of the test-tube. The bunsen is turned off. When the residue has cooled, the gas supply is completely turned off and the whole tube is allowed to cool. The test-tube and the residue are now weighed. If, by accident, a small piece of the end of the gas supply tube has been broken off during the experiment, it may be 'hooked out' of the test-tube by means of a length of copper wire with the end bent into a small hook, before the tube is weighed.

Your results should be set out as follows:

Mass of empty test-tube   = a grams
Mass of test-tube + copper oxide = b grams
∴ Mass of copper oxide   = $(b - a)$ grams
Mass of test-tube + copper   = c grams
∴ Mass of copper   = $(c - a)$ grams

82

Now, $(b - a)$ grams of copper oxide contain $(c - a)$ grams of copper

$\therefore$ Percentage of copper is $\dfrac{(c-a)}{(b-a)} \times 100 = z\%$

$\therefore$ Percentage of oxygen is $(100 - z)\%$

During the last fifty years, it has been shown that there are some solid lattice compounds whose compositions are *not* always the same. Iron oxide and iron sulphide are two examples. In honour of Berthollet, compounds such as these, which have a variable composition, are now known as **Berthollide compounds**.

It was these two laws of chemical combination that led John Dalton to put forward the first modern atomic theory (see Section 3.2). As a result of this theory, Dalton was led to see that, as a logical deduction, there ought to be another fundamental law of chemical combination and, in 1803 or 1804, he put this new law forward.

8.3. The law of multiple proportions

Dalton stated the law formally in the following way:
If two elements combine to form more than one compound, then the different masses of one of those elements that combine with a fixed mass of the other element are in a simple ratio to one another.

In order to explain the meaning of this law more clearly, we can use the figures for the compositions of the two oxides of copper: as well as the black oxide (whose percentage composition was found in Investigation 8b), copper forms a red oxide.

Experiment has shown that in 100 g of the black oxide, 79.89 g of copper are combined with 20.11 g of oxygen, while in the red oxide, the figures are 88.82 g and 11.18 g respectively.

From these figures we may calculate the weights of copper that are combined with, say, 16 g of oxygen in each case.

$$\text{Black oxide } \frac{79.89}{20.11} \times 16 \text{ or } 63.5 \text{ g}$$

$$\text{Red oxide } \frac{88.82}{11.18} \times 16 \text{ or } 127.0 \text{ g}$$

Thus the masses of copper that combine with 16 g of oxygen are in the ratio of 127.0 to 63.5, or 2 to 1, which is a simple ratio.

As the relative atomic mass of copper is 63.5, then one mole has a mass of 63.5 g (see Section 3.7).

One mole of oxygen atoms has a mass of 16 g and as moles contain the same number of atoms it is clear that, in black copper oxide, one copper atom is present for each oxygen atom and the empirical formula (see Section 8.5) is CuO. In the red oxide, it is equally clear that there are two copper atoms present for each oxygen atom and the empirical formula is Cu_2O. The agreement of these results with the law of multiple proportions is plain.

8.4. The formulae of covalent compounds

As we saw in Section 3.5, each element is given a symbol to represent one mole of the element. By joining the symbols in a 'formula' we can show the moles of each sort of atom in one mole of the compound, and even their arrangement. Some simple 'rules' for writing formulae are now given:

a. Two separate moles of hydrogen atoms (for example) are shown by putting the figure 2 in front of the symbol – 2H. A mole of hydrogen gas in which the atoms go round in pairs is shown as H_2. Three moles of hydrogen gas are shown as $3H_2$.
b. Water has a molecule made from two hydrogen atoms and one oxygen atom. The mole is shown as H_2O. Two moles of water are shown as $2H_2O$.
c. When a compound contains more than one 'group' of the same atoms in its molecule, the group is put in a bracket and a small subscript number after the bracket shows the number of groups. For example, calcium hydrogencarbonate, which has one calcium ion joined to two hydrogencarbonate ions, is shown as $Ca(HCO_3)_2$. Ammonium sulphate (VI), with two ammonium ions to each sulphate(VI) ion, is written as $(NH_4)_2SO_4$.

8.5. Empirical formulae

The empirical formula of a compound is the simplest formula which correctly expresses the relative numbers of atoms of each element present in the compound. For example, the benzene molecule contains six carbon and six hydrogen atoms – that is the formula of the molecule, its molecular formula, is C_6H_6. For each carbon atom, there is one hydrogen atom. Thus the empirical formula is CH. Table 8.2 makes this idea clearer.

Two points will be noted about Table 8.2. First, it will be seen that several compounds may have the same empirical formula and, second, that ionic compounds do have an empirical formula, though they do not have molecules. The empirical formulae of ionic compounds are often used in chemical equations to represent the substances.

TABLE 8.2.   MOLECULAR AND EMPIRICAL FORMULAE

Compound	Molecular Formula	Empirical Formula
Benzene	C_6H_6	CH
Ethene	C_2H_4	CH_2
Propene	C_3H_6	CH_2
Butene	C_4H_8	CH_2
Glucose	$C_6H_{12}O_6$	CH_2O
Methanal	CH_2O	CH_2O
Sodium chloride	—	Na^+Cl^-
Calcium carbonate	—	$Ca^{2+}CO_3^{2-}$
Lead(II) nitrate(V)	—	$Pb^{2+}(NO_3^-)_2$
Aluminium sulphate(VI)	—	$Al_2^{3+}(SO_4^{2-})_3$

8.6. Calculating the empirical formula of a compound

The empirical formula of any compound may be found very easily if its composition by mass is known. The following examples show how the calculation is carried out.

Example A. Calcium carbonate

Experiment shows that in 100 g of calcium carbonate there are 40 g of calcium, 12 g of carbon and 48 g of oxygen. The relative atomic masses of calcium, carbon and oxygen are 40, 12 and 16 respectively.

We can thus calculate the number of moles of the elements in 100 g of the compound:

$$\text{for calcium } \frac{40}{40} \text{ or } 1$$

$$\text{for carbon } \frac{12}{12} \text{ or } 1$$

$$\text{for oxygen } \frac{48}{16} \text{ or } 3.$$

As we saw when calculating the empirical formulae of the oxides of copper (see Section 8.3), one mole of any element always contains the same number of atoms. Thus for each one calcium atom in calcium carbonate there is one carbon atom and three oxygen atoms and the empirical formula is $CaCO_3$.

Example B. Sodium thiosulphate(VI) ('hypo')

Each 100 g of this compound contains 18.55 g of sodium, 25.80 g of sulphur, 19.35 g of oxygen and 36.30 g of water. We can, as before, calculate the number of moles of the elements, sodium, sulphur and oxygen. We can also calculate the number of moles of water by dividing the actual mass of water by its mole mass. The

85

mole mass of a compound is the sum of the mole masses of the atoms in one molecule of the compound:

$$\text{for sodium } \frac{18.55}{23} \text{ or } 0.808$$

$$\text{for sulphur } \frac{25.80}{32} \text{ or } 0.808$$

$$\text{for oxygen } \frac{19.35}{16} \text{ or } 1.210$$

$$\text{for water } \frac{36.30}{18} \text{ or } 2.018.$$

If we take each of these results and divide by the smallest of the numbers, then we should get the simplest ratio:

$$\text{for sodium } \frac{0.808}{0.808} \text{ or } 1.0$$

$$\text{for sulphur } \frac{0.808}{0.808} \text{ or } 1.0$$

$$\text{for oxygen } \frac{1.210}{0.808} \text{ or } 1.5$$

$$\text{for water } \frac{2.018}{0.808} \text{ or } 2.5.$$

But, it is impossible to have half an atom, or half a molecule! So the simplest *whole number* ratio is clearly 2, 2, 3 and 5 and the empirical formula of sodium thiosulphate(VI) is $Na_2S_2O_3.5H_2O$.

8.7. The chemical equation

A chemical equation is simply a shorthand way of indicating the changes that take place in the combinations of the atoms, ions and molecules in the course of a chemical reaction. We use the word 'equation' because, as follows from the law of conservation of matter, exactly the same numbers of atoms of each element present in the **reactants** (reacting substances) must also be present in the products of the reaction. The law of constant composition enables us to write – in most cases – the formulae of the substances with certainty.

It is very important to realize what information may be conveyed by a chemical equation and, perhaps, even more important to realize what information is *not* conveyed. For example, while an equation can show the formulae of the reactants and the products, it does not tell us, unless we so indicate by special means, the conditions under which the reaction takes place – whether heat or light is needed to activate the change, whether increased pressure is used or whether or not a catalyst (see Section 14.3) is required.

We can, by using suitable letters, show the physical states of the substances, and this is often very useful. For example, $H_2O(g)$ means water as a gas or vapour; $H_2O(l)$ means liquid water and $H_2O(s)$ means solid water (ice). The letters (aq), from the Latin *aqua* (water), after a formula mean that the substance is dissolved in water. Thus $Na^+Cl^-(aq)$ means sodium chloride dissolved in water. We can see also from this last example that it is very easy to indicate an ionic compound by showing the electric charges that the ions carry and by using the empirical formula.

It is also important to understand that **an equation can only be used to represent what actually happens in an actual chemical change**. Just because it is possible to write an equation, it does not follow that the change represented can actually take place.

8.8. Writing chemical equations

It is necessary to know what happens in the change concerned and first to express it in words. The formulae for *one* molecule (or one empirical formula, if the compound is ionic) of each substance should then be written down. Charges on ions may be put in if useful, but this is not essential. Again, the physical states of the substances are sometimes left out. In order to write the correct formulae of the substances, a knowledge of the symbols concerned and the valencies of the atoms or groups of atoms is needed (see Section 7.5).

Example A. The action of heat on chalk

Chalk is calcium carbonate and, when it is heated, it breaks up to give lime (calcium oxide) and carbon dioxide gas. These facts should be put in the form of a word equation:

calcium carbonate → calcium oxide + carbon dioxide

The formula (or empirical formula) for each substance is now written:

$$Ca^{2+}CO_3^{2-}(s) \rightarrow Ca^{2+}O^{2-}(s) + CO_2(g)$$

or, more simply:

$$CaCO_3(s) \rightarrow CaO(s) + CO_2(g)$$

The arrow is used to indicate in which direction the change is taking place. It used to be common to use an 'equals' symbol in place of the arrow, but the arrow is greatly to be preferred. In the case of reactions that can go in either direction according to changes in the conditions, that is **reversible reactions**, a double arrow may be used: $\rightleftharpoons$. The 'plus' symbol between formulae signifies 'together with'.

When the formula for each substance has been written, the next step is to count up all the atoms of each kind on either side of the arrow, to see if the numbers are equal. In the example given, this is so, and the equation is correct as it stands.

Example B. The action of heat on potassium chlorate(V)

When potassium chlorate(V) is heated, it breaks up to give solid potassium chloride and oxygen gas is liberated:

$$\text{potassium chlorate(V)} \rightarrow \text{potassium chloride} + \text{oxygen}$$
$$KClO_3(s) \quad \rightarrow \quad KCl(s) \quad + O_2(g)$$

If we now count up the numbers of each kind of atom on either side of the arrow, we find that we have one potassium atom and one chlorine atom on either side, but, whereas there are three oxygen atoms on the left-hand side, there are only two on the right. The reason for putting O_2, and not just 'O' on the right-hand side is that oxygen atoms always go in pairs when they are not combined with other elements.

The balance of the numbers of oxygen atoms can be put right by doubling the number of potassium chlorate(V) units and by trebling the number of oxygen molecules:

$$2KClO_3(s) \rightarrow KCl(s) + 3O_2(g)$$

If we now check again, we find that while the oxygen atoms are now in balance we now have two potassium atoms and two chlorine atoms on the left-hand side, but only one of each on the right-hand side. Doubling the number of potassium chloride units puts this right:

$$2KClO_3(s) \rightarrow 2KCl(s) + 3O_2(g)$$

The equation is now correctly balanced.

Example C. The burning of ethane gas in air

Ethane is a hydrocarbon (see Chapter 33). When it is burned in air, the products are water and carbon dioxide gas:

$$\text{ethane} \quad + \quad \text{oxygen} \rightarrow \text{water} \quad + \quad \text{carbon dioxide}$$
$$C_2H_6(g) \quad + \quad O_2(g) \quad \rightarrow \quad H_2O(l) \quad + \quad CO_2(g)$$

As there are two carbon atoms and six hydrogen atoms on the left-hand side, there must be at least three water molecules and two carbon dioxide molecules on the right-hand side:

$$C_2H_6(g) + O_2(g) \rightarrow 3H_2O(l) + 2CO_2(g)$$

A count of the number of oxygen atoms now shows two on the left and seven on the right. As oxygen atoms always go around in pairs, it is necessary to multiply the number of oxygen molecules on the left-hand side by 3.5:

$$C_2H_6(g) + 3.5O_2(g) \rightarrow 3H_2O(l) + 2CO_2(g)$$

The equation in this form is quite correct, and there is no real need to go further, but it may be considered desirable to get rid of the decimal on the left-hand side. This is done by 'doubling-up' the whole equation:

$$2C_2H_6(g) + 7O_2(g) \rightarrow 6H_2O(l) + 4CO_2(g)$$

Practice makes perfect! You may care to try your hand at balancing the following equations:

a. $NaOH(aq) + H_2SO_4(aq) \rightarrow Na_2SO_4(aq) + H_2O(l)$
b. $CuO(s) + HCl(aq) \rightarrow CuCl_2(aq) + H_2O(l)$
c. $NH_4Cl(aq) + NaOH(aq) \rightarrow NaCl(aq) + NH_3(g) + H_2O(l)$
d. $NaHCO_3(s) \rightarrow Na_2CO_3(s) + CO_2(g) + H_2O(g)$
e. $Pb(NO_3)_2(s) \rightarrow PbO(s) + NO_2(g) + O_2(g)$
f. $Fe_2O_3(s) + CO(g) \rightarrow Fe(s) + CO_2(g)$
g. $CO_2(g) + C(s) \rightarrow CO(g)$
h. $C_2H_4(g) + O_2(g) \rightarrow CO_2(g) + H_2O(l)$
i. $NH_3(g) + CuO(s) \rightarrow Cu(s) + N_2(g) + H_2O(g)$
j. $NH_3(g) + O_2(g) \rightarrow NO(g) + H_2O(g)$
k. $FeSO_4(aq) + Al(s) \rightarrow Al_2(SO_4)_3(aq) + Fe(s)$

Test your understanding

1. State the laws of conservation of matter and constant composition.
2. Describe an experiment by which the law of conservation of matter might be tested.
3. State the law of multiple proportions.
4. Two oxides of tin have the following percentage compositions:
   oxide 1 tin 88.1%, oxygen 11.9%
   oxide 2 tin 78.8%, oxygen 21.2%
   By calculating the masses of tin that are combined with 16 g of oxygen in each of the two oxides, show that these figures illustrate the law of multiple proportions. (Relative atomic masses are: tin = 119, oxygen = 16.)
5. Nitrogen forms three oxides. The percentage compositions are:
   oxide 1 nitrogen 63.7%, oxygen 36.3%
   oxide 2 nitrogen 46.7%, oxygen 53.3%
   oxide 3 nitrogen 30.5%, oxygen 69.5%
   By calculating the masses of oxygen that are combined with 14 g of nitrogen in each of these oxides separately, show that the figures support the law of multiple proportions. (Relative atomic masses are: nitrogen = 14, oxygen = 16.)
6. Calculate the empirical formulae for the following compounds whose compositions are given below for 100 g of each compound:
   a. sulphur 40.05 g, oxygen 59.95 g
   b. calcium 36.19 g, chlorine 63.81 g

c. nitrogen 22.23 g, oxygen 76.17 g, hydrogen 1.60 g

d. carbon 12.15 g, oxygen 16.17 g, chlorine 71.68 g

e. carbon 37.48 g, calcium 62.52 g

f. carbon 7.81 g, chlorine 92.19 g

g. nitrogen 35.00 g, oxygen 59.96 g, hydrogen 5.04 g

h. sodium 20.70 g, oxygen 50.40 g, sulphur 28.90 g

(Relative atomic masses are: hydrogen $= 1$, carbon $= 12$, nitrogen $= 14$, oxygen $= 16$, sodium $= 23$, sulphur $= 32$, chlorine $= 35.5$, calcium $= 40$.)

7. Explain the meaning of each of the following in terms of the numbers of moles involved. For example, $2HCl$ would be explained as 'two moles, each containing one mole of hydrogen atoms and one mole of chlorine atoms'.

a. $2Na$, b. Cl_2, c. $3NH_3$, d. Na_2SO_4, e. $6NaCl$, f. NH_4Cl, g. $C_{12}H_{22}O_{11}$, h. $Ca(HCO_3)_2$, i. $2Na_2CO_3.10H_2O$

8. Explain exactly what information is conveyed by the following equations:

(i) $CaCO_3(s) + 2HCl(aq) \rightarrow CaCl_2(aq) + CO_2(g) + H_2O(l)$

(ii) $Fe(s) + CuSO_4(aq) \rightarrow Cu(s) + FeSO_4(aq)$

(iii) $2Pb(NO_3)_2(s) \xrightarrow{heat} 2PbO(s) + 4NO_2(g) + O_2(g)$

(iv) $H_2O_2(aq) \xrightarrow{MnO_2} H_2O(l) + \frac{1}{2}O_2(g)$

(v) $CuSO_4(s) + 5H_2O(l) \rightarrow CuSO_4.5H_2O(s)$

(vi) $NH_4Cl(s) \underset{cool}{\overset{heat}{\rightleftharpoons}} NH_3(g) + HCl(g)$

Chapter 9

Air and Oxygen

The ancient Greek philosophers believed that all material substances were made up of four '*elements*'. These elements are not, of course, to be confused with our modern chemical elements. The four elements of the Greeks were **fire, air, earth and water**.

For some 2 000 years, air was thought to be a single substance and it was not until the seventeenth century that, largely due to the work of the Englishman, John Mayow, the single nature of air was questioned.

9.1. Oxygen in the air

Investigation 9a

Take a glass tube, about 1 m long and 25 mm in diameter, and seal one end. This is best done by heating the glass in a bunsen flame until it softens and then sealing the end off. However, a well-fitting rubber bung will do to block the end if sealing by heating is not found convenient.

Fill the tube with water and then empty the water out again. This is done in order to make sure that the tube is filled with fresh air before the investigation starts. While the walls of the tube are still wet, sprinkle iron filings into it so that some of them stick to the walls of the tube near the sealed end. Invert the tube over some water in a bowl and adjust the levels of the water inside and outside the tube so that they are equal. This is best done by inserting a length of rubber tubing into the open end and carefully sucking out some air (see Figure 9.1).

Now, measure the length of the air column above the water and write it down. Leave the tube to stand for three or four days, each day measuring the length of the air column above the water surface in the tube. Continue doing this until there is no further change. Then calculate the percentage rise of the water in the tube as follows:

$$\text{Length of air column at start} = a \text{ cm}$$
$$\text{Length of air column at end} = b \text{ cm}$$
$$\text{Rise of water} = (a - b) \text{ cm}$$

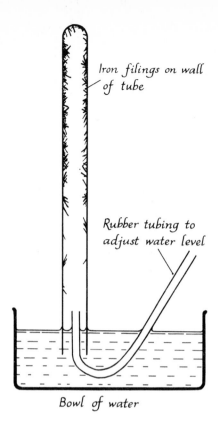

Iron filings on wall of tube

Rubber tubing to adjust water level

Bowl of water

Figure 9.1 Setting up the apparatus for Investigation 9a

Percentage rise is $\dfrac{(a-b)}{a} \times 100$

$$= \quad \%$$

As the tube has the same diameter all through its length (neglecting the small rounded part at the closed end, if the tube has been sealed by heating), the percentage rise of the water is exactly the same as the percentage of the air that has 'disappeared' during the course of the experiment. Examine the iron filings carefully. What do you notice?

Now, carefully put a cork, or rubber bung, into the open end of the tube while the open end is still under the water surface. Remove the tube from the water and invert it. Light a wooden splint, remove the cork and carefully put the lighted splint into the tube. What happens?

Experiments on these lines led scientists to realize that air contains at least two different substances, one of which allows substances to burn – that is it supports combustion – while the other does not support combustion.

The part of the air which supports combustion is the gas called **oxygen**; the remaining part of the air, which does not support combustion, is largely the gas **nitrogen**.

Investigation 9b. Another way of determining the percentage of
oxygen in the air

Take a glass tube of the same type as that used in Investigation
9a and make sure that it is filled with fresh air. Place about 30 mm
of water in the tube and add four or five pellets of sodium hydroxide. Now, add enough benzene-1,2,3-triol to make a heap on a new
halfpenny piece and *quickly* and *firmly* insert a rubber bung. Stand
the tube on the bench with the bunged end uppermost and
measure the length AB (see Figure 9.2). This is best done by using
the surface of the bench as a reference level and measuring from
the bench surface separately to A and B in order to find the
distance AB by subtraction. Then allow the liquid in the tube to
run backwards and forwards along the length of the tube for about
five minutes. What happens to the liquid?

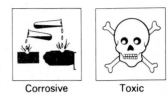

Corrosive   Toxic

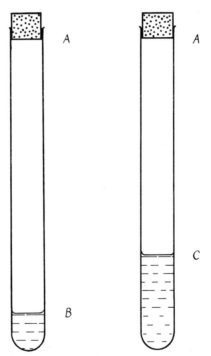

Figure 9.2 Making the
measurements for
Investigation 9b

When all the oxygen has been absorbed by the liquid, place
the stoppered end of the tube in a large bowl of water and gently
ease out the bung. You may find that this is quite hard to do. Why
is this so? What happens when the bung is removed? Take care
that, while you are removing the bung, the end of the tube does
not come out of the water in the bowl. Now, with the bung out of
the tube, level-up the water inside and outside the tube (see
Figure 9.3). Replace the bung. Stand the tube on the bench as
you did when taking the first readings and measure the length AC.

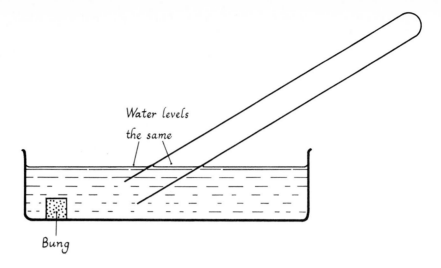

Figure 9.3 Levelling-up   Bung

The percentage of oxygen in the air is given by the expression:

$$\frac{(AB - AC)}{AB} \times 100$$

9.2. The composition of the air

Today we believe that air is a mixture of many gases and Table 9.1 gives the approximate composition by volume of *dry* air.

TABLE 9.1. THE COMPOSITION OF THE AIR

Gas	Volume Percentage
Nitrogen	78.08
Oxygen	20.95
Argon	0.93
Carbon dioxide	0.03
Neon, helium, krypton, xenon	0.002

In addition to these gases, there are traces of many others. In particular, near towns there are sulphur dioxide and hydrogen sulphide, together with some carbon monoxide. Near the sea, traces of salt are to be found. Dust is always present, the amount varying from small quantities in moist air near the sea or other areas of water, to very large quantities near industrial towns.

Water is always present in large or small amounts, but, as the quantity is variable, it is normally omitted from the percentage composition. The amount of water vapour actually present in the air at any given time, compared with the maximum amount that the air *could* hold, is known as the **relative humidity**.

94

Living organisms – bacteria, both helpful and harmful, and the spores of fungi – are also present.

9.3. The industrial extraction of oxygen from the air

A very large, and ever increasing, tonnage of oxygen is extracted from the air. The process used is to change the air to a liquid by cooling it and then to separate the substances present by a process of fractional distillation (see Section 20.6).

When a gas is compressed, it becomes hot. You will probably have noticed this fact if you have ever pumped up a bicycle tyre. On the other hand, when a compressed gas is allowed to expand, it becomes cool.

Investigation 9c. The cooling of a compressed gas on expansion

If your school has a cylinder of compressed gas (oxygen, nitrogen or carbon dioxide), the cooling effect can be shown by ,opening the valve for a short time and allowing some of the gas to escape freely. The metal end of the cylinder near the valve will become very cold.

★ *Warning – this should only be done by the teacher.*

It is possible to obtain solid carbon dioxide from a cylinder of the gas. A bag of loosely woven cloth is held round the outlet of the cylinder and the tap is fully opened for a few seconds. If the cylinder is of the normal type, it should be inverted; if it is of the syphon type, it should be kept upright.

In the liquefaction of air, air is compressed to about 150 times its normal pressure. Carbon dioxide is removed by absorbing it in sodium hydroxide solution. The air is then dried. The gases are cooled back to normal temperature by passing them through pipes over which cold water is flowing, and are cooled still further by meeting some of the cooled gases from the apparatus. The cooled gases are then allowed to expand and, in so doing, are made to work an 'expansion engine'. By these means, the temperature falls to about 110 K (about $-160\,°C$) at a pressure of about six times normal. Final expansion causes the gases to liquefy. The nitrogen, which boils at 77 K ($-196\,°C$), is boiled off. This leaves the oxygen, which boils at 90 K ($-183\,°C$), as a liquid.

9.4. The large-scale uses of oxygen

One of the important uses of oxygen today is in the manufacture of steel. For this purpose, 'tonnage' oxygen is produced at a low cost per tonne. Other uses include the treatment of patients suffering from diseases of the lung, such as pneumonia, and in heart disease in order to reduce the strain. Oxygen is also used by pilots flying at high altitudes where there is little oxygen, and by astronauts where there is no atmosphere at all. Again, oxygen is used to obtain the high temperatures needed for melting and welding metals. This is done by burning hydrogen or ethyne in oxygen (the oxy-hydrogen and oxy-ethyne flames).

9.5. Oxygen in the laboratory

Investigation 9d. The laboratory preparation of oxygen

There are many ways in which oxygen gas may be prepared in the laboratory, and two such methods are given in detail. Other methods will be indicated briefly.

Method 1

By far the simplest – and safest – way to prepare oxygen is by liberating the gas from hydrogen peroxide solution. Hydrogen peroxide is a relatively unstable substance and easily breaks down to give oxygen and water:

$$2H_2O_2(l) \rightarrow 2H_2O(l) + O_2(g)$$

The breakdown of the compound can be slowed by the addition of certain substances, such as propane-1,2,3-triol. Substances that behave in this manner and slow down the rate of a chemical change are called **inhibitors**.

On the other hand, it is possible to find substances that speed up chemical changes. In the case of the decomposition of hydrogen peroxide, the black solid, manganese(IV) oxide, is a particularly powerful agent. Substances acting like manganese(IV) oxide in speeding up chemical changes are known as **catalysts.** We shall learn more about catalysts and inhibitors in Chapter 14.

Oxidising

Harmful

Set up the apparatus shown in Figure 9.4. Put two or three grams (the actual amount is not important) of manganese(IV) oxide into the flask. Add '20 volume' hydrogen peroxide solution through the thistle funnel so that oxygen is liberated at the rate you require. There is no need for heating. The gas given off is collected in the gas jar 'over water' – oxygen does not dissolve to a great extent in water. The term '20 volume' as applied to the hydrogen peroxide solution means that each unit of volume of the solution can give twenty times that volume of oxygen, the volume being measured at 273 K (0 °C) and at a pressure of one atmosphere.

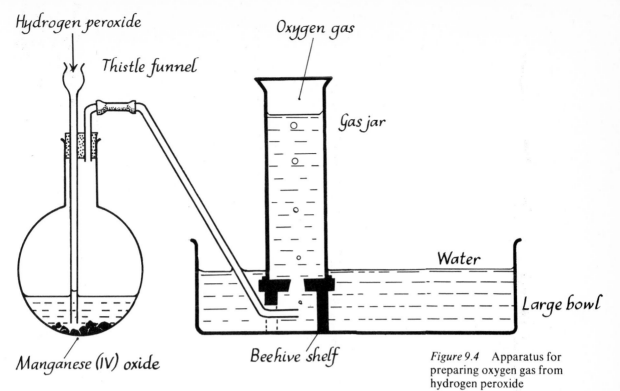

Hydrogen peroxide

Thistle funnel

Oxygen gas

Gas jar

Water

Large bowl

Beehive shelf

Manganese (IV) oxide

Figure 9.4 Apparatus for preparing oxygen gas from hydrogen peroxide

Each gas jar is filled with water in the bowl and is then turned upside-down over the beehive shelf. Greased glass discs may be slipped over the open end of each jar when it is full of the gas so that the jar may be removed from the bowl for the purpose of testing the properties of the gas.

Method 2

In this method, oxygen is prepared by the action of heat on potassium chlorate(V):

$$2KClO_3(s) \rightarrow 2KCl(s) + 3O_2(g)$$

Again, a catalyst is used. It is manganese(IV) oxide. About one part of this substance is used for each eight parts of potassium chlorate(V). The apparatus is shown in Figure 9.5, the method for collecting the gas being exactly the same as for Method 1.

Oxygen may also be prepared in the laboratory by heating sodium nitrate(V) or potassium nitrate(V):

$$2NaNO_3(s) \rightarrow 2NaNO_2(s) + O_2(g)$$
$$2KNO_3(s) \rightarrow 2KNO_2(s) + O_2(g)$$

Oxygen was first prepared in Great Britain by Joseph Priestley in 1774 by heating mercury(II) oxide:

$$2HgO(s) \rightarrow 2Hg(l) + O_2(g)$$

Oxidising

Explosive

★ *Warning – mixtures of potassium chlorate(V) with other substances are dangerous.*

★ *Warning – on no account should this method be tried in your own laboratory – the fumes from mercury are extremely poisonous.*

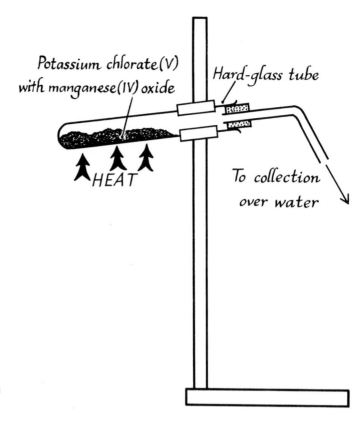

Figure 9.5 The preparation of oxygen from potassium chlorate(V)

Investigation 9e. The properties of oxygen

When considering the properties of a gas, it is convenient to consider the physical properties and the chemical properties. There is a very simple way by which we can remember eight of the most important properties about which we should always ask. If we take the word COWSLIPS, we may use the letters of the word to remind us of the initial letters of the eight properties:

1. C for **colour**. Does the gas have any colour?
2. O for **odour**. Does the gas have any smell?
3. W for **weight**. What is the density of the gas?
4. S for **solubility**. Does the gas dissolve in water?
5. L for **litmus**. Does the gas have any action on litmus solution?
6. I for **inflammability**. Does the gas burn in air?
7. P for **poisonous**. Is the gas poisonous?
8. S for **supporter of combustion**. Will other substances burn in the gas?

The first four of these properties are physical in character; the second four are chemical.

1. *Colour*. The colour of a gas is easiest to see if the greatest possible depth is observed. For example, if the gas is in a test-tube, it is better to observe the colour by looking down the length of the tube rather than by looking across the breadth of the tube.

2. *Odour*. When testing for the smell of a gas, great care must be taken not to get too much of the gas into the nostrils or the lungs. The test-tube or jar containing the gas should be held about 100 mm from the nose and the hand should be used to impel a small quantity of the gas towards the nose while the mouth is kept shut. Very gentle sniffing is sufficient. It is important that the smells of the different gases should be learned; the smell is one of the most useful ways to identify a gas.

3. *Weight*. There is no *simple* way of determining the density of a gas. We normally use **relative density** for the purpose of comparing gas densities. The relative density of a gas is its density compared with the density of the least dense gas known, which is hydrogen. On this scale, the density of hydrogen is taken as unity. It is useful to remember that the relative density of air is 14.4.

4. *Solubility*. To test a gas for solubility, take a boiling tube (150 mm × 25 mm) of the gas, with the end closed with a cork or bung. Open the tube under some water in a bowl and move the tube about so that the water can 'slop up' inside it, taking care not to let the end of the tube come out of the water. Note how quickly, if at all, the gas is dissolved in the water.

5. *Litmus*. Use litmus paper for this test. Take a small piece of the paper, damp it well with water, and hold it in the gas. It is best to use a pair of tongs to hold the paper while the test is being carried out, so that the fingers do not come into close contact with the gas. Note the effect of the gas on the paper. Both blue and red papers may be used, or neutral papers (purple) may be obtained. It is best to use blue litmus paper first. Note any change in the colour of the paper. A change from blue to red shows an acid gas; red to blue shows an alkaline gas. Some gases **bleach** the litmus paper, that is they remove the colour from it.

6. *Inflammability*. Take a test-tube of the gas and bring a lighted wood splint up to the open end. Take care that the gas does not have time to escape from the tube while you are bringing up the splint!

7. *Poisonous character*. It is, of course, not possible to make any simple test for this. Again, it is very important to learn which gases are dangerous.

8. *Supporter of combustion*. This is best tested by taking a lighted splint and placing it in a test-tube full of the gas. In the case

of very active supporters of combustion, a *glowing* splint may re-light on being put into the gas.

Carry out these tests on samples of oxygen gas collected in Investigation 9d. Pay particular attention to Test 8.

9.6. Oxides

Investigation 9f. Burning different substances in oxygen

Collect a number of gas-jars full of oxygen gas and carry out the following experiments by burning the substances referred to in Table 9.2. Have greased glass discs ready to close the jars when the burning is completed, so that the products of combustion cannot escape. Then add about 10 cm³ of water to the jar, and make sure that the products come into close contact with the water. Now, add some litmus solution (or some universal indicator solution) to the jar and note any colour changes. For the interpretation of all the indicator colours, see Chapter 13 on 'Acids', but, for the purposes of this investigation, it is only necessary to note whether the litmus goes red or blue, or whether the universal indicator is red/orange/yellow (showing an acid), or green/blue/purple (showing an alkali). Tabulate your results as shown in Table 9.2.

TABLE 9.2.   HOW TO SET OUT THE RESULTS FOR INVESTIGATION 9f

Substance	Indicator Colour	Acid/Alkali
Sulphur		
Magnesium		
Iron		
Carbon		
Phosphorus (*red*)		
Zinc		

Harmful

You will find it convenient to use powdered materials, and to heat them in deflagrating spoons. The spoons should be cleaned thoroughly between each test by heating in the bunsen flame, but care is needed to see that the end of the spoon is not loosened by over-heating.

Try also to burn a candle in oxygen and to test the gaseous products with indicator. The candle can be supported on a bent piece of copper wire (see Figure 9.6).

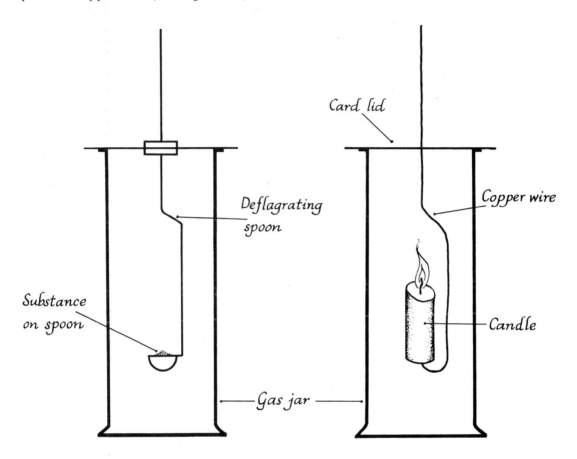

Figure 9.6 Burning substances in oxygen

The products formed when elements burn in oxygen are called **oxides**. An oxide which, when dissolved in water, turns litmus from blue to red, is called an **acidic oxide.** The solution of the oxide in water produces an acid (see Chapter 13). Which elements in Investigation 9f have formed acidic oxides? The equations are:

$$S(s) + O_2(g) \rightarrow SO_2(g)$$
$$2Mg(s) + O_2(g) \rightarrow 2MgO(s)$$
$$3Fe(s) + 2O_2(g) \rightarrow Fe_3O_4(s)$$
$$C(s) + O_2(g) \rightarrow CO_2(g)$$
$$4P(s) + 5O_2(g) \rightarrow 2P_2O_5(s)$$
$$2Zn(s) + O_2(g) \rightarrow 2ZnO(s)$$

You will notice that oxides contain only one other element in addition to the oxygen.

9.7. The classification of oxides

As we have just seen, an acidic oxide is one which dissolves in water to form an acid. The opposite of an acidic oxide is a **basic oxide**. Most basic oxides are insoluble in water, but they all have the power of **neutralizing** acids.

Investigation 9g. The neutralization of an acid

Harmful

★ *Warning – these compounds are dangerous and need careful handling.*

About one-third fill a boiling-tube (150 mm × 25 mm) with water and add a drop or two of litmus solution, followed by *one* drop of dilute hydrochloric acid so that the solution is red in colour. Now, sift a little magnesium oxide powder into the liquid and continue sifting in the oxide until there is no further change in colour. Repeat the experiment, using zinc oxide. What colour changes can be noticed in each case?

Only a few basic oxides are soluble in water, and these first react with water to form **hydroxides**. Such soluble substances are known as **alkalis**, and their solutions turn litmus from red to blue. The chief alkalis are the oxides (or hydroxides) of sodium, potassium and calcium:

$$Ca^{2+}O^{2-}(s) + H_2O(l) \rightarrow Ca^{2+}(OH^-)_2(s)$$
$$Na^+_2O^{2-}(s) + H_2O(l) \rightarrow 2Na^+OH^-(s)$$
$$K^+_2O^{2-}(s) + H_2O(l) \rightarrow 2K^+OH^-(s)$$

In addition to the acidic and basic oxides, there are some oxides which, under the right conditions, can behave either like acidic oxides or like basic oxides. They are known as **amphoteric oxides**. Zinc oxide and aluminium oxide are of this type. In the next section, we shall also meet **super-oxides**.

9.8. Some important oxides

Further study of the acidic oxides will be carried out in Chapter 13. Let us now investigate the properties of the metallic oxides, including those of calcium, magnesium, zinc, iron, lead and copper.

Investigation 9h

Toxic

Carry out the following tests, using calcium oxide, magnesium oxide, zinc oxide, iron(III) oxide, lead(II) oxide, lead(IV) oxide and copper(II) oxide.

1. Heat a small quantity of each oxide in a 75 mm × 10 mm test-tube. Note any colour changes and the colour of the cold residue. In each case, test for the evolution of oxygen gas (glowing splint). Set out your results as shown in Table 9.3.
2. Test a little of each oxide with some dilute (3M) hydrochloric acid. Fill about one-third of a test-tube (150 mm × 19 mm) with the acid and add a very small amount of the oxide. Shake well

102

and gently warm the tube. Does the oxide dissolve? List your results, making a table on the lines shown in Table 9.3 for Test 1.

TABLE 9.3. HOW TO SET OUT THE RESULTS FOR INVESTIGATION 9h

Oxide	Colour Hot	Colour Cold	Oxygen Evolved?
Calcium oxide			
Magnesium oxide			
Zinc oxide			
Iron(III) oxide			
Lead(II) oxide			
Lead(IV) oxide			
Copper(II) oxide			

3. Test a little of each oxide with some dilute (3M) sodium hydroxide solution, using the same technique as for the dilute acid. Take care, when warming the solution, as sodium hydroxide solution is very liable to 'bump' when it is heated. Note whether the oxide is soluble in the sodium hydroxide solution. Again, tabulate your results.

Corrosive   Harmful

4. Put about 1 cm³ of *concentrated* hydrochloric acid into a small test-tube. Add a small quantity of the oxide and warm very gently. *Cautiously* smell the gas evolved (if any) and hold a piece of moistened, 'starch-iodide' paper in the fumes. Is any chlorine gas evolved? (If chlorine gas is evolved, the starch-iodide paper will rapidly turn dark blue. The smell of the gas is also very distinctive). Tabulate your results.

★ *Warning – care is needed.*
★ *Warning – chlorine is a very poisonous gas.*

Oxides which give off oxygen when heated, and which can liberate chlorine from hydrochloric acid, are rich in oxygen. They are called **oxidants** and may be classed as super-oxides. Those oxides which dissolve in dilute hydrochloric acid, but do not dissolve in sodium hydroxide solution, are basic oxides. Those oxides which dissolve both in dilute hydrochloric acid and in sodium hydroxide solution are amphoteric oxides.

As a result of your investigations, to which of these classes do the oxides you have been testing belong? What type of element forms acidic oxides? What type of element forms basic oxides?

Another interesting point is that metals may often be identified by the colours of their oxides. The difference in colour (if any)

103

between the colour of the oxide when it is hot and when it is cold is an important additional help here.

Test your understanding

1. Describe in detail how you would show that air contains at least two gases.
2. How would you find the percentage of oxygen gas in air? Give a fully labelled diagram of the apparatus you would use.
3. What is the normal composition by volume of dry air? Suggest TWO places where the composition might differ somewhat from the normal.
4. What impurities are likely to be found in air in and around towns?
5. What do you understand by the term 'relative humidity'?
6. Explain in outline how oxygen gas may be obtained from air on a large scale.
7. What are the chief large-scale uses of oxygen?
8. 'The relative density of oxygen is 16, while that of nitrogen is 14.' Explain the meaning of this statement.
9. Explain the meaning of the terms: (i) catalyst, (ii) inhibitor.
10. How would you prepare and collect a sample of oxygen gas in the laboratory? Give a fully labelled diagram of the apparatus you would use, name the chemicals you would need and give the equation for the reaction.
11. Name FOUR substances which give off oxygen gas when they are heated.
12. You are given a sample of a white powder which you suspect to be a basic oxide. Explain how you would test the sample to see if your suspicion is correct.
13. A chocolate-brown powder gives off oxygen gas when heated in an ignition tube. When warmed with a little concentrated hydrochloric acid, a gas is evolved with a typical smell. The gas turns moistened 'starch-iodide' paper a very dark blue. To what class of oxide does the substance belong? What is the gas given off with the hydrochloric acid?
14. Name, and give the formulae of, THREE acidic oxides.
15. What is an 'alkali'? Name THREE alkalis and give their formulae.
16. An oxide is found to dissolve in dilute hydrochloric acid and also to dissolve, on warming, in sodium hydroxide solution. To what class of oxide does it belong? Name TWO oxides that it could be.

Chapter 10

Corrosion

10.1. The rusting of iron

In Investigation 9a, we used the reaction of iron with oxygen in order to determine the percentage of oxygen in the air. On examination, the iron filings, originally a grey colour, were found to be covered with a brownish deposit. This deposit is known as **rust**. We must now investigate this reaction of iron in much greater detail.

Investigation 10a. The conditions for the rusting of iron

Take four test-tubes (150 mm × 25 mm) and a supply of clean bright iron nails.

Tube 1. Put a few small lumps of anhydrous calcium chloride into the bottom of the tube. On top of these, place a small pad of cotton wool. Then put in three or four of the nails. Finally, close the tube tightly with a rubber bung or a good cork.

Tube 2. Put a small volume of water into this tube and stand three or four nails in the water so that part of each nail is submerged and part is above the surface.

Tube 3. Fill this tube almost completely with water, after first having put three or four nails in the tube. Make sure that the nails are well below the surface of the water.

Tube 4. Put some pure water (distilled or de-ionized) into this tube, together with three or four nails, so that the nails are completely submerged. Now, boil the water gently for about five minutes. Make sure that when the boiling is completed the nails are still well under the surface. Cool the tube by allowing cold water to run over the outside, and immediately pour about 10 mm depth of medicinal paraffin on to the surface of the water in the tube. Then close the tube tightly with a rubber bung or a good cork. If no medicinal paraffin is available, the tube should be closed while the water in it is still very hot, care being taken to leave a space of about 10 mm between the water surface and the bottom of the stopper. Figure 10.1 shows the arrangement of the tubes.

Leave the tubes to stand in the rack for several days, examining them at regular intervals. Note all the changes that take place. Note where the greatest rusting has taken place on the nails. What conclusions can you draw from this investigation? (Hint – the water in tube 4 was boiled in order to remove any dissolved air.)

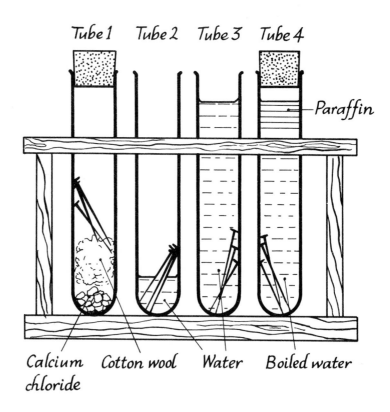

Figure 10.1 The tubes for Investigation 10a

10.2. The results of Investigation 10a

Your observations of the results of Investigation 10a will probably have been along the following lines:

Tube 1. There are no signs of rusting and the nails remain bright.

Tube 2. There is considerable rusting, particularly where the water surface cuts across the nails.

Tube 3. There are signs of considerable rusting having taken place.

Tube 4. Here the results will depend on how well you managed to remove all the air from the water. If the work has been done well, there will be little, if any, signs of rusting. If you failed to remove all the air, then there will be some signs of rusting, but the total amount will be far less than that seen in tube 3.

The clear conclusion to be drawn from this investigation is that *both air and water* are necessary for the rusting of iron to occur.

10.3. What is iron rust?

Investigation 9a showed that, when rust appears on iron, oxygen disappears from the air. Investigation 10a has shown that water also is needed.

Investigation 10b. The nature of iron rust

Scrape a quantity of the brownish substance from the surface of a piece of rusted iron into a small hard-glass test-tube. Heat the powder quite strongly and observe what happens. Test any liquid drops that form on the cooler parts of the tube with some anhydrous copper(II) sulphate(VI). If the white powder is turned blue, then the drops are shown to contain water (see page 55).

Harmful

Iron rust consists of a compound of iron and oxygen – iron(III) oxide – combined with some water. Its composition is variable, depending on the conditions under which it is formed.

10.4. The cost of the rusting of iron

The rusting of iron costs this country alone many millions of pounds each year, and the protection of iron against rusting is of great importance. There are a number of ways in which this can be done.

   a. *The use of paint*. This is the most widely used method and simply involves the exclusion of both air and water from the metal. It is necessary to renew the paint as it becomes worn away. It is a well-known fact that the Forth Rail Bridge, for example, requires the continual attention of a number of painters in order to keep it in good condition.

   Iron which has already rusted has to receive special treatment before it can be painted successfully. If this special treatment is not given, the rust will spread under the paint and the paint will flake off. Treatment may consist of first scraping off the loose rust and then of painting the iron with aluminium paint or with solutions containing phosphoric(V) acid. The aluminium acts by displacing iron from its oxide (see Section 16.2), while the phosphoric(V) acid converts the iron oxide of the rust into iron(III) phosphate(V). This compound forms a hard, compact mass and does not flake off. Indeed, one of the main problems with iron rust is that it so easily comes away from the surface and so exposes fresh metal to the air and water. Other metals, as we shall see, do rust, but in most cases the rust remains firmly attached to the metal and forms a protective layer.

b. *Greasing and oiling.* Paint protects iron from rusting, but, if the iron has to slide in contact with other metal parts, for example in an engine, the paint layer would soon wear off. In this situation, something else has to be used which will not wear off. Where the parts are large, grease is used. Where parts are smaller, oil will do. The grease and oil also reduce friction and allow parts to move more smoothly.

Grease is also used to protect the surfaces of new iron objects that have to be stored. This is done, for example, in the case of cooking pots. It is easier to remove the grease than it would be to remove paint. In the same way, oil can be used to protect the parts of a bicycle from rusting if it has to be put away in a damp place.

c. *Galvanizing and tin-plating.* Paint, oil and grease will all wear off in time and have to be replaced. If the iron is covered with a thin layer of another metal, it is possible to protect it for a longer time. Two metals which are used for this purpose are zinc and tin.

In **galvanizing**, the iron is first thoroughly cleaned. It is then dipped in a bath of molten zinc. When it is withdrawn, the surface is covered with a thin layer of zinc (see Figure 10.2). **Tin-plating** is carried out in the same way, using tin for the protective coating.

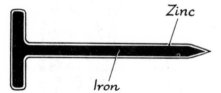

Figure 10.2 A nail that has been galvanized

If tin-plated iron is scratched, and a little tin is removed, the iron around the scratch will rust very rapidly. The rust spreads underneath the tin layer until all the iron has rusted. In the case of galvanized iron, even if the zinc is scratched off in parts, the iron is still protected and will not rust.

The reason for the difference will be understood if Section 16.2 is read: zinc comes before iron in the electrochemical or displacement series, whereas tin comes after iron. This means that any iron which might start to react with air and water to form iron rust will immediately be replaced by zinc. Tin, on the other hand, would be replaced by iron, and the rusting is thus accelerated. Figure 10.3 shows the difference in effect of scratches on galvanized and tin-plated iron.

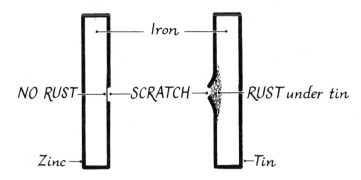

Figure 10.3 The difference between galvanized and tin-plated iron.

Tin does not, however, dissolve in food juices, and so it is used for covering 'tins' that are going to be used for storing food. Zinc, which is better for protection, is used for outdoor work, for example in galvanized sheeting and buckets.

d. *Electroplating.* It is also possible to deposit thin layers of other metals on iron by the use of an electric current (see Section 16.6). For this purpose, chromium, nickel, silver and cadmium are often used. It is possible to produce a very bright shiny finish, which is also wear-resistant and rustproof, by using the correct conditions for the electroplating.

e. *Alloying.* The whole topic of alloys is dealt with in Chapter 18. An alloy is a mixture of metals. By mixing in about 12 or 15 per cent of chromium with the iron, an alloy is produced which is very resistant to rusting. Alloys of this type are known as '**stainless steel**'. Some nickel is also added to improve the anti-rusting properties.

f. *The use of vapour phase inhibitors.* It has been found that certain compounds containing the element phosphorus will give off vapours that will prevent the rusting of any iron in the vicinity. These compounds are known as **vapour phase inhibitors** (or **V.P.I**s for short). Paper soaked in solutions of these substances and then dried out is used to give off the vapour; objects wrapped in such paper will not rust. Pieces of the paper placed near iron objects have the same effect; for example a piece of the paper placed in a tool-box or in a camera case will act as a rust preventer.

Investigation 10c. The effect of air and water on iron that has been treated in various ways

Take eight test-tubes (150 mm × 25 mm) and put about 30 mm

109

of tap water in the bottom of each. Place the tubes in a suitable rack with a piece of wood above the tubes. Hang the objects listed below, one in each tube, so that they are close to, but not touching, the surface of the water (see Figure 10.4).

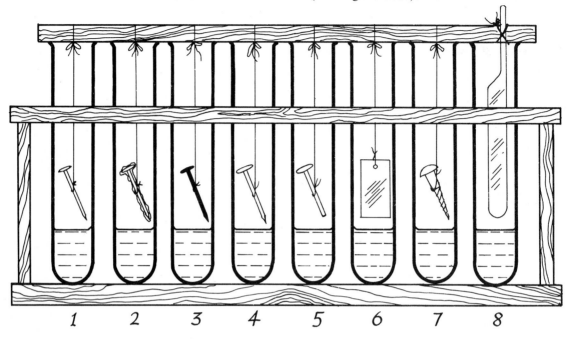

Figure 10.4 The arrangement of the tubes for Investigation 10c

Tube 1. A bright iron nail.
Tube 2. A nail that has been well greased.
Tube 3. A nail that has been painted.
Tube 4. A galvanized nail.
Tube 5. A galvanized nail with the point cut off.
Tube 6. A small piece of tin-plate, cut from a 'tin'.
Tube 7. A chromium-plated screw.
Tube 8. Support a small, stainless steel knife-blade in this tube.

Observe the changes (if any) that take place in the objects over the course of about two weeks.

10.5. The rusting of metals other than iron

We have seen that if iron is exposed to moist air it will rust and that the iron rust flakes off. If the same thing happened with zinc and tin, there would be little point in using these metals to cover the surface of the iron. We must now investigate how some other metals behave when they are placed in moist air.

Investigation 10d. The rusting of aluminium, zinc, lead and copper

Take four test-tubes (150 mm × 25 mm) and put about 30 mm of

110

tap water in the bottom of each. Stand the tubes in a rack. Cut strips of aluminium (milk bottle top, plain 'silver' colour), zinc, lead and copper, each about 50 mm × 10 mm. Polish them, so that the surfaces are shiny, and then place one strip in each tube so that it stands partly in and partly out of the water. Observe the changes that take place over a period of about one or two weeks.

10.6. Survey of the results of Investigations 10c and 10d

In Investigation 10c, the protection against rusting is good in tubes 2, 3, 4, 7 and 8. In tube 5, no rusting takes place where the iron has been exposed by cutting off the point of the nail. In tube 6, however, it can be seen that the tin-plated metal has rusted around the edges where the iron is exposed. The nail in tube 1 is used as a **control** to see how much rusting takes place in untreated iron during the time of the investigation.

In Investigation 10d, you will have noticed that the metal surfaces are no longer bright; they have become tarnished. If you rub the surfaces with your finger, you will find that the tarnish will not rub off. This is the way in which the rusting of these metals is different from the rusting of iron.

10.7. The use of aluminium, zinc, lead and copper in building

These metals have been seen to tarnish when exposed to moist air, but the surface film formed does not flake off. Because of this property, it is possible to use these metals without special treatment in the building industry. The film of rust, once it is formed, protects the underlying metal from further attack.

Aluminium

Because of its low density, aluminium is used for roofing. When alloyed with other metals, it is used for constructional work.

Zinc

We have seen that zinc is used to protect iron from rusting by the process of galvanizing. Galvanized iron is often used for roofing temporary buildings, barns, garden sheds, etc. Zinc, by itself, is often used for **flashings**. Figure 10.5 shows what a flashing is.

Lead

Lead has, in the past, been used for making roofs, but it is now so expensive that it is not often used for this purpose. Instead, it can be used, like zinc, for flashings. It is also used for 'leaded' lights in making decorative windows.

FLASHING

Figure 10.5 A zinc or lead flashing to a chimney stack

Copper

Copper is used for making water-pipes and for the roofs of buildings, particularly those with 'flat' roofs. Its rust has a pleasing pale-green shade.

10.8. Rusting, oxidation and reduction

When a substance, such as iron, combines with oxygen, it is said to be **oxidized**. When a substance which contains oxygen has some, or all, of its oxygen taken away, it is said to be **reduced**.

Thus, rusting is oxidation. Many metals (we have seen some above) undergo similar reactions to iron when placed under similar conditions. The change is always such that the metal goes back to the state in which it is found naturally.

Some metals – gold is an example – do not undergo this process of rusting; they may be found naturally as the free metal. It is for this reason, among others, that gold is used for making ornaments and trinkets. It retains its appearance without becoming tarnished or corroded. We use the word **corrosion** to describe the dulling and eating away of the surface of a metal brought about by the sort of changes we have been investigating in this chapter.

When any metal is extracted from one of its natural compounds, that compound has to be reduced. Thus, the extraction of any metal is a reduction process.

We saw in Section 5.1 that when sodium combines with chlorine the sodium loses an electron to the chlorine and becomes a

112

positively charged ion. In the same way, when iron reacts with oxygen, it gives up electrons to the oxygen:

$$2Fe(s) + 1\tfrac{1}{2}O_2(g) \rightarrow Fe_2{}^{3+}O_3{}^{2-}(s)$$

Thus, the oxidation of iron involves the iron losing electrons and this, indeed, is what happens in *all* oxidations – **the substance which is oxidized loses electrons**. The adding of oxygen to a substance is therefore only one sort of oxidation process.

In the same way, when oxygen is taken away from iron(III) oxide, the iron picks up electrons:

$$Fe_2{}^{3+}O_3{}^{2-}(s) \rightarrow 2Fe(s) + 1\tfrac{1}{2}O_2(g)$$

The term 'reduction' thus covers all cases in which a substance picks up electrons and is not limited just to the case where a substance loses oxygen.

It should be realized that an oxidation cannot take place without a corresponding reduction. The electrons which are removed have to find a home somewhere!

Test your understanding

1. What conditions are required for rusting to take place?
2. Describe an experiment to find out the conditions needed for iron to rust.
3. How would you show that iron rust contains water?
4. Give THREE methods of preventing moist air from getting to the surface of iron.
5. What is 'galvanized iron'? Why is it preferred to tin-plate for outside use?
6. What is tin-plate? For what purposes is it preferred to galvanized iron, and why?
7. What is 'stainless steel'?
8. Name FOUR metals that are often used to protect iron and may be deposited on the iron by using an electric current.
9. What is a 'vapour phase inhibitor'?
10. How would you first treat an iron surface that had started to rust before you painted it?
11. What important difference is there between iron rust and the rust of other metals, such as copper?
12. Outline the main uses of lead and zinc in building.
13. What is 'oxidation'? Give one example of a chemical change which is an oxidation.
14. What is the meaning of the word 'reduction'?
15. Why must every case of oxidation be accompanied by a corresponding reduction?

Chapter 11

Water and Metals

Water is one of the commonest substances. It occurs in vast quantities in seas and oceans; it falls as rain or snow. In this chapter, we shall make some preliminary investigations to try to discover something about the properties of water and its reactions with other common substances. You should remember that we have already seen that it is a very good solvent.

11.1. The action of cold water on metals

Whilst we expect water to dissolve sugar or salt, we should be very surprised to see a nail dissolved by water. However, let us examine the action of some metals on water.

Investigation 11a. The action of sodium on cold water

★ Warning – this investigation should be carried out by a teacher. A safety screen must be placed between the apparatus and the class, and safety spectacles should be worn.

Explosive

Flammable

Corrosive

Take a large bowl, glass if possible, at least 200 mm in diameter, and half fill it with water. *Using a pair of tongs*, carefully remove a piece of sodium from beneath the oil in the bottle. Dry the oil from the surface of the sodium with a filter paper, and cut off a small cube, of about 2 mm side. Look quickly at the freshly cut side of the cube and, at once, return the remainder of the sodium to the bottle. Take the small cube in the tongs and drop it on to the surface of the water in the bowl. Observe all that happens.

Make sure that the tongs are dry and then cut off another piece of sodium. Place the sodium on the water surface and this time hold a lighted splint over it. This will be rather difficult to do. *Your face should be kept well away from the water.*

When all the action has ceased, put a few drops of universal indicator into the water, or dip in a piece of red and a piece of blue litmus paper.

Answer these questions

1. Was the surface of the sodium shiny after cutting?
2. Was the surface of the large piece of sodium from the bottle shiny?
3. Why is the sodium kept under oil?
4. What gas is given off when sodium is put into water?
5. What colour does the indicator change to?
6. What does this colour show? (See Investigation 9f.)

Your teacher may repeat the experiment using a piece of potassium instead of a piece of sodium.

We should have seen that, like other metals, sodium has a shiny surface just after cutting, but that it soon becomes dull when exposed to the air. This is because it combines very readily with the oxygen in the air. It is for this reason that it is protected from the air by being kept under oil. When dropped on to water, it rushes over the surface at a great speed and becomes so hot that it melts. During the reaction, hydrogen gas is given off and the remaining liquid is a solution that gives an alkaline reaction with indicators.

If you were fortunate enough to see a piece of potassium put on to water, you will have noticed that it becomes so hot that a lilac coloured flame is produced. This lilac colour is characteristic of potassium and can be used as a test for potassium.

The new substance that is formed dissolves in water to give an alkaline reaction to indicators; it is called sodium hydroxide. A **hydroxide** is an oxide that has combined with some water. Hence we could write:

$$\text{sodium } + \text{ water } \rightarrow \text{ sodium hydroxide } + \text{ hydrogen}$$
$$2Na(s) + 2H_2O(l) \rightarrow 2NaOH(aq) + H_2(g)$$

You will remember that we found that some metals combine with oxygen to form oxides that, when dissolved in water, give an alkaline reaction with indicators.

Investigation 11b. The action of calcium on water

Take a glass beaker, at least 250 cm³, and half fill it with water. Now take a small test-tube and fill it with water. Hold your thumb firmly over the open end of the tube and turn it upside-down; place the open end of the tube under the water in the beaker and then remove your thumb. The tube should remain full of water. Drop two small pieces of calcium metal into the beaker and collect, in the tube, some of the gas produced. When the tube is full of gas, remove it from the beaker and, keeping the open end downwards, apply a lighted splint to the mouth of the tube. If the gas burns it is hydrogen. Test the water left in the beaker for its action on indicators.

Harmful

Flammable

Explosive

Notice that the hydroxide formed in this experiment, calcium hydroxide, appears as a white solid, because it is not very soluble in water:

$$\text{calcium } + \text{ water } \rightarrow \text{ calcium hydroxide } + \text{ hydrogen}$$
$$Ca(s) + 2H_2O(l) \rightarrow Ca(OH)_2(s) + H_2(g)$$

This investigation should be repeated with some other metals to see if they behave in the same way. Try to judge the reactivity of the metals by observing which react most readily.

11.2. The action of steam on metals

If the metals will not react readily with cold water, then we might try the effect of heating the metals in a flow of steam. Since we have seen that sodium, potassium and calcium react readily with cold water, we shall start by trying a different metal, magnesium.

Investigation 11c. The action of steam on magnesium

★ *Warning – this investigation should be carried out by the teacher.*

Take a soft-glass test-tube and heat a small part at the sealed end strongly in a flame, using a blowpipe if possible. When the end is hot remove the tube from the flame and gently blow, from the open end, so that a small hole is formed in the end of the test-tube, as shown in Figure 11.1. Then heat the tube in a cool flame for a few minutes, in order to prevent the glass cooling too quickly and hence cracking.

SMALL HOLE

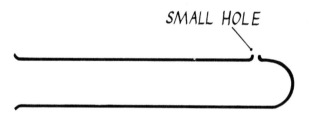

Figure 11.1   The test-tube needed for investigation 11c

Fix the prepared test-tube to a flask containing water, as shown in Figure 11.2. Adjust the clamps so that the tube is well supported.

Remove the test-tube. Scrape both sides of a 200 mm length of magnesium ribbon, to remove the dull surface, and loosely coil the ribbon around a pencil. Fit the coiled magnesium ribbon into the test-tube so that it is about half-way down the tube. Replace the test-tube at the end of the piece of glass tubing. Heat the water in the flask until it boils steadily. Then transfer the heat to the test-tube, so that the magnesium ribbon is heated. What is observed?

When the test-tube is cool, shake some of the residue into water and test with indicator.

The reaction between magnesium and steam is similar to that between calcium and water; an insoluble oxide is produced and

116

the hydrogen burns at the end of the tube. The oxide becomes a hydroxide when mixed with water. Try to write the equation for the action of steam on magnesium. (The valency of magnesium is two, the same as that of calcium.)

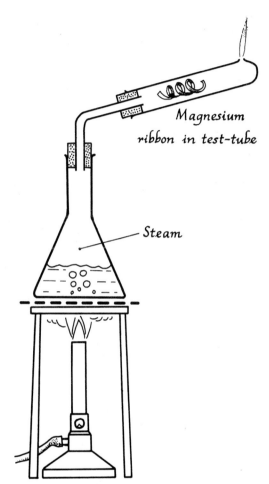

Magnesium ribbon in test-tube

Steam

Figure 11.2 The action of steam on magnesium

Iron will also react with steam, but only if it is heated until it is red-hot. A diagram of the apparatus needed to cause iron to react with steam is shown in Figure 11.3.

$$\text{steam} + \text{iron} \rightarrow \text{hydrogen} + \text{iron oxide}$$
$$4H_2O(g) + 3Fe(s) \rightarrow 4H_2(g) + Fe_3O_4(s)$$

If you examine the panel pins at the end of the investigation in which iron is heated in a stream of steam, you will find that they have a *black* oxide coating. What colour is the other oxide of iron that you have seen?

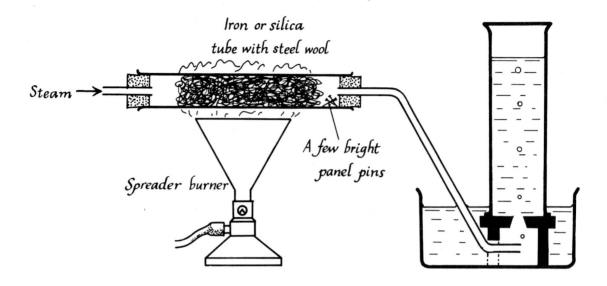

Steam →

*Iron or silica
tube with steel wool*

*A few bright
panel pins*

Spreader burner

Figure 11.3 The action
of steam on iron

Zinc also gives some reaction with steam, but other metals are less reactive.

11.3. The activity series

We have seen that some metals react with cold water and that some metals react with steam, whereas others do not react with either. From the experiments that we have seen, we could place the metals in an order in which the most reactive is placed first. If we do this we obtain Table 11.1.

TABLE 11.1.   ACTION OF METALS ON WATER

Potassium	Very active, melts and burns	With
Sodium	Quite reactive, melts	Cold
Calcium	Quite reactive, a steady stream of hydrogen is evolved	Water
Magnesium	Burns when heated in steam	With
Iron	Liberates hydrogen from steam when red-hot	Steam

This list is known as an activity series, because it places the metals in an order of reactivity. We shall see from other experiments how more metals can be added to the list below iron.

Compare the metals in Table 11.1 with the list of metals in Chapter 7.

11.4. Hydroxides and oxides

When metals are burned in air or oxygen, they form oxides which have an akaline reaction with indicators; such oxides are said to be **basic**. We have seen that when metals react with water they form either oxides or hydroxides that have an alkaline reaction. If carried out at high temperatures, as with magnesium and steam, the oxide will always form. At lower temperatures, the compound (oxide or hydroxide) formed is the one which forms with the greatest evolution of heat.

Investigation 11d. The conversion of calcium oxide to hydroxide

Take a lump of freshly roasted calcium oxide, about 10 cm³ in volume, and place it in a basin. Arrange the basin so that water drops steadily on to the lump of calcium oxide at a rate of about two drops per second. What do you observe?

Calcium oxide, like other strongly basic metal oxides, combines with water to form a hydroxide and evolves heat in the process. It requires strong heating to convert the hydroxide back to the oxide. This process in which water is added is known as **slaking**, and will be examined in more detail in Chapter 22. If calcium oxide is left in a badly sealed bottle, it will take water from the air; hence it can be used to dry substances. Sodium oxide combines so readily with water that it is very difficult to obtain a pure sample of this oxide.

Harmful

Metals that form hydroxides that are readily soluble in water are known as **alkali metals**, for example sodium and potassium. Metals that form hydroxides that are only slightly soluble in water, giving an alkaline reaction, are known as the **alkaline earth metals**, for example magnesium and calcium. Hydroxides that are soluble in water are known as **alkalis**. We shall see later that other hydroxides are formed that are not soluble in water, and that these hydroxides lose water very easily to form oxides.

11.5. Properties of alkalis

Alkalis are formed by metals that are high in the activity series. These alkalis are able to displace other metals, as hydroxides, from their salts. In addition to calcium, sodium and potassium hydroxides, alkaline properties are found in a solution of ammonia gas in water; it is, therefore, known as ammonium hydroxide. (More details will be given about this substance in Chapter 25.)

Investigation 11e. The action of alkalis on solutions of salts

For this investigation, solutions of sodium and ammonium

Corrosive

Harmful

hydroxides will be needed that should contain approximately one mole per cubic decimetre. (Further information about salts will be found in Chapter 15.) The effect of adding alkali solution, *a little at a time*, to about 10 cm³ of the salt solution should be observed. The following salts will be found useful: copper(II) sulphate(VI), zinc sulphate(VI), iron(III) chloride, iron(II) sulphate(VI). The solutions of salts should be nearly saturated.

Notice that in some cases the reaction seems to go in two stages. The first, the formation of a precipitate and then, second, the disappearance of the precipitate. The second stage of the reaction occurs when more alkali is added than is necessary to complete the first stage of the reaction. The alkali is then said to be present **in excess**.

On adding excess sodium hydroxide to zinc sulphate(VI) solution, the white jelly-like precipitate that had formed dissolves. This is because the hydroxide of zinc behaves like an acidic oxide in the presence of a strongly alkaline substance.

When excess ammonium hydroxide is added to copper(II) sulphate(VI) solution or to zinc sulphate(VI) solution, the precipitate that formed initially redissolves. In the case of the copper(II) sulphate(VI) solution a very dark-blue solution is obtained. This is because the ammonia from the ammonium hydroxide solution forms a complex compound with the salt.

Investigation 11f. The action of heat on a suspension of copper(II) hydroxide

Corrosive

Harmful

Take the tube which contains the copper(II) hydroxide precipitate (obtained by adding sodium hydroxide solution to copper(II) sulphate(VI) solution) and transfer the precipitate and solution to a boiling tube. Hold the tube in a suitable holder and gently heat in a bunsen flame until the water begins to boil. What do you see? What do you think has happened to the copper hydroxide?

Compare the change with that noticed when calcium hydroxide was heated in a tube.

Strong alkalis precipitate hydroxides of metals from solutions of their salts:

sodium hydroxide + copper(II) sulphate(VI) → sodium sulphate(VI) + copper(II) hydroxide
$$2NaOH(aq) \quad + \quad CuSO_4(aq) \quad \rightarrow \quad Na_2SO_4(aq) \quad + \quad Cu(OH)_2(s)$$

Metals that are high in the activity series form hydroxides with the evolution of heat; their hydroxides only lose water to form oxides if they are heated strongly. Metals that are low in the activity series form hydroxides when solutions of their salts are treated with alkalis. These hydroxides lose their water to form oxides more easily. For example, if light-blue coloured copper(II) hydroxide is boiled in water it forms copper(II) oxide:

$$\text{copper(II) hydroxide} \rightarrow \text{copper(II) oxide} + \text{water}$$
$$\text{Cu(OH)}_2(s) \quad \rightarrow \quad \text{CuO(s)} \quad + \text{H}_2\text{O(l)}$$

Test your understanding

1. What is formed when sodium is dropped into water?
2. What colour does litmus become if it is placed in the water which has reacted with sodium?
3. Name the products formed when calcium is dropped into water.
4. What happens to calcium oxide when you add water to it?
5. Make a list of the metals that you know in order of activity.
6. What do you think will happen when a solution of sodium hydroxide is added to a solution of lead(II) nitrate(V)?
7. Write an equation for the reaction of potassium with water.
8. Two new elements are given symbols X and Y. When X is dropped into water, it sinks and slowly gives off hydrogen. When Y is heated to red-heat in a stream of steam, no action is observed. Where would you place elements X and Y in Table 11.1?

Chapter 12

Hydrogen

In Chapter 11 we saw how the action of certain metals on water produced a gas – hydrogen. In this chapter, we are going to study some other methods of preparing the gas. We shall then investigate some of its properties and consider its large-scale production and uses.

Explosive

Harmful

Figure 12.1 Apparatus for the preparation and collection of hydrogen gas

12.1. The preparation of hydrogen in the laboratory

Investigation 12a. The liberation of hydrogen from dilute sulphuric(VI) acid by the action of zinc

Set up the apparatus shown in Figure 12.1. The flask contains zinc in granulated form. Granulated metals are made by melting the metal and then by dropping small quantities into a cooling liquid, usually water. The sulphuric(VI) acid should be about 3/2 M. Make sure that the end of the thistle funnel is very close to the bottom of the flask so that when the acid is added through the

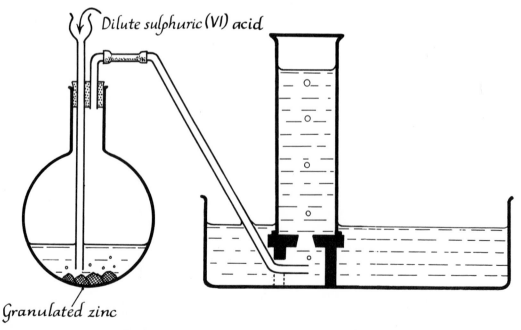

Dilute sulphuric (VI) acid

Granulated zinc

funnel, the end of the funnel will be under the surface of the liquid in the flask. If this is not the case, the gas will escape up the thistle funnel and so will not force its way through the water in the bowl and into the gas-jar.

You will probably find that when you add the sulphuric(VI) acid to the zinc, very little seems to happen. If this is the case, make a solution of a little copper(II) sulphate(VI) in water and pour it down the funnel. Note the difference that the addition of the copper(II) sulphate(VI) makes to the rate of the reaction. The copper in the copper(II) sulphate(VI) is acting as a catalyst in the reaction. The copper is displaced from the copper(II) sulphate(VI) by the action of the zinc:

Harmful

$$Zn(s) + Cu^{2+}(aq) \rightarrow Zn^{2+}(aq) + Cu(s)$$

The solid copper displaced in this way is deposited on the surface of the zinc and acts as an impurity. The presence of impurities is a most important factor in determining the rate at which a chemical reaction can take place (see Chapter 14).

Collect several gas-jars full of the gas. Test its properties (see Investigation 9e – COWSLIPS) as far as you can; in particular test the inflammability. You may notice that the first jars collected ignite with a 'squeaky pop', as compared with the jars collected later on, which burn quietly. The explosion in the case of the jars collected earlier is due to the fact that the gas is mixed with air from the apparatus and forms an explosive mixture.

Explosive

Flammable

For this reason, it is most important that small quantities of hydrogen gas are tested very carefully before any larger supply of the gas is ignited at its source.

★ *Warning.*

The same warning applies to any inflammable gas, for example, methane or hydrogen sulphide.

Make a list of the properties of hydrogen gas as suggested by the word COWSLIPS.

Investigation 12b. The preparation of hydrogen by the electrolysis of water

On a larger scale, water may be electrolysed in a Hofmann or Worcester pattern water voltameter (see Figure 12.2). This method may be done as a demonstration.

Take enough water to fill the voltameter and add a small quantity of sulphuric(VI) acid to it (enough to make it about 1 M). Fill the voltameter and carefully mark the level of the liquid, either in the centre tube in the case of the Hofmann apparatus, or on the jar if the Worcester pattern is used. Apply a d.c. voltage of between 12 and 24 volts across the platinum electrodes.

Corrosive

Harmful

Note that the gases are liberated *at the electrodes*. Compare the

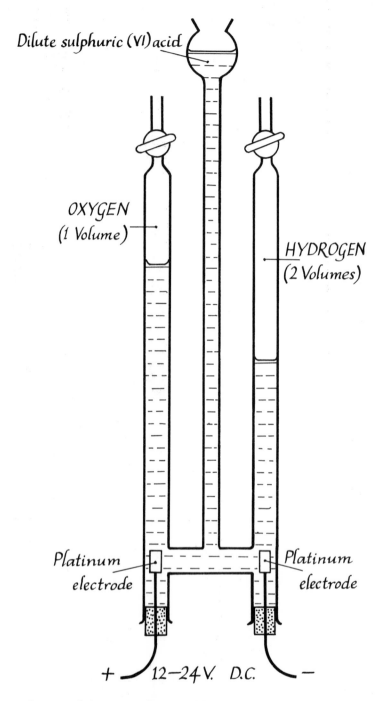

Dilute sulphuric (VI) acid

OXYGEN
(1 Volume)

HYDROGEN
(2 Volumes)

Platinum
electrode

Platinum
electrode

$+$　12–24 V.　D.C.　$-$

Figure 12.2a The
Hofmann water voltameter

volumes of the gases that are liberated at the positive and negative electrodes. When sufficient gas has collected, switch off the current, carefully open the tap at the top of the collecting tube on the positive side of the apparatus and apply a *glowing* splint. What happens? Open the tap at the top of the collecting tube on the

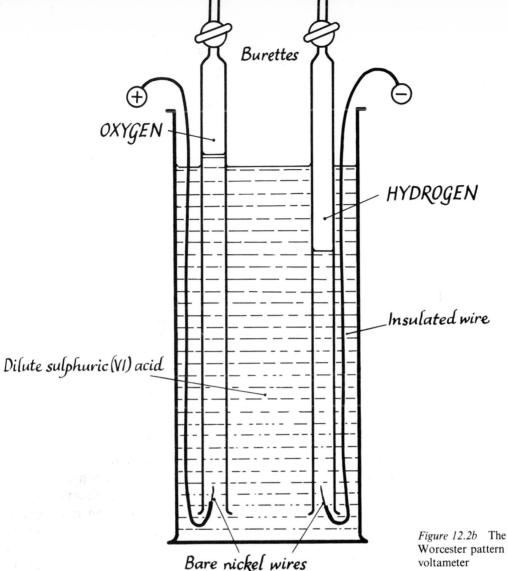

Burettes

OXYGEN

HYDROGEN

Insulated wire

Dilute sulphuric (VI) acid

Bare nickel wires

Figure 12.2b The Worcester pattern water voltameter

negative side in a similar way and apply a *lighted* splint. What happens?

When your tests are completed, allow the water in both collecting tubes to rise to the top. Compare the level of the liquid in the centre tube, or the jar, with what it was at the start of the experiment. Do you notice any change?

On the small scale, the apparatus shown in Figure 12.3 may be used for the electrolysis of water.

The apparatus is filled with 1 M sulphuric(VI) acid and a d.c. voltage of about 6 volts is applied across the electrodes. The gases which collect in the test-tubes may be tested with lighted and glowing splints. Care must be taken to see that the gases do not escape from the tubes before the tests can be applied.

These experiments on the electrolysis of water show that when a direct current is applied across the electrodes, gases are evolved

Harmful

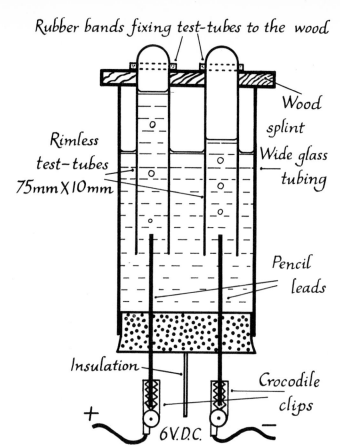

Rubber bands fixing test-tubes to the wood

Rimless
test–tubes
75mm X 10mm

Wood
splint

Wide glass
tubing

Pencil
leads

Insulation

Crocodile
clips

+   6V. D.C.   −

Figure 12.3 Class
apparatus for the
electrolysis of water

at the negative and positive electrodes in the proportions of two to one by volume. Tests show that these gases are hydrogen and oxygen respectively. An analysis of the liquid left after the electrolysis shows that the quantity of sulphuric(VI) acid remaining is exactly the same as the quantity of acid used at the start of the experiment. The hydrogen and the oxygen come entirely from the water. The function of the acid is to lower the resistance of the water to the passage of the electric current. In this sense, the acid acts as a catalyst. Pure water will conduct an electric current, but the voltage needed to force the current through the water is very high.

12.2. The industrial production of hydrogen

Hydrogen used to be produced on the large scale from water-gas (see Section 4.6) by the Bosch process. This process consisted of mixing the water-gas with more steam and passing the mixture over iron(III) oxide which acted as a catalyst, converting the carbon monoxide in the water-gas into carbon dioxide and producing more hydrogen:

$$CO(g) + H_2(g) + H_2O(g) \rightarrow CO_2(g) + 2H_2(g)$$

The carbon dioxide was removed from the mixture by 'scrubbing', i.e. washing the gases with water under high pressure, whence the carbon dioxide dissolved in the water:

$$CO_2(g) + H_2O(l) \rightarrow H_2CO_3(aq)$$

In countries where electric power is cheap, that is, in countries which have plenty of hydro-electric power available – the Scandinavian countries, for example – hydrogen can be produced by a large-scale electrolysis of water. Today, hydrogen is being produced from petroleum (see Chapter 20).

12.3. The large-scale uses of hydrogen

The most important use of hydrogen on the large scale is in the manufacture of ammonia by the Haber process (see Section 24.9). This involves the reaction between hydrogen and nitrogen:

$$N_2(g) + 3H_2(g) \rightleftharpoons 2NH_3(g)$$

Another big use is in the manufacture of **margarine** from vegetable oils. Vegetable oils are compounds of hydrogen and carbon-**hydrocarbons** – which are 'unsaturated'; that is, they do not contain as much hydrogen as the valency of the carbon atoms allows. Treatment of these oils with more hydrogen under the correct conditions results in the addition of more hydrogen to the oil molecules:

$$oil + hydrogen \rightarrow fat$$

The temperature of the oil is raised to about 450 K (about 180 °C), and the hydrogen is forced in under a pressure about four times that of the normal air pressure. A catalyst is used; this is metallic nickel, and it is afterwards filtered off from the hot liquid mixture. The process is known as **hardening**, as the liquid vegetable oil is converted into a solid fat.

Another use for hydrogen is found in the **oxy-hydrogen flame**. The high temperature produced by burning hydrogen in oxygen is used for the cutting and welding of metals.

At one time, hydrogen was used to fill the gas-bags of balloons and airships, but, because of the dangerously inflammable nature of the gas and the large number of accidents that took place, its use for this purpose was abandoned. Small quantities of hydrogen are still used to fill meteorological balloons and balloons used to carry scientific instruments to high altitudes for the purposes of research.

12.4. Hydrogen as a reducing agent

In Investigation 8b we made use of methanol, a compound of hydrogen, to remove the oxygen from the copper in copper oxide. We have also seen (Section 10.8) that the removal of oxygen from a substance is known as reduction. Thus, hydrogen is a **reducing agent** or **reductant**.

We shall see later (Chapter 17) that other substances, for example carbon and carbon monoxide, can be used as reductants. The use of reductants is of great importance in the extraction of metals from their compounds.

12.5. Hydrogen and water

In Investigation 12b we saw that when water is electrolysed hydrogen and oxygen are evolved in the proportions by volume of two to one respectively.

Explosive   Flammable

★ *Warning – this investigation must be carried out only by an experienced teacher. A safety screen must be used.*

★ *Warning – the gas must not be ignited until it has been tested for freedom from air.*

Investigation 12c. The burning of hydrogen in air

Set up the apparatus shown in Figure 12.4, using a hydrogen cylinder.

The use of an ordinary hydrogen preparation apparatus is not advisable as there is more danger of getting an explosive mixture. Before igniting the hydrogen at the jet, it is essential to collect small quantities in test-tubes over water and to burn them in order to check that all air has been displaced from the apparatus. This will be the case when a test-tube of hydrogen burns quietly and there is no 'pop' or explosion when a flame is applied.

When tests have shown that the gas is safe to ignite, light it at the jet and keep a small flame playing against the cooled flask. Arrange the watch glass to catch any liquid product that may be formed. Continue the experiment until about 2 cm³ of liquid have been collected.

The following tests may be carried out on portions of the liquid collected.

a. Taste a small drop.
b. Find the boiling-point of the liquid, using a small-scale method (Investigation 1i).
c. Find the freezing-point of the liquid. This is easily done by putting a few drops in a small test-tube (75 mm × 10 mm) and putting a − 10° to 110 °C thermometer in the liquid. Prepare a freezing-mixture by crushing up some ice and adding

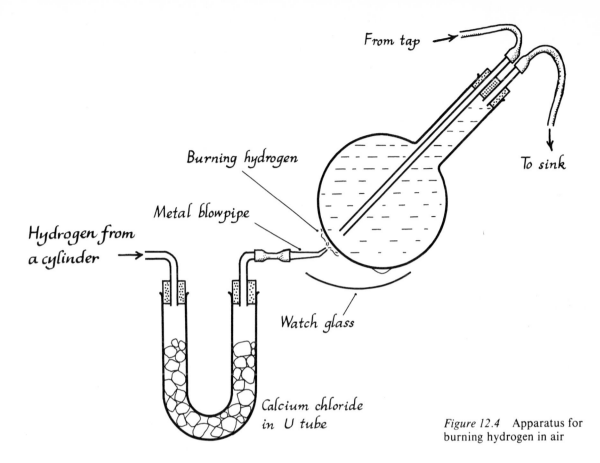

From tap →

To sink

Burning hydrogen

Metal blowpipe

Hydrogen from a cylinder →

Watch glass

Calcium chloride in U tube

Figure 12.4 Apparatus for burning hydrogen in air

to it about one-quarter as much salt. Mix well, and stand the test-tube in the mixture. Stir the liquid with the thermometer. Note the temperature at which the liquid starts to solidify.

d. Add some of the liquid to a little anhydrous copper(II) sulphate(VI). What happens to the colour of the white solid?

As a result of these tests, what is the liquid formed when hydrogen is burned in air? What other tests might you carry out to confirm your ideas?

Harmful

12.6   The composition of water by mass

We have seen that when water is electrolysed, two volumes of hydrogen and one volume of oxygen are evolved; let us now investigate the masses of the two gases that combine to form water.

Investigation 12d

Set up the apparatus shown in Figure 12.5. The combustion tube should be about 300 mm long and should be made of silica.

About 20 g of copper(II) oxide, preferably in wire form, are

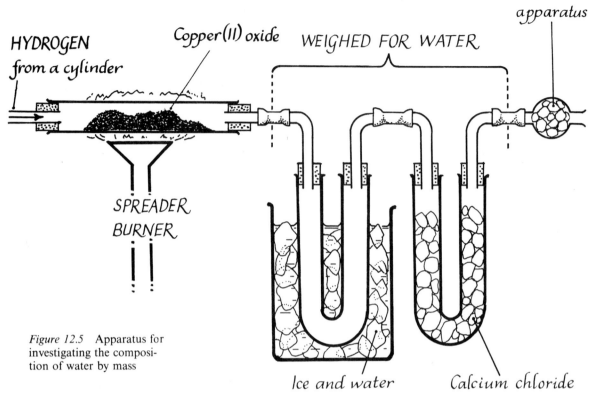

HYDROGEN *from a cylinder*

Copper(II) oxide

WEIGHED FOR WATER

CALCIUM CHLORIDE in TUBE *to stop water from the air getting into the apparatus*

SPREADER BURNER

Figure 12.5 Apparatus for investigating the composition of water by mass

Ice and water

Calcium chloride

★ *Warning – this investigation should only be carried out by an experienced teacher. A safety screen must be used and it is absolutely essential to test the hydrogen coming out of the apparatus for purity before heating is begun.*

Explosive

Flammable

heated to redness and then allowed to cool in a desiccator. These are then placed in the combustion tube. The ends of the tube are closed and the whole tube is weighed as accurately as possible. The two U-tubes are also weighed together, the one empty and the other containing anhydrous calcium chloride. The apparatus is then re-assembled, care being taken to retain the corks with which the combustion tube was fitted while the weighing was being carried out.

Hydrogen from a cylinder is passed through the *cold* apparatus until all the air has been displaced. The purity of the issuing hydrogen is checked in the usual way by collecting small test-tubes of the issuing gas over water and igniting them. Only when the collected gas burns quietly is it safe to start heating the combustion tube. Great care must also be taken to see that the bungs used in the apparatus fit tightly and that no air can leak into the apparatus during the experiment.

The burner is now lit under the combustion tube and a steady stream of hydrogen is passed through the apparatus. The copper(II) oxide is reduced to copper, and the water which is formed in the reaction either condenses in the first, empty U-tube or is absorbed by the calcium chloride in the second tube. When all the copper(II) oxide has been reduced, the heating is stopped and a slow stream of hydrogen is kept passing through the apparatus until it is quite cool.

Explosive

Flammable

When the apparatus is cool, the combustion tube is removed, the original corks are re-fitted, and it is again weighed. The loss of mass of this tube gives the mass of oxygen that is in the water that has been collected in the U-tubes.

The U-tubes are also removed, carefully dried on the outside and weighed again. The increase in mass of these tubes gives the mass of water that has been formed.

By subtracting the mass of oxygen used up from the mass of water formed, the mass of hydrogen that has reacted with the oxygen is found. Tabulate your results as follows:

Mass of combustion tube at start $= a$ grams

Mass of combustion tube at end $\quad = b$ grams

Mass of oxygen used $\quad\quad\quad\quad = (a - b)$ grams

Mass of U-tubes at start $\quad\quad\quad = c$ grams

Mass of U-tubes at end $\quad\quad\quad = d$ grams

Mass of water formed $\quad\quad\quad = (d - c)$ grams

Mass of hydrogen used $\quad\quad\quad = (d - c) - (a - b)$ grams

$\quad\quad\quad\quad\quad\quad\quad\quad\quad\quad\quad = z$ grams

We are now in a position to calculate from our results the mass of hydrogen that has combined with 16 g of oxygen, that is with one mole of oxygen atoms. Again, remembering that the mole mass of hydrogen atoms is 1 g, we are able to determine the empirical formula of water.

12.7. The explosion of hydrogen with oxygen

When hydrogen is mixed with air in certain proportions, the mixture is explosive and is very dangerous. When hydrogen is mixed with oxygen, the danger is considerably increased. A mixture of two volumes of hydrogen with one volume of oxygen is the most dangerous mixture of all. As these are the proportions that are obtained when water is electrolysed, the mixture is known as **electrolytic gas**.

★ *Warning – this experiment should be carried out only by the teacher.*

Explosive

Harmful

Fill a Hofmann-type water voltameter (see Investigation 12b) with dilute sulphuric(VI) acid, making sure that both collecting tubes are full and that the level of the liquid in the third tube is just up to the level of the taps. Fix a short piece of rubber tubing to each tap and join them by a T- or Y-piece, as shown in Figure 12.6. Join a longer piece of rubber tubing to the T- or Y-piece and fit it with a clip. Finally, fit a short piece of glass tubing to the end of the rubber tube.

Nearly fill a plastic coffee cup (or plastic carton of similar type) with a detergent solution. Have a second plastic container full of water.

Open *both* taps of the voltameter, leaving the clip on the rubber tubing closed. Have the glass outlet tube under the surface of the water in the second plastic container. When the liquid level in the collecting tubes of the voltameter has fallen about half-way

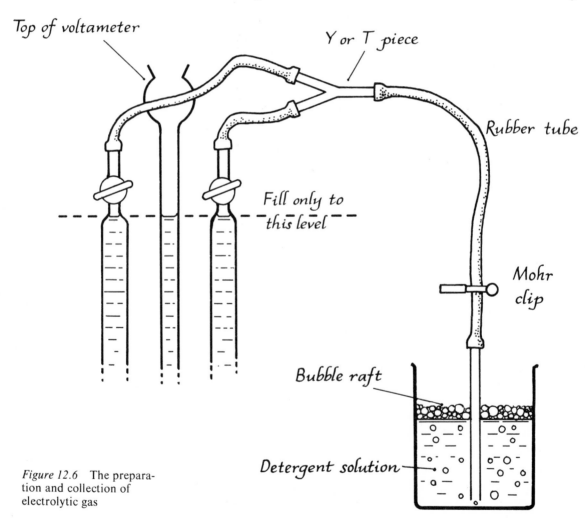

Top of voltameter

Y or T piece

Rubber tube

Fill only to this level

Mohr clip

Bubble raft

Detergent solution

Figure 12.6 The preparation and collection of electrolytic gas

down, squeeze the clip and allow the gas to escape to the air. Repeat this process twice more. This is done in order to clear any air from the apparatus.

Now, allow some more gas to collect in the apparatus and transfer the glass outlet tube to the plastic container with the detergent solution. Squeeze the clip and blow a 'raft' of gas bubbles on the surface of the detergent solution. Transfer the glass outlet tube back to the water and move the container with the detergent well away from the water voltameter. Apply a lighted splint to the bubble raft, taking care to use a long splint so that the hand does not come too close to the container. A really deafening explosion should occur.

The use of bubbles to contain the explosive mixture removes the danger of fragments of apparatus being thrown out. Provided that a light is not allowed anywhere near the voltameter, the experiment is quite safe.

When the experiment is completed, immediately remove the rubber connecting tubes from the voltameter and make sure that all traces of gas are removed from them by running water through them. Make sure also that no gas is left in the voltameter.

The very loud sound produced is caused by the air outside the bubbles rushing together and meeting when liquid water is formed from the gas. The liquid occupies a negligible volume compared to the volume of the gases from which it is formed. This is really an **implosion** – not an explosion. The same effect occurs when an evacuated container, such as a cathode-ray tube, is broken. Implosions of pieces of apparatus such as this can be very dangerous.

Liquid mixtures of hydrogen and oxygen have been used for blasting in enclosed spaces – for example, in the making of tunnels – where the poisonous fumes from more usual explosives would be hazardous.

Explosive

* *Warning – a safety screen must be used. The head should be turned away as the light is applied.*

12.8. Fuel cells

It is clear that a lot of energy is available when two volumes of hydrogen are combined with one volume of oxygen to form water. If this energy is harnessed, then it can be used. It can, of course, be used to produce an explosion; it can be used in the oxy-hydrogen flame; but, by far the most efficient way of harnessing the energy is to reverse the process of electrolysis, that is, to obtain electricity *directly* from the reaction. Other methods are wasteful and only a relatively small proportion of the energy available is usefully obtained.

The first fuel cell was made as long ago as 1839, using hydrogen and oxygen with sulphuric(VI) acid as electrolyte and finely divided platinum as a catalyst. More recent work by F. T.

Bacon in this country has resulted in a renewed interest, and in 1959 a cell of about five kilowatts output was demonstrated. Other interesting information about fuel cells is to be found in issue No. 157 of the *School Science Review*, on pages 521–6.

Test your understanding

1. Describe, giving a fully labelled diagram of the apparatus you would use, how you would prepare and collect a sample of hydrogen gas in the laboratory.
2. When dilute sulphuric(VI) acid is added to pure zinc, very little reaction, if any, occurs. If a few drops of copper(II) sulphate(VI) solution are then added, a vigorous reaction begins. Why is this?
3. It is very dangerous to try and light hydrogen gas coming from any apparatus without first testing a very small quantity of the gas. Why is this?
4. What are the chief large-scale uses of hydrogen?
5. Under what conditions may hydrogen be used to convert a vegetable oil into a fat? What is the importance of this change?
6. How would you show that when hydrogen gas is burned in air a liquid is formed?
7. What tests would you apply to the liquid formed when hydrogen burns in air to show that the liquid contains water?
8. Explain, giving a fully labelled diagram of the apparatus, how you could determine the composition of water by mass.
9. What is 'electrolytic gas'?
10. 'Hydrogen is a good reductant.' Explain what is meant by this statement and give an actual example of its use in this way.
11. Explain the meaning of each of the following: (i) implosion; (ii) electrode; (iii) voltameter.
12. Outline, giving the equations for the reactions, how hydrogen can be obtained from water-gas.
13. What is a 'fuel cell'? Why would the invention of a really efficient device of this nature be important?

Chapter 13

Acids

In Chapter 9, we saw that when non-metals are burned in air they produce oxides that turn blue litmus paper red. These oxides were said to be acidic. We have already made a detailed study of basic oxides in Chapter 11; in this chapter we shall make a further study of substances that give an acid reaction to litmus.

13.1. What is an acid?

The word acid is associated with a sour taste, but obviously this is not a suitable test to use. We have, however, seen that it is possible to use a specially prepared paper to test for the acidic nature of substances; we shall now try to find out which substances give an acid reaction.

Investigation 13a. To find out which substances have an acid reaction to litmus

As an introduction to this investigation, place a little lemon juice on the tip of your tongue. What sensation do you get in your mouth? Now place a drop of the lemon juice on a piece of blue litmus paper. What happens to the litmus paper? You could also try this with a drop of vinegar or with an acid drop. *You should not taste any other substances.*

The simple introductory experiment gave you a chance to see if a sour taste is to be connected with substances that turn litmus paper red.

For the main part of this investigation, we shall need some blue litmus paper and a teat pipette. The substances listed below should be examined in the following way:

a. *If a gas*, squeeze the teat on the pipette, so that the air is displaced, hold the jet of the pipette in the gas to be tested and release the pressure on the teat. This will cause the pipette to fill up with gas. Hold a piece of *dry* litmus paper near the jet of the pipette and gently squeeze the teat of the pipette, so that the gas flows on to the paper. Record any change in colour of the blue litmus paper. Repeat the experiment using a piece of moistened litmus paper. Again record any change in colour. Do you notice any difference in the results with dry and moist litmus paper?

b. *If a liquid*, use the teat pipette to put a drop of the liquid on to the dry litmus paper. Again record any colour change and repeat the experiment with a piece of moist litmus paper.

c. *If a solid*, place a little of the solid on a dry piece of blue litmus paper and examine the area around the solid. If no change in colour is observed, moisten the solid with a little water. Again examine the area around the solid.

Harmful

The substances that should be examined are dilute sulphuric(VI) acid, carbon dioxide, water, dilute hydrochloric acid, sulphur dioxide, dilute nitric(V) acid, lemon juice and vinegar (if these were not used earlier).

Notice that although some of the substances tested turn litmus paper red at once, some substances will only turn litmus paper red if either the substance or the litmus paper is moist. From this observation, we may suggest that water is connected with the acid behaviour. In Section 7.2, we learned about the action between the dipole of hydrogen chloride and the dipole of the water molecule to produce $H^+(aq)$ ions.

13.2. What substances are acids?

We have observed that an acid reaction is obtained if some oxides are dissolved in water. This is, however, not the only way in which acids may be prepared. In fact, not all substances that give an acid reaction in water contain oxygen. A very good example of a substance that gives an acid reaction in water is hydrogen chloride. As we have seen, when hydrogen chloride is dissolved in water, it produces $H^+(aq)$ ions, in the same way that sulphur dioxide dissolves in water to give an acid reaction. The three common laboratory acids, hydrochloric, sulphuric(VI) and nitric(V), are made by dissolving gases in water. Further details of the preparation of these acids will be found in Chapters 30 (hydrochloric acid), 28 (sulphuric(VI) acid) and 26 (nitric(V) acid). They are often called mineral acids.

Investigation 13b. The properties of mineral acids

In this investigation, we shall examine the reactions of the three acids mentioned earlier, and later we shall compare these reactions with those shown by some other acids.

For this investigation, you should have some dilute solutions of these three acids: about 1M sulphuric(VI) acid and about 2M hydrochloric acid and nitric(V) acid.

Harmful

[*A solution that contains one mole dissolved in one cubic decimetre of* **solution** *is said to be a* **one molar** *(1M) solution.*]

Place a little sodium carbonate in the bottom of a test-tube (about as much as would cover the end of a small spatula) and pour in about 5 cm³ of the acid being investigated. Examine the reaction of each of the three acids with some sodium carbonate. Test the gas evolved for carbon dioxide as explained below.

Test for carbon dioxide

Using a teat pipette, as described in Investigation 13a, remove some gas from the test-tube. In another clean test-tube place about 2 cm³ of lime-water. Lower the teat pipette into the second tube and gently squeeze the teat, so that the gas in the pipette slowly bubbles through the lime-water. Keep the teat of the pipette squeezed and remove the teat pipette from the test-tube. If the lime-water has not turned cloudy, bubble in some more carbon dioxide.

We shall now examine the reaction of these acids with a metal, zinc. Place a piece of technical grade granulated zinc at the bottom of a test-tube. (Technical grade will give better results than pure grade zinc. If only pure zinc is available, this may be used, but a few drops of copper(II) sulphate(VI) solution should be added to the test-tube.) Pour about 5 cm³ of the hydrochloric acid into the tube. Observe what happens and collect some of the gas produced in a boiling tube as shown in Figure 13.2.

Harmful

After some time, remove the boiling tube, keeping the open end downwards, and apply a lighted splint to the open end. What happens? What is the gas evolved from the acid by the metal?

Repeat this investigation with the other two acids. Is the same gas evolved from all the acids when they react with zinc?

★ *Warning – if a brown-coloured gas is obtained at any time, be careful not to inhale any.*

These three mineral acids are said to be **strong acids**, because of the reactions that we have observed. They are called strong acids because of *their properties*, not because the solution contains a large amount of the acid. The amount of acid in a solution is known as its concentration. For example, a solution of hydrochloric acid containing one-thousandth of a mole in one cubic decimetre of water is a dilute solution of a strong acid; whereas soda water is a concentrated solution of a weak acid, carbonic acid (carbon dioxide dissolved in water is known as carbonic acid).

From this investigation, we see that a strong acid will react with a carbonate, liberating carbon dioxide. Notice that carbon dioxide dissolved in water is an acid. Hence the acids we have used are strong enough to displace the weak acid, carbonic acid, from its salts:

$$Na_2CO_3(s) + 2HCl(aq) \rightarrow 2NaCl(aq) + H_2O(l) + CO_2(g)$$

137

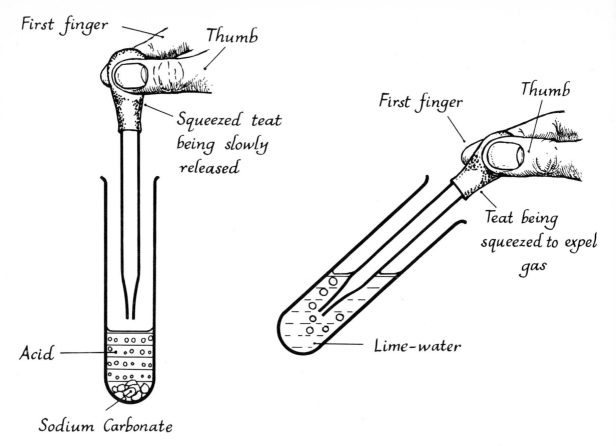

First finger

Thumb

Squeezed teat being slowly released

First finger

Thumb

Teat being squeezed to expel gas

Acid

Lime-water

Sodium Carbonate

Figure 13.1 Testing for carbon dioxide

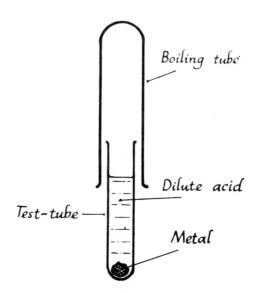

Boiling tube

Dilute acid

Test-tube

Metal

Figure 13.2 The action of acid on zinc

We have also seen that a strong acid will attack metals. In most cases the acids evolve hydrogen when treated with a metal:

$$Zn(s) + 2HCl(aq) \rightarrow ZnCl_2(aq) + H_2(g)$$
$$Zn(s) + H_2SO_4(aq) \rightarrow ZnSO_4(aq) + H_2(g)$$

Nitric(V) acid is an exception to this because it is a good source of oxygen, and the hydrogen that is liberated is converted to water before it can escape. Only magnesium is able to liberate hydrogen from nitric(V) acid, and then only from cold dilute acid.

In the reaction between hydrochloric acid and zinc, one atom per formula (the single hydrogen atom) is displaced by the zinc, but in the reaction between sulphuric(VI) acid and zinc, two hydrogen atoms per formula are displaced by the zinc. The number of hydrogen atoms in a formula of an acid that can be displaced by a metal is known as the **basicity** of the acid. Hence hydrochloric acid is said to be **monobasic** and sulphuric(VI) acid is said to be **dibasic.**

In addition to the three acids that we have examined, there is another acid in common use that could be called a strong acid. This is phosphoric(V) acid, H_3PO_4. This acid is used to convert rust into a hard phosphate that will not flake, before painting rusted steel.

13.3. Some weaker acids

In the previous section, we examined the properties of the three common laboratory acids; these were all strong acids. There are also many other acids in everyday use in our homes, but these are weaker acids.

Investigation 13c. Comparison of weak acids with strong acids

Repeat Investigation 13b using the following acids instead of the mineral acids: *acetic* acid, *carbonic* acid (made by dissolving carbon dioxide in water), *tartaric* acid and *citric* acid.

Compare the results of this investigation with the results obtained in Investigation 13b. Many of these weaker acids occur naturally (names in italics are the traditional names):
lactic acid is found in sour milk,
butyric acid is found in rancid butter,
citric acid is found in the juice of most fruit, especially in lemons,
tartaric acid is found in over-ripe grapes,
malic acid is found in rotting apples, and
acetic acid is formed when ethanol, for example in wine, is left exposed to the air for some time.

Notice that all these acids are formed when air is in contact with the original substance for some time. The change is caused by the action of oxygen from the air on one or more chemicals in the substance. This addition of oxygen is usually catalysed (see Chapter 14) by an enzyme.

Many of these weaker acids contain carbon and hydrogen chains (see Chapter 33), but not all the hydrogen atoms are acidic.

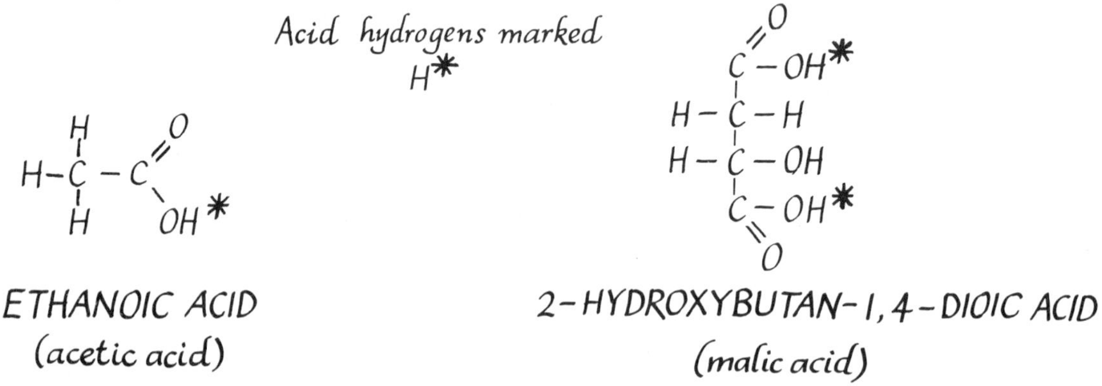

Acid hydrogens marked H✳

ETHANOIC ACID
(acetic acid)

2-HYDROXYBUTAN-1,4-DIOIC ACID
(malic acid)

Figure 13.3 The structures of two acids

13.4. Indicators

In Investigation 13a, we used litmus paper to examine the acidic nature of some substances, and we have said that acids are substances that produce $H^+(aq)$ ions when dissolved in water. Hence, we are using litmus to show the presence of $H^+(aq)$ ions. Substances that are used to do this are called indicators. Indicators, then, show whether the concentration of $H^+(aq)$ ions is high or low; for example, litmus turns blue when the concentration of H^+ (aq) ions is low and red when it is high.

Investigation 13d. Properties of some indicators

In this investigation, we shall use the three indicators that are commonly found in the laboratory: litmus, phenolphthalein and screened methyl orange (if screened methyl orange is not available, then methyl orange may be used).

Take three test-tubes and pour into each about 2 cm³ of dilute hydrochloric acid. Now add one drop of an indicator to the acid in each tube, using a different indicator for each tube. Make a note of the colour of the liquid in the tubes. Pour away the hydrochloric acid and wash the tubes thoroughly with water, preferably distilled or de-ionized.

Repeat the experiment with dilute ethanoic acid, dilute sodium

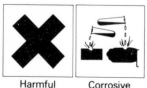

Harmful   Corrosive

140

hydroxide solution and dilute sodium carbonate solution. From this investigation, draw a conclusion about the colours shown by these three indicators in acidic and basic solutions.

Investigation 13e. The action of indicators on sodium hydrogencarbonate

Take three clean test-tubes and pour into each about 5 cm³ of dilute sodium hydrogencarbonate solution. Put one drop of litmus solution into one tube, one drop of screened methyl orange into the second tube and one drop of phenolphthalein into the third tube. What do you observe? Compare the results of this investigation with those of your previous investigation. Is sodium hydrogencarbonate acidic or basic?

Investigation 13d should have given you results that agree with the information in Table 13.1.

TABLE 13.1 THE COLOUR SHOWN BY INDICATORS IN ACIDIC AND BASIC SOLUTIONS

Indicator	Colour in Acidic Solution	Colour in Basic Solution
Litmus	Red	Blue
Phenolphthalein	Colourless	Red
Screened methyl orange	Red	Green
(Methyl orange)	Red	Yellow

Investigation 13e gives us a rather surprising result. Sodium hydrogencarbonate solution appears to be acid to phenolphthalein and basic to screened methyl orange. This is because the indicators change colours at different $H^+(aq)$ ion concentrations. Phenolphthalein changes colour (colourless to red) on the basic (that is low $H^+(aq)$ ion concentration) side and screened methyl orange changes colour (green to red) on the acidic side (that is high $H^+(aq)$ ion concentration). By using different indicators, we can draw up a table of acid strength. Strong acid character is given a low number and strong basic character is given a high number; water lies in the middle of this scale and is said to be **neutral**. This scale is known as the **pH scale** and runs from 0 (strongly acid) to 14 (strongly basic). Figure 13.4 shows the pH values and the indicator colour changes.

If we wish to find out the pH value for a particular substance, we can do this quickly by using a mixture of indicators known as universal indicator (or a piece of universal indicator paper). Figure 13.5 shows the colours shown by a common universal indicator at different pH values.

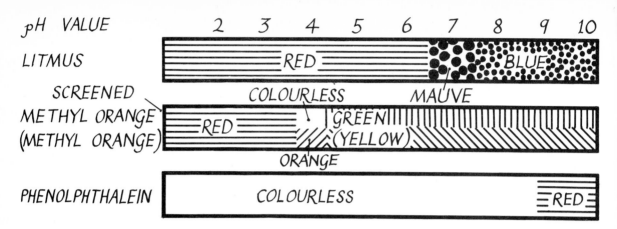

Figure 13.4 The pH scale and indicators

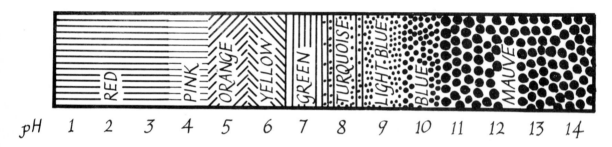

Figure 13.5 Universal indicator colours

Investigation 13f. The neutralization of acids and alkalis

When an acid is reacted with an alkali so that the mixture has neither acidic nor basic properties, the acid is said to be **neutralized**. This neutral point is given the value 7 on the pH scale.

Harmful

Take a burette and make sure that the tap operates properly by putting a little water into the burette and opening and closing the tap. Now close the tap and pour into the burette some M/10 hydrochloric acid until the burette is about one-quarter full. Open the tap and allow some of the acid to run out into the sink and then close the tap. Now tilt the burette so that the rest of the acid runs out through the open end. As it runs down the burette rotate the burette so that all the walls are rinsed with the acid. Repeat this rinsing process so that the water left in the burette from washing is removed.

Fill a burette with M/10 hydrochloric acid solution. Place the burette in a burette stand and put a white tile, if available, on the base of the stand. Place a beaker under the burette jet and open the tap fully. The acid will run rapidly into the beaker. As it does so, any air that is trapped in the tap of the burette will be expelled. When about 10 cm³ of acid has run out of the burette, close the tap and refill the burette. Then adjust the level until the bottom of the meniscus is on the zero mark.

Take a 10 cm³ pipette fitted with a syringe as shown in Figure 13.6. Cover the hole in the side of the syringe with a finger and hold the pipette so that the jet is below the surface of an M/10 solution of sodium hydroxide in a beaker. Carefully and slowly, pull

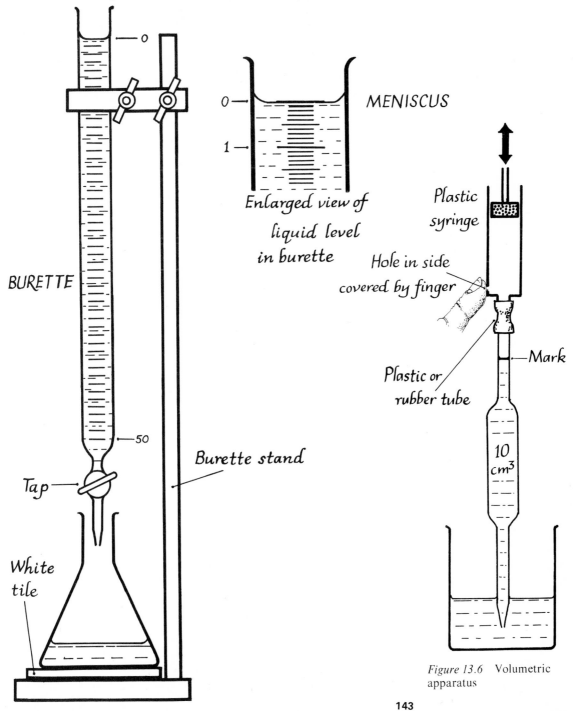

Enlarged view of liquid level in burette

MENISCUS

BURETTE

Burette stand

Tap

White tile

Plastic syringe

Hole in side covered by finger

Plastic or rubber tube

Mark

10 cm³

Figure 13.6   Volumetric apparatus

up the piston of the syringe. The solution from the beaker should rise up the tube of the pipette. Continue until the solution just enters the bulb of the pipette. Then remove the pipette from the beaker and tilt it to a horizontal position. Gently ease your finger off the hole and allow the solution to run up until it is just above the mark and then replace your finger over the hole. Hold the pipette with the jet over a sink and remove your finger so that the solution can run out; rotate the pipette as the solution runs out. Repeat this rinsing process.

Now place the jet of the pipette back into the beaker of M/10 sodium hydroxide solution and make sure that the piston of the syringe is fully down. Cover the hole with your finger and slowly pull up the piston until the solution is just above the mark. By carefully removing your finger, allow the level of the liquid to fall until the meniscus is again on the mark. Put the jet of the pipette into a conical flask and take your finger away from the hole. When all the liquid has run into the flask, touch the surface of the liquid in the flask with the jet of the pipette. You have now put exactly 10 cm³ of sodium hydroxide solution into the flask. Add about five drops of universal indicator to the sodium hydroxide solution and swirl the flask. Note the colour of the solution.

Place the flask under the jet of the burette and open the tap to allow 1 cm³ of the acid to run into the flask. Swirl the flask and note the colour of the solution. Repeat this, adding 1 cm³ of the acid at a time, until you have added a total of 20 cm³ of acid. The process that you have carried out is known as **titration**.

Compare the colours that you have recorded with the colour code on the bottle of indicator or with Figure 13.5, and note down the pH value for each stage in the experiment. Make a copy of the graph shown in Figure 13.7 (on graph paper if

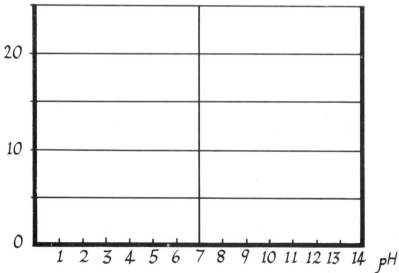

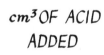

cm³ OF ACID
ADDED

Figure 13.7 The neutral-
ization of acids and bases

possible), and mark a cross to correspond with the volume of acid added and the pH value of the solution at that stage. When you have entered every result, draw a smooth curve through all the points. Write *hydrochloric acid against sodium hydroxide* on this line.

Clean the pipette with distilled water and repeat the experiment using M/10 ammonium hydroxide solution instead of sodium hydroxide solution. Then clean the burette and refill it with M/10 ethanoic acid and repeat the titration against sodium hydroxide and ammonium hydroxide solutions. Plot the results on the same graph paper as before and label each curve.

From your graph, make a note of the volume of acid that had to be added to reach pH 7 (the neutral point). Notice that with a strong acid (hydrochloric acid) and with a strong base (sodium hydroxide) the change in pH is sudden, but with weak acids (ethanoic acid) and weak bases (ammonium hydroxide) the change in pH is gradual.

13.5. Acids and metals

We have seen that one of the properties of acids is that they react with metals. This property can be used to give us more information about the reactivity of the metals.

Investigation 13g. The action of hydrochloric acid on different metals

For this investigation, you will need samples of different metals. These samples should be all of the same type, if possible, for example foil or powder. Foil will be found best, since the powdered metals soon become covered with a film of protective oxide that cannot be removed. Samples of foil about 100 mm² in area should be used. The following metals should be investigated: magnesium, aluminium, zinc, iron, lead, tin and copper. Take as many test-tubes as you have metal samples and put about 5 cm³ of 2M hydrochloric acid into each tube.

Harmful

If any of the metals have dull surfaces, scrape the surface with a penknife blade or rub the surface with some emery cloth. When all the metals have been cleaned, drop them into the tubes of acid. Make a note of which metals you put in which tube. Watch the reactions and try to place the metals in an order, with the most active first. You can judge the activity of the metals by the rate at which the hydrogen bubbles are given off.

We saw, in Chapter 11, that magnesium gives off hydrogen from steam; in the last investigation, we saw that magnesium evolves hydrogen rapidly from an acid. Hence the list of metals, in order of reactivity with acids, can be added to the list of metals in order of reactivity with water. Compare this combined list with the order of the metals given in Table 7.2.

13.6. Acids and soils

The acid content of the soil has a considerable effect on the growth of plants; for example, potatoes will grow quite well in an acid soil but wheat grows better in a basic soil. Most soils in Great Britain have pH values between 4 and 9, although values outside this range do occur. Soils tend to become more acid as time passes. There are two reasons for this. First, rain is acidic, since carbon dioxide and other acidic gases in the air dissolve in the rain as it falls. Secondly, the rain water causes the basic substances in the soil to be washed away. The acid nature of rain water further helps this washing-out or **leaching** process, since it dissolves calcium carbonate (in the form of limestone and chalk). Acid soils prevent the decomposition of dead plant material, that is humus, and as a result the soil becomes waterlogged. Peat bogs are good examples of very acid soil conditions.

The acidic nature of a soil is usually neutralized by the use of calcium hydroxide (lime) or calcium carbonate (chalk).

Investigation 13h. To examine the acid nature of the soil

Take a sample of soil and place it in a large beaker, so that the beaker is about half full. Now add water, distilled if possible, until the beaker is about three-quarters full. Stir thoroughly and then filter the suspension. Add universal indicator to the filtrate or dip in a piece of universal indicator paper. If the filtrate is acid, neutralize it by slowly adding powdered calcium carbonate to the filtrate until the indicator shows that the pH is 7.

If possible, repeat this investigation with samples of soils from other areas. It will be found that sandy soils tend to be acid and that clay soils tend to be basic. The difference is due, in part, to the ease with which water percolates through sandy soils, carrying soluble basic substances with it.

Test your understanding

1. What simple test can we use for an acid?
2. Many substances only give an acid reaction in the presence of water. Why is this?
3. How can you test for carbon dioxide?
4. Why are some acids said to be strong acids?
5. What is the basicity of (a) ethanoic acid, (b) 2-hydroxybutan-1,4-dioic acid?

6. What colour is phenolphthalein when it is added to sodium hydroxide solution?

7. When some screened methyl orange is added to a liquid the indicator changes from green to red. What can you say about the liquid?

8. A solution produces the following colours with the indicators listed:
   litmus turns red,
   screened methyl orange turns green,
   phenolphthalein turns colourless.
   Suggest a pH value for the solution.

9. An element is burnt in air. The residue is added to water and some universal indicator. The indicator turns mauve. What type of element was burnt?

10. An acid is placed in a flask with some universal indicator. An alkaline solution is then slowly run in from a burette, the solutions being mixed as the alkaline solution runs in. The colour of the indicator is seen to change in the following sequence: red, pink, orange, yellow, green and mauve. What can you say about the type of acid and alkaline solutions being used?

11. Place the metals potassium, magnesium, copper, zinc and tin in an order of reactivity.

12. Why do farmers often add calcium hydroxide to their land?

Chapter 14

Fast and Slow Reactions

14.1. Fast reactions

When an explosive charge is set off, then it is clear that the chemical reaction involved is instantaneous – or, at least, very rapid indeed. In the same way, when a lighted match is applied to the gas from a bunsen burner, the gas ignites immediately and continues to burn as it comes out of the burner. Again, the reaction is very fast.

Investigation 14a. Some rapid reactions

Toxic

Harmful

★ *Warning – do not look directly at the flame.*

a. Take a few crystals of sodium sulphate(VI) and dissolve them in about 10 cm³ of water in a test-tube. Take a few crystals of barium chloride (*poisonous*) and dissolve them in water in a second test-tube, again using about 10 cm³. Alternatively, put about 10 cm³ of barium chloride solution from the laboratory reagent bottle into a test-tube. Now, pour the contents of one of the test-tubes into the other test-tube – it does not matter which into which. What happens?

b. Dissolve a few crystals of potassium chromate(VI) in about 10 cm³ of water in a test-tube. Add an equal volume of silver nitrate(V) solution from the laboratory reagent bottle. What happens?

c. Place some powdered magnesium (not too much) in a piece of glass tubing. Adjust a bunsen burner to give a fairly strong (but not roaring) blue flame. Blow the powdered magnesium through the flame. What happens?

Investigation 14b. A slower reaction

Harmful

Make a *cold* solution of sodium thiosulphate(VI) ('hypo') by dissolving about 1 g of the crystals in 50 cm³ of pure water. Pour a little of this solution into a test-tube and add about 3 cm³ of dilute hydrochloric acid. Watch the tube carefully for some time. What do you see happen?

Put a fresh portion of the sodium thiosulphate(VI) solution in another test-tube and warm it over the bunsen flame. Now, add the same volume of dilute hydrochloric acid as before. What happens this time?

Investigation 14c. The effect of concentration on the rate of a chemical change

The reaction between sodium thiosulphate(VI) and dilute hydrochloric acid is ideal for a simple study of the effect of concentration on reaction rate.

Prepare a solution of sodium thiosulphate(VI) which contains 50 g dm^{-3}. Prepare also a 1M solution of hydrochloric acid. Using a measuring cylinder, put 50 cm^3 of the sodium thiosulphate(VI) solution into a 100 cm^3 conical flask or a 100 cm^3 beaker. Now, write your name on a piece of paper and stand the flask or the beaker over the writing. Measure 10 cm^3 of the acid into a test-tube. Have a stop-clock ready, or use a watch with a seconds hand.

When you are ready, pour the acid into the sodium thiosulphate(VI) solution and give the container a quick swirl, at the same instant starting the stop-clock, or noting the time with the seconds hand. Look directly down from above the flask or beaker at your name written on the paper. As the precipitate of sulphur forms, the writing will become less distinct and will eventually vanish. When this happens, stop the clock or note the time.

Repeat the experiment, this time using 40 cm^3 of the sodium thiosulphate(VI) solution and adding 10 cm^3 of pure water in order to make the total volume 50 cm^3. Again, add 10 cm^3 of the acid. Repeat again, using 30 cm^3, 20 cm^3, 10 cm^3 of the sodium thiosulphate(VI) solution in turn, in each case adding enough pure water to make a total volume of 50 cm^3. In all cases, 10 cm^3 of acid is used. Record your results as shown in Table 14.1.

Harmful

TABLE 14.1. HOW TO RECORD THE RESULTS IN INVESTIGATION 14c

Volume of Sodium Thiosulphate(VI)	Volume of Water	Time Taken in Seconds	Reciprocal of Time Taken × 1000
40 cm^3	10 cm^3		
30 cm^3	20 cm^3		
20 cm^3	30 cm^3		
10 cm^3	40 cm^3		

Plot a graph of the volume of sodium thiosulphate(VI) solution used against the times taken for the writing to vanish. Plot also a graph of the volumes of sodium thiosulphate(VI) solution used against the reciprocals of the times – that is against 1/time (multiplied by 1 000). The reciprocal of the time taken represents the **rate** of the reaction. What do your graphs show you?

Investigation 14d. A more detailed study of the effect of temperature on reaction rate

You will have noticed (Investigation 14b) that, when the acid was added, the heated solution of sodium thiosulphate(VI) deposited sulphur far more quickly than the cold solution. We may use the techniques described in Investigation 14c to make a more detailed study of the temperature effect.

Harmful

Take 10 cm³ of the sodium thiosulphate(VI) solution and dilute it with pure water to make 50 cm³. Put a thermometer into the solution, and warm it to a few degrees above the temperature you want. Now, carrying out the same technique as before, add 10 cm³ of the hydrochloric acid. Measure the time needed for your name to vanish, and note the temperature of the mixture. Repeat the experiment, using exactly the same volumes of sodium thiosulphate(VI) solution, water and hydrochloric acid each time, but altering the temperature. Suitable temperatures would be about 290 K, 300 K, 310 K, 320 K and 330 K (about 20 °C, 30 °C, 40 °C, 50 °C and 60 °C respectively). Record your results as shown in Table 14.2.

TABLE 14.2 HOW TO RECORD THE RESULTS IN INVESTIGATION 14d

Temperature	Time Taken in Seconds	Reciprocal of Time Taken × 1 000

Plot graphs of the times taken against the temperatures at which the experiments were carried out, and also of the reciprocals of the times taken against the temperatures. How do these graphs compare with the graphs from Investigation 14c?

Investigation 14e. The effect of surface area on the rate of a chemical change

    a. Take a length of magnesium *ribbon*, hold it in a pair of crucible tongs, and put the other end in a bunsen flame. As in

Investigation 14a, put some *powdered* magnesium into a glass tube, and blow it through a bunsen flame. Which sample ignites more easily?

★ *Warning – do not look directly at the flame.*

b. Take a large *marble chip* and put it into a 100 cm³ beaker. Take, as nearly as you can estimate, an equal mass of *powdered chalk* and put it into another 100 cm³ beaker. At the same instant add about 60 cm³ of hydrochloric acid (3M) to each beaker. Compare what happens in the two beakers.

c. Take three test-tubes (150 mm × 25 mm). Into the first, put a length of magnesium *ribbon*; into the second, put some magnesium *turnings* (or some *very* tiny pieces of magnesium cut from a length of magnesium ribbon); and into the third tube, put some *powdered* magnesium. Now, take some very dilute (0.3 M) hydrochloric acid and add about 40 cm³ of it to each of the three tubes, as quickly as you can, starting with the tube containing the ribbon and ending with the tube containing the powdered magnesium. Compare the reactions in the three tubes.

Harmful

Investigations such as these show that the more finely divided a reacting substance is, the quicker it reacts. It is for this reason that finely powdered substances, such as coal, flour, wood powder or dust, when mixed with air, can be very dangerous and may lead to explosions if a spark is brought near them.

Investigation 14f. The effect of adding a catalyst

Take three small test-tubes (75 mm × 10 mm) made of hard glass (Pyrex). Into the first put some potassium chlorate(V); into the second, put some manganese(IV) oxide; and into the third, put a mixture of one part of manganese(IV) oxide and four parts of potassium chlorate(V). Arrange the tubes in a triangle (see Figure 14.1) so that they may be heated with a bunsen burner. Adjust the burner to a fairly small, blue flame.

Oxidising   Explosive

While heating the tubes, test them periodically for the evolution of oxygen gas, using glowing splints. Which tube gives off oxygen first? Does any oxygen come off from the manganese(IV) oxide alone? Does any oxygen come from the potassium chlorate(V) alone?

14.2. The factors which affect the rate of a chemical change

All the investigations in this chapter have had the aim of finding out something about the rates at which chemical changes take place and the factors that can affect the rates. The chief factors which can alter the rates of chemical changes are concentration, temperature, physical size and the presence of a catalyst.

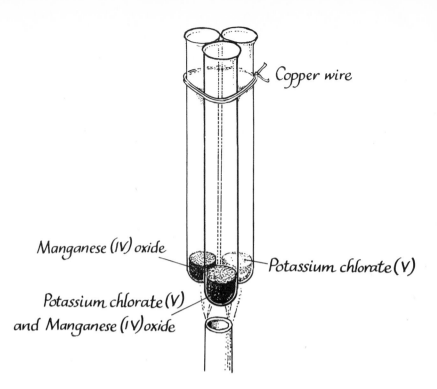

Concentration

In every case, an increase in concentration brings about an increase in the rate of reaction.

Temperature

Again, in every case, an increase in the temperature brings about an increase in the rate of reaction.

Physical size

We have seen that the more finely divided the substances are, the more rapidly they react. This is because a solid substance can only react at its surface. Powdering a substance enormously increases the surface area. Try to calculate the increase in surface area brought about when a piece of solid in the form of a 10 mm cube is divided into 1 000 cubes, each of side 1 mm, and 1 000 000 cubes, each of side 0.1 mm.

Catalysts

The effect of catalysts is considered in more detail below.

14.3. Catalysts and inhibitors

Catalysts are so important in increasing the rates at which chemical reactions take place that it is not an exaggeration to say that the whole of our modern chemical industry depends on the use

152

of suitable catalytic materials. Some important industrial processes which use catalysts are listed below:

   a. The Haber process for the production of ammonia gas from nitrogen and hydrogen (see Section 24.9);
   b. The Ostwald process for producing nitric(V) acid from ammonia gas and air (see Section 26.1);
   c. The Contact process for the manufacture of sulphuric(VI) acid (see Section 28.2);
   d. The production of ethanol from ethene (see Section 34.2); and
   e. The manufacture of many plastics (see Chapter 36).
      Details of all these processes will be found in the relevant sections as shown above.

Substances which slow down the rates of chemical changes are known as inhibitors, as we saw in Chapter 9. Inhibitors are also important on the industrial scale. For example, an inhibitor is added to the rubber used in the manufacture of tyres in order to slow down the rate at which the rubber is attacked by the oxygen in the air. Attack by oxygen causes the rubber to **perish** and become brittle. Another inhibitor is used to prevent the too rapid decomposition of hydrogen peroxide solutions during storage.

14.4. The effect of pressure on the rates of reactions between gases

Our investigations have shown us that an increase in concentration is followed by an increase in the rate of a chemical reaction. In the case of gases, an increase in concentration can be brought about by an increase in pressure. This makes the volume of the gases smaller, and so brings the particles of the gases closer together. Thus, an increase in pressure results in an increase in reaction rate. This fact is made use of in many industrial processes where gases are concerned.

14.5. Reversible chemical reactions

Not all chemical reactions 'go to completion' – that is go on until *all* the reactants are used up and have been converted into the products. For example, when nitrogen and hydrogen react together, under the normal conditions of the Haber process (see Section 24.9), only about 12 per cent of the gases put into the process are converted into ammonia. Under other conditions of temperature and pressure, different percentages of ammonia are obtained. Again, when iron and steam are allowed to react together in a closed container, only a proportion of the steam is converted into hydrogen, an equivalent amount of iron oxide being formed at the same time.

Reactions such as these are known as **reversible**, because if, in the latter case for example, hydrogen is heated with iron oxide in a closed container, iron and steam are formed. The final proportions of the substances present are exactly the same as when iron and steam are reacted together, provided that the pressure and the temperature are the same in both cases.

By a careful choice of pressure and temperature, it is possible to ensure that the maximum possible conversion of the reactants into the products is obtained. In the case of exothermic reactions (those where heat is evolved), a bigger yield is obtained at lower temperatures; where the reaction is endothermic (heat being absorbed), a bigger yield is obtained at higher temperatures.

Pressure is only important in the case of reactions where a gas or gases are involved, and then only when the number of particles (molecules) of the gaseous products is different from that of the gaseous reactants. If the number of particles of the gaseous products is *less* than that of the gaseous reactants, a pressure increase will bring about a better yield; if the number of gaseous product particles is *greater*, a better yield can be obtained at lower pressures.

It is most important not to confuse the degree of conversion that is obtained in a reversible reaction with the *rate* at which the conversion takes place. A rise in temperature, for example, will *always* increase the rate of reaction, but it may well decrease the yield. Likewise, in the case of reactions between gases, an increase of pressure will *always* increase the rate of the reaction, but it may well decrease the yield.

All industrial processes are a compromise between the yield obtainable under a given set of conditions, and the rate at which the conversion can be brought about. The controlling factor is one of *cost*. It would clearly be useless to get 100 per cent conversion, if the time needed is several months. On the other hand, a yield of some 0.1 per cent would be of little value, although the time taken might be very short. A compromise has to be made.

Test your understanding

1. Zinc will dissolve in dilute hydrochloric acid, giving off hydrogen gas. If you wanted to collect the largest possible volume of hydrogen in the shortest possible time, would you: (i) use zinc filings or granulated zinc or zinc powder; (ii) use 1 M, 2 M or $\frac{1}{2}$ M hydrochloric acid; (iii) keep the apparatus cool, or warm it? In each case, explain the reason for your choice of answer.

2. Give TWO examples of chemical reactions that are so fast that they

may be considered to be instantaneous, and TWO examples of chemical reactions that take an appreciable period of time.

3. A lump of coal will burn quietly in the grate, but there have been many serious explosions in coal-mines. These explosions have been thought to be due to coal-dust in the air. Why should this be dangerous?

4. In the manufacture of hydrogen from water-gas by the Bosch process, iron(III) oxide is used as a 'catalyst'. What is the function of a catalyst in a chemical reaction?

5. Copper(II) oxide is said to act as a catalyst in the decomposition of potassium chlorate(V) by heat. Describe, giving full experimental details, how you would check this statement.

6. Styrene (phenylethene), if kept for too long in its bottle, changes from a liquid into a glass-like solid. When a trace of an acid compound is added, this change takes place much more slowly. What name is given to substances that slow down chemical changes in this way? Give TWO other examples of chemical changes that can be affected in the same way.

7. Why does an increase in pressure bring about an increase in the rate of reaction where gases are concerned?

8. What is a 'reversible' reaction?

9. Some hydrogen peroxide solution is placed in a small flask attached to a gas syringe. At zero time, a small quantity of manganese(IV) oxide is added to the flask and readings of the volumes of oxygen gas given off are taken at time intervals as shown below:

Time/s	0	5	10	15	20	30	40	50	60	70	80	90	100
Volume/cm^3	0	11	20	27	32	42	50	57	63	67	69	70	70

Plot a graph showing the volume of oxygen evolved (y-axis) against the time in seconds (x-axis). On the same axes, sketch the sort of curve you would expect to get if you repeated the experiment with everything exactly the same as before except that an equal volume of pure water was added to the hydrogen peroxide before the manganese(IV) oxide was added.

10. When ammonia is manufactured from hydrogen and nitrogen by the Haber process, the proportion of ammonia obtained gets less as the temperature is raised. Nevertheless, the process is carried out at some 550 °C rather than at room temperature. Why is this?

Chapter 15

Salts

We have seen in earlier chapters that when an alkali is dissolved in water it produced metal ions and hydroxyl ions, and that an acid produces $H^+(aq)$ ions and acid radical ions. We can represent what happens in these cases by equations:

$$alkali + water \rightarrow M^+(aq) + OH^-(aq)$$
$$acid + water \rightarrow H^+(aq) + A^-(aq)$$

If an acid is mixed with an alkali, the oppositely charged $H^+(aq)$ and $OH^-(aq)$ ions react with each other to form water:

$$H^+(aq) + OH^-(aq) \rightarrow water$$

The acid radical and metal ions remain dissolved in the water. They may be separated from the water by evaporation. The substance formed by these ions is called a **salt**.

15.1. The formation of salts by the action of an acid on an alkali

We have seen that a salt may be prepared by mixing together an acid and an alkali. For our first investigation, we could use sodium hydroxide and hydrochloric acid.

Investigation 15a. The preparation of sodium chloride

Corrosive

Harmful

Put about 30 cm³ of 2M sodium hydroxide solution into a 100 cm³ beaker. Drop in a small piece of litmus paper. From a burette, run in 2M hydrochloric acid until the litmus paper just turns red. (If you add too much acid, add a little more of the sodium hydroxide solution to turn the litmus paper blue again.) Pour the solution into an evaporating basin and place this on a gauze over a bunsen burner.

Light the bunsen burner and adjust it so that it gives a small semi-luminous flame. After a short while, turn up the gas and open the air holes on the bunsen to obtain a hot flame. Continue heating the solution until most of the water has been evaporated and the solid in the basin begins to 'spit'. As soon as this happens, turn off the bunsen and allow the remainder of the water to evaporate without further heating. When the basin is cool, scrape out some of the solid on to a filter paper and allow it to dry off in a warm place.

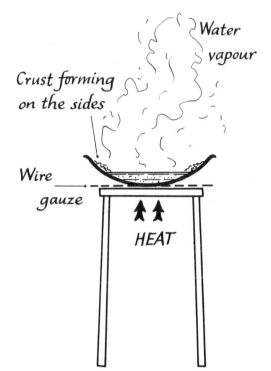

Figure 15.1 Evaporation of water from salt solution

The salt that you have made is sodium chloride (or common salt):

$$NaOH(aq) + HCl(aq) \rightarrow NaCl(aq) + H_2O(l)$$

In the reaction, water was formed from the $H^+(aq)$ ions and the $OH^-(aq)$ ions, leaving the sodium ions and the chloride ions. When the water is removed, these crystallize into a simple cube-shaped solid (see Figure 5.3). Examine some of the solid that you have made and compare the shape of the solid with that shown in Figure 2.5.

A similar investigation may be made using potassium hydroxide instead of sodium hydroxide and either sulphuric(VI) or nitric(V) acid in place of hydrochloric acid: for example, potassium hydroxide with nitric (V) acid will form potassium nitrate(V).

When evaporating solutions, tests should be carried out periodically to see if the solution is concentrated enough to crystallize. To do this, carefully pour a little of the solution into

Corrosive

Harmful

157

a test-tube and cool it under cold water running from a tap. If crystals form in the tube, the evaporation should be stopped and the solution in the basin allowed to cool slowly. If crystals do not form, the solution in the test-tube should be returned to the basin and evaporation should be continued for a little longer. These tests are not needed with sodium chloride, since the solubility of sodium chloride does not change very much with temperature changes (see Figure 2.4). To obtain crystals of salts that have similar solubilities in hot and cold water, it is necessary to evaporate nearly to dryness.

Many salts that are obtained by crystallization from water, especially if they are formed by slow cooling of hot solutions, contain water of crystallization (see Section 15.7).

Investigation 15b. The formation of ammonium chloride

Harmful

In the last investigation, we took care to see that the solution was almost exactly neutral before evaporation of the water was started. The work can be repeated by running ammonium hydroxide solution into hydrochloric acid until a piece of litmus paper turns blue:

$$NH_4OH(aq) + HCl(aq) \rightarrow NH_4Cl(aq) + H_2O(l)$$

In this case, however, there is no need to carefully add more acid. Pour the solution into an evaporating basin and place it on a tripod and gauze, if possible in a fume cupboard. Boil the solution and the extra ammonium hydroxide will be removed from the solution as ammonia gas:

$$NH_4OH(aq) \rightarrow NH_3(g) + H_2O(l)$$

Stop heating as soon as white fumes appear to be coming from the evaporating basin. You will soon realize why it is preferable to carry this experiment out in a fume cupboard, if one is available.

15.2. Neutralization of acids and bases

In the investigations that you have just finished, you probably noticed that the solution became hot when the acid and the alkali were mixed. We shall now find out how much heat is liberated when a neutralization occurs.

Harmful

Corrosive

Investigation 15c. The heat of neutralization

For this investigation, you will need some 3M hydrochloric acid, 3M nitric(V) acid, 3M sodium hydroxide solution and 3M potassium hydroxide solution, and some 3/2 M sulphuric(VI) acid.

158

(The sulphuric(VI) acid is of a different strength to the other solutions because it is dibasic – see Chapter 13.)

Use a test-tube (100 mm × 16 mm is a suitable size) to measure some of the acid into a polythene (or other plastic) beaker. Carefully stir the acid with a −10 °C to 110 °C thermometer. Meanwhile, use another test-tube of the same size to measure out a similar volume of the alkali to be used. Allow the alkali to stand in the test-tube in a rack for about five minutes (so that it is at room temperature). When the temperature of the acid in the beaker is constant, record the temperature. Pour the alkali from the test-tube into the beaker and stir the mixture with the thermometer. Record the new steady temperature shown by the thermometer.

Make a copy on Table 15.1 and enter in the spaces the rise in temperature for each combination of acid and alkali.

TABLE 15.1. RISE IN TEMPERATURE WHEN AN ACID IS NEUTRALIZED BY AN ALKALI

Alkali	Acid		
	Hydrochloric	Nitric(V)	Sulphuric(VI)
Sodium hydroxide			
Potassium hydroxide			

You may also try using ethanoic acid and ammonium hydroxide solution. These also should be 3M solutions.

Can you suggest why several of the results are nearly the same?

Harmful

15.3. The action of acids on metals

We have already seen that one of the properties of acids is their corrosive action on metals. When an acid acts on a metal, hydrogen gas is liberated and a salt is formed. We should now test this general principle to see if it holds good for all metals and acids.

Investigation 15d

For this investigation you will need a number of test-tubes and some boiling tubes. You will also need some dilute (about 2M) hydrochloric, nitric(V) and sulphuric(VI) acids. We can then investigate the action of these acids on some metals. Draw up a

Harmful

table similar to Table 15.2 and fill in the names of the metals that you use as you carry out each stage in the investigation.

TABLE 15.2   THE ACTION OF METALS ON ACIDS

Metal	Acid		
	Hydrochloric	Nitric(V)	Sulphuric(VI)
Copper			
Zinc			
Iron			
Lead			
Tin			
Magnesium			

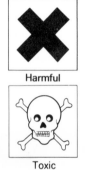

Harmful

Toxic

Metals that you should use are copper, zinc, iron, lead, tin and magnesium.

About half fill a test-tube with the dilute acid and then drop in the sample of the metal (note: if the metal is dull in appearance, it should be scraped with a knife blade or rubbed with emery cloth before use). Cover the open end with a boiling tube as shown in Figure 15.2.

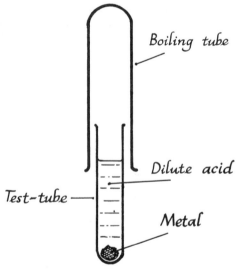

Figure 15.2 Examining the products of the reaction between a metal and an acid.

Record, on your copy of Table 15.2, the amount of action that occurs and the nature of any gas that is evolved. A gas is evolved if you are able to observe bubbles coming from the surface of the metal. Observe the colour of the gas. If it is colourless, test it for hydrogen; if the gas is brown, stop the reaction by pouring the contents of the tube into a sink full of water.

Make a list of metals in order of their activity towards the acid. This list is an extension of the series that was started in Table 11.1. Add hydrogen to the list so that all the metals above hydrogen react with acids or water giving off hydrogen and all those below hydrogen do not give off hydrogen when they are dropped into acid.

Only one metal is able to evolve hydrogen from nitric(V) acid – that metal is magnesium; even this metal will only liberate hydrogen if the acid is cold and very dilute. Most other metals cause a brown gas, called nitrogen dioxide, to be given off. This occurs because nitric(V) acid is a good source of oxygen.

Now take the solution of zinc and/or magnesium in sulphuric(VI) acid and add more metal until no more will dissolve. Transfer the mixture to a small beaker and evaporate off the water. On warming, any small pieces of undissolved metal should dissolve; if they do not, add a little more sulphuric(VI) acid. As the water is evaporated, test for crystallization as described in Investigation 15a. The crystals that form are magnesium or zinc sulphate(VI). When the solution is ready to crystallize, set it aside and allow it to cool. When you have a good collection of crystals, lay them on a piece of paper and try to draw the shape of the crystal.

Harmful

15.4. Other ways to make a salt from an acid

In Investigation 15a, we prepared salts by the action of an acid on an alkali; we saw in Chapter 11 that an alkali was a soluble base. We can now examine the action of an acid on an insoluble basic oxide.

Investigation 15e. The action of sulphuric(VI) acid on a metal oxide

Take a 100 cm³ beaker and about one-third fill it with 2M sulphuric(VI) acid. Now stir the acid, adding, a little at a time, some copper(II) oxide, zinc oxide or magnesium oxide. When there is a little undissolved oxide left in the beaker, place the beaker on a tripod and gauze and warm the mixture. Add more acid or more oxide, as necessary, so that the final hot solution does not contain any undissolved oxide and so that it contains a slight excess of acid. Now evaporate off some of the water and test for crystallization as before. When crystals begin to form, set the beaker aside and allow the crystals to settle at the bottom of the beaker. Again make a sketch of the crystal shape and compare this with your other results and with the shape shown in Figure 2.5.

Harmful

The action of copper(II) oxide with sulphuric(VI) acid has an equation of the following form:

$$CuO(s) + H_2SO_4(aq) \rightarrow CuSO_4(aq) + H_2O(l)$$

In Investigation 13b, we saw that acids evolved carbon dioxide from carbonates. This occurs because carbonic acid, that is the acid from which carbonates are made, is a weaker acid than the acids that we usually meet in the laboratory.

Investigation 15f. The action of sulphuric(VI) acid on metal carbonates

Harmful

We can now examine the result of the action of an acid on carbonates, by repeating Investigation 15e using copper(II) carbonate, magnesium carbonate or zinc carbonate in place of the oxides. Again allow the solutions to crystallize and compare the shape of the crystals with those made before.

The equation for the action of sulphuric(VI) acid on magnesium carbonate is:

$$MgCO_3(s) + H_2SO_4(aq) \rightarrow MgSO_4(aq) + CO_2(g) + H_2O(l)$$

15.5. Direct combination

In addition to using acids, salts can be prepared by combining two elements together directly; in fact, this is the only way by which some salts can be obtained. This direct combination is known as **synthesis**.

Investigation 15g. The synthesis of a chloride of iron

Iron is an element that has two different valency states, both of which are fairly common. These are two valent, when the compounds were called *ferrous*, and three valent, when the compounds were called *ferric*. We shall prepare some iron(III) chloride.

Toxic

Arrange a test-tube with a side arm and a length of glass tubing, about 200 mm long and 8–9 mm in bore, as shown in Figure 15.3(a). If a test-tube with a side arm is not available, use a test-tube fitted with a rubber bung as shown in Figure 15.3(b). Slide three or four granules of calcium chloride into the length of glass tubing as shown. Now take two or three pieces of iron wire, about 100 mm long, and scrape them clean with a knife blade or emery cloth. Remove any grease from the surface by dipping them in a suitable solvent, tetrachloromethane for example. After this, handle them at one end only. Allow the iron wire to dry in the air and then slide it into the glass tube. Fit a teat pipette to the bung for the test-tube and place a small pile of potassium manganate(VII) crystals into the test-tube. *Carefully* fill the teat pipette, keeping the jet pointing downwards, with concentrated hydrochloric acid and replace the bung in the test-tube, so that the acid will drop on to the potassium manganate(VII) crystals.

Corrosive

Figure 15.3 Synthesis of iron(III) chloride

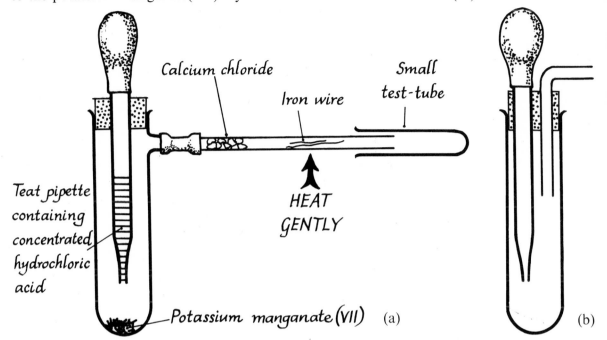

Calcium chloride

Iron wire

Small test-tube

Teat pipette containing concentrated hydrochloric acid

HEAT GENTLY

Potassium manganate (VII)   (a)

(b)

Gently squeeze the teat of the pipette so that *two or three* drops of the acid fall on to the potassium manganate(VII). A greenish coloured gas, chlorine, will be produced. Add a *few* more drops of acid until you can see the chlorine in the glass tube near the iron wire. Now, using a small burner if possible, heat the glass tube near the iron wire until the reaction starts. Once it has started, remove the burner and add more acid, *a drop at a time*, to the potassium manganate(VII). The chlorine combines directly with the iron and black coloured crystals of iron(III) chloride are collected in the small test-tube at the end of the apparatus:

$$2Fe(s) + 3Cl_2(g) \rightarrow 2FeCl_3(s)$$

★ *Warning – the gas produced in this investigation is poisonous. Take care not to inhale any of it.*

Toxic

Some salts (iron(III) chloride is one of these) have to be prepared in this way, since if they are made in solution the water reacts with the salt to prevent the formation of the salt as a solid.

15.6. The preparation of insoluble salts

We have learned that acids, bases and salts are ionic in their nature and that solutions of these will contain ions. We will now examine the effect of mixing some salt solutions together.

Investigation 15h. Mixing solutions of salts

Toxic

Take a small test-tube and place a little barium chloride solution in the tube; add to this some copper(II) sulphate(VI) solution. The solutions will turn milky and if allowed to stand, a white solid will settle out at the bottom of the tube. This solid is called a **precipitate**. Now make a copy of Table 15.3.

TABLE 15.3. FORMATION OF PRECIPITATES

Salt Solution	Metal Ion \ Acid Ion	AgNO₃ NO₃⁻	Na₂CO₃ CO₃²⁻	CuSO₄ SO₄²⁻	NaCl Cl⁻	Pb(NO₃)₂ NO₃⁻	K₂SO₄ SO₄²⁻	BaCl₂ Cl⁻
AgNO₃	Ag⁺	X						
Na₂CO₃	Na⁺		X					
CuSO₄	Cu²⁺			X				P
NaCl	Na⁺				X			
Pb(NO₃)₂	Pb²⁺					X		
K₂SO₄	K⁺						X	
BaCl₂	Ba²⁺			P				X

Repeat the previous experiment using different combinations of solutions. When a precipitate is formed put a P in the appropriate boxes: the table shows the result of the experiment with copper(II) sulphate(VI) and barium chloride solutions.

From your table of results try to decide the answers to these questions:

a. Which is the most soluble acidic ion?

b. Which is the least soluble acidic ion?

c. Which is the least soluble metal ion?

From these experiments we see that salts that are insoluble may be prepared by mixing solutions containing their ions. Hence, we can represent the formation from solutions of copper(II) sulphate(VI) and barium chloride as:

$$Ba^{2+}(aq) + SO_4^{2-}(aq) \rightarrow BaSO_4(s)$$

or

$$BaCl_2(aq) + CuSO_4(aq) \rightarrow CuCl_2(aq) + BaSO_4(s)$$

164

15.7. Water and salts

Many of the salts that we have made owe their special crystalline shape to the presence of water in the ionic lattice. If this water is removed the crystal loses its characteristic shape. This water is known as **water of crystallization**.

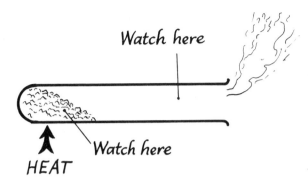

Figure 15.4 Heating copper(II) sulphate(VI) crystals

Investigation 15i. Removing water of crystallization

Take a small test-tube and place a few crystals of copper(II) sulphate(VI) in the tube. Hold the tube in a horizontal position and gently warm it. Carefully watch both the substance and the colder walls of the tube near to the mouth (see Figure 15.4).

Harmful

As the water is driven out, the crystals lose their shape and blue colour. The hydrated blue copper(II) sulphate(VI) has been changed into **anhydrous** white copper(II) sulphate(VI) powder:

$$CuSO_4 . 5H_2O(s) \rightarrow CuSO_4(s) + 5H_2O(l)$$

When crystalline, copper(II) sulphate(VI) contains five moles of water of crystallization to each mole of copper ion.

When the residue is cold, tip it out on to a watch glass. Now allow a few drops of water to fall on to the anhydrous copper(II) sulphate(VI). What happens?

Investigation 15j. The atmosphere and salts

Take two watch glasses: on one place a few washing soda crystals and leave this watch glass in a dry place; on the other place some granules of calcium chloride and leave this one uncovered in the laboratory.

Some substances which contain water of crystallization lose water to the air on standing: such substances are said to be **efflorescent**. Substances that are efflorescent do not always lose water – whether they do or not depends upon the amount of water already in the air. A salt may effloresce in the dry air of a laboratory, but may retain its water of crystallization in the damper air of a bathroom or kitchen.

Other substances become damp when exposed to the air: these substances are said to be **hygroscopic**. If a hygroscopic substance absorbs so much water that it dissolves in this water, it is said to be **deliquescent**. What sort of substances are washing soda and calcium chloride?

Substances that are able to absorb water from the air are used as **drying agents**. The apparatus that is used to contain the drying agent and the substance to be dried is called a **desiccator** (see Figure 15.5).

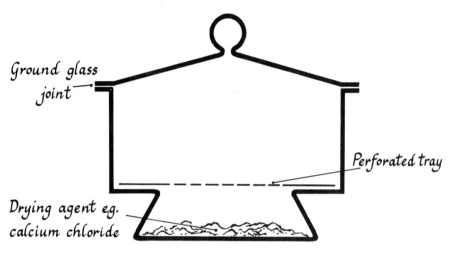

Figure 15.5   A desiccator

Washing soda crystals are a hydrated form of sodium carbonate, $Na_2CO_3.10H_2O$. In dry air, the washing soda loses some of its water of crystallization to the air and becomes a powder. Another good example of a substance that, like calcium chloride, is deliquescent, is sodium hydroxide.

15.8. Salts of dibasic acids

We saw, in Chapter 13, that an acid is formed when a substance dissolves in water producing $H^+(aq)$ ions. We also learned that some acids can produce more than one $H^+(aq)$ ion from each molecule. Such acids include sulphuric(VI) acid, which can produce two $H^+(aq)$ ions in dilute solution, and phosphoric(V) acid, which can produce three $H^+(aq)$ ions in dilute solution. We will now examine the result of dissolving a salt of sulphuric(VI) acid in water.

Investigation 15k. A solution of sodium hydrogensulphate(VI)

Take a test-tube and half fill it with distilled water. Drop in a piece of blue litmus paper or a few drops of universal indicator solution. Now add a little sodium hydrogensulphate(VI) and shake the tube. What happens?

Sodium hydrogensulphate(VI) contains an ion which is itself an acid, the hydrogensulphate(VI) HSO_4^- ion. When dissolved in water this ion produces more H^+ (aq) ions:

$$HSO_4^-(s) + aq \rightarrow H^+(aq) + SO_4^{2-}(aq)$$

Test your understanding

1. If solutions of potassium hydroxide and nitric(V) acid are mixed, which ions are removed?
2. Write the equation for the action of potassium hydroxide on nitric(V) acid.
3. Which metals lie below hydrogen in the activity or electrochemical series?
4. Why is nitric(V) acid not used in the preparation of hydrogen?
5. Write the equation for the reaction of nitric(V) acid with zinc oxide.
6. Write the equation for the reaction of copper(II) carbonate with hydrochloric acid.
7. What is the formula for iron(II) chloride?
8. Why do you think that you had to put some calcium chloride into the tube when you made some iron(III) chloride? Why can you use calcium chloride for this purpose?
9. Name TWO metals whose common salts are all soluble.
10. Why do some salts have to be prepared by direct synthesis from the elements when others may be made from solution?

Chapter 16

Displacement of Metals

We have seen that strong acids are able to displace weak acids from salts; for example, when hydrochloric acid is dropped on to a carbonate the carbonic acid is displaced and carbon dioxide is liberated. In this chapter, we shall examine the behaviour of metals to see if they show a similar reaction.

16.1. Replacement by metals

We have already learned that some metals are able to exist in a free state; we use these metals for coinage.

Investigation 16a. The displacement of copper from solution

Harmful

For this investigation, take seven test-tubes and half fill them with a saturated solution of copper(II) sulphate(VI). You will also need some magnesium, aluminium, zinc, iron, lead, tin and copper. These metals should be in strip form (either wire or ribbon or thin strips cut from foil). Arrange the strips of metals so that they hang in the test-tubes dipping into the copper(II) sulphate(VI) solution. Set these tubes aside for about fifteen minutes and then examine the tubes. What do you see at the bottom of the tubes?

Take out a metal strip. What has happened to it? Do all the metals behave in the same way?

Compare the colour of the solution remaining in the tubes with some of the original solution. Has it changed?

If the solution has changed colour, copper has been removed from the solution and its place has been taken by the other metals. You should have seen a coating of copper on the metal or possibly some copper at the bottom of the tube. You may have observed that the solution was not such a bright blue as the original solution:

$$Cu^{2+}(aq) + Fe(s) \rightarrow Cu(s) + Fe^{2+}(aq)$$

The ionic equation given is for the experiment with iron. The copper ions in the solution of copper(II) sulphate(VI) have lost their charge and have been deposited as copper atoms, whilst the iron atoms in the strip of iron have lost electrons to become ions.

This is a chemical change and so we should expect that the energy of the system has changed also.

Investigation 16b. Is heat evolved or absorbed when copper ions are removed from solution?

This experiment is best done in polythene test-tubes or small polythene bottles (about 30 cm³ capacity), but glass test-tubes could also be used. Measure 10 cm³ of one molar copper(II) sulphate(VI) solution into a tube or bottle. (The meaning of molar is explained in Section 13.2.) Stir this with a thermometer and record the temperature. Add a *spatula-load* of powdered zinc to the test-tube and stir with the thermometer. Record the highest temperature reached.

Harmful

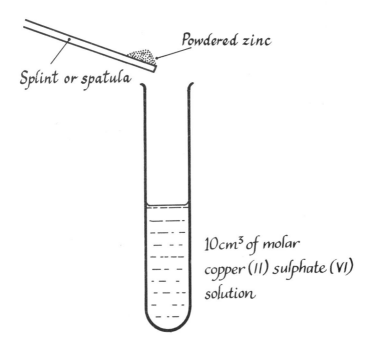

Powdered zinc

Splint or spatula

10cm³ of molar copper (II) sulphate (VI) solution

Figure 16.1 The addition of zinc to copper(II) sulphate(VI) solution

The exact quantity of metal added to the tube is not important, but should be sufficient to decolourize the copper(II) sulphate(VI) solution. A wood splint may be used instead of a spatula; a heap of powdered metal on the 5 mm at the end would be sufficient.

This investigation should be repeated with powdered iron, tin, magnesium and lead.

Make a list of the metals that you use, placing the one that gives the greatest rise in temperature first and the one that gives the least rise in temperature last. Do you notice anything familiar about this order?

16.2. The electrical properties of metals

We observed, in Investigation 16a, that copper(II) ions may be displaced from solution by other metals. That is, the copper(II) ions gain electrons to become atoms:

$$Cu^{2+}(aq) + 2e^- \rightarrow Cu(s)$$

whilst other metals lose electrons to become ions:

$$M(s) \rightarrow M^+(aq) + e^-$$

or

$$M(s) \rightarrow M^{2+}(aq) + 2e^-$$

This constitutes a flow of electrons. We can examine this flow if we place two metals in solution.

Investigation 16c. Electricity between two metals

For this investigation, you will need a copper block, a copper rod and a voltmeter reading up to 3 volts, together with some connecting wires. The copper block may be made by screwing a piece of copper sheet (about 50 mm square) to a block of wood and fixing a terminal, preferably by soldering, to one corner of it. The arrangement is shown in Figure 16.2.

Figure 16.2 Investigating the flow of electricity between metals

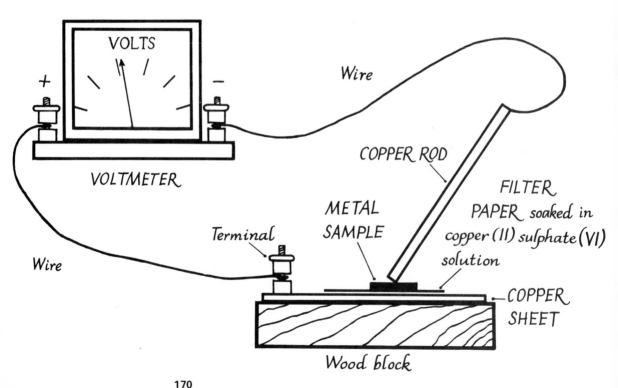

Before use, the copper block and the copper rod should be thoroughly cleaned with steel wool. Connect the copper block to the positive terminal of the voltmeter and the copper rod to the negative terminal. Take a piece of filter paper (a small piece may be used) and soak it in copper(II) sulphate(VI) solution. Lay this piece of filter paper on the copper block, place a piece of zinc foil on top of the filter paper and press down hard with the copper rod. Record the reading on the voltmeter.

Repeat the experiment, using small pieces of other metals. If the meter deflects in the opposite direction, reverse the connections and record the result as a negative value.

Your teacher may repeat this experiment using *small* pieces of sodium and potassium metals.

Make a list of the results, placing the highest value first and the lowest positive (or highest negative) value last.

In this investigation, you have made a simple electric cell by using two different metals separated by a solution of an electrolyte (see Section 5.2).

Compare the order of the metals that you obtained with that given in Table 16.1.

Harmful

Explosive

Flammable

★ *Warning – this should only be done by a teacher. It is advisable to place a safety screen between the class and the apparatus.*

TABLE 16.1. THE ELECTROCHEMICAL SERIES

Potassium
Calcium
Sodium
Magnesium
Aluminium
Zinc
Iron
Lead
(Hydrogen)
Copper
Mercury
Silver
Gold

Hydrogen is placed in this series since, in some ways, it resembles metals. You will notice that the metals listed before hydrogen may liberate the gas from dilute hydrochloric acid. Those placed after hydrogen in the list do not react with dilute hydrochloric acid. You should also note that metals that are high in the series will displace metals below them from solutions of their salts.

16.3. Sources of metals

Only the metals that are low in the electrochemical series (that is gold, silver, and a little mercury and copper) occur in the uncombined state. All the other metals must be obtained from compounds that they have formed with other elements; these compounds are known as **ores**. Generally speaking, the more reactive metals, like sodium, are obtained from chlorides and the less reactive metals, like iron, from oxides or sulphides.

In all these cases, the metal ore has to be converted into a pure metal: in other words, metal ions have to be changed to metal atoms, for example:

$$Na^+(l) + e^- \rightarrow Na(l)$$

This process, the addition of electrons, is known as **reduction**. Table 17.1 gives a list of the ores of some metals and the methods used to obtain the metals from their ores.

16.4. Reduction of metal oxides

Where a metal is combined with oxygen, we need to use a substance that will remove the oxygen from the metal. A suitable element for this purpose, in some cases, is carbon. Carbon has the advantage that its compound with oxygen, carbon dioxide, is a gas. This gas will escape and leave the uncombined metal.

Investigation 16d. Reduction of some metal oxides

Toxic

Take a charcoal block (70 mm × 30 mm × 30 mm is a suitable size) and, by rotating a penny, make a shallow depression at one end. Place a little lead(II) oxide (litharge) in this depression and, to prevent it being blown out, allow one drop of water to fall on to the oxide. We must now heat the oxide on the charcoal block with a very hot flame. This requires quite a lot of practice.

Take a mouth blowpipe and place the wider end in your mouth; blow down this end and you should feel a jet of air coming out of the small hole at the other end. Practise this until you can keep up this stream of air for about a minute. Now take a bunsen burner and set it to give a luminous flame about 100 mm high. Place the narrow end of the blowpipe near (or just on) the barrel at the top of the bunsen burner. Turn the blowpipe so that the jet of air will blow across the flame and blow steadily. When you can do this, hold the charcoal block at one end so that the lead(II) oxide is in the hot jet of flame. This is illustrated in Figure 16.3.

When you have heated the lead(II) oxide for about two minutes, remove the charcoal block from the flame and see if there has been any change. If not, repeat the heating process for a little longer. What change can you see?

172

When the charcoal is cool, scrape out the residue and look at it carefully. Is it shiny? Can you cut it with a knife? Does it mark paper?

MAKING THE DEPRESSION

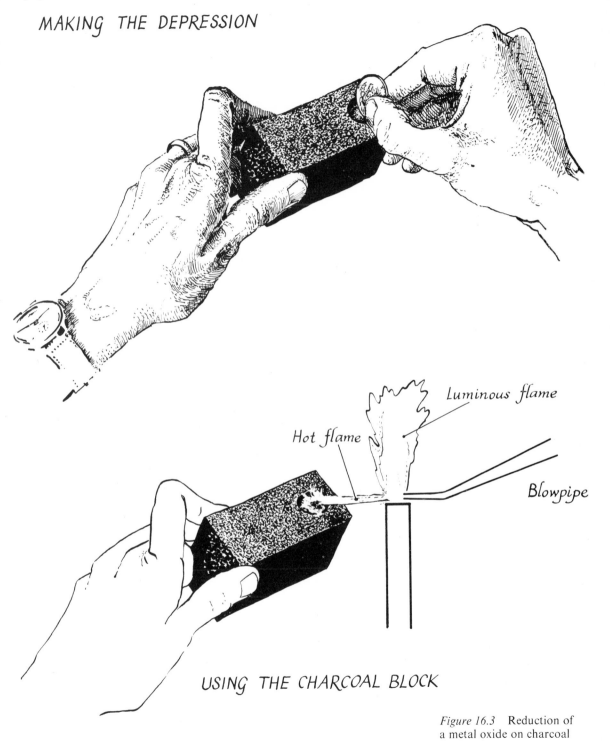

USING THE CHARCOAL BLOCK

Figure 16.3 Reduction of a metal oxide on charcoal

If the answer to these three questions is *yes*, you have reduced the lead(II) oxide to lead:

$$2PbO(s) + C(s) \rightarrow 2Pb(s) + CO_2(g)$$

This is one of the easiest charcoal block reductions to carry out. You should now try to reduce some black copper(II) oxide and some red iron(III) oxide in the same way.

How could you examine the residues that you have obtained to see if you have copper and iron?

Other methods can be used to reduce metal oxides to metals. Gases which are capable of combining with oxygen in the oxide are particularly suitable. In Investigation 8b, black copper(II) oxide is reduced to copper metal by using a gas. One of the gases that will reduce some metal oxides is hydrogen. Another gas which will reduce metal oxides to metals is carbon monoxide. Both hydrogen and carbon monoxide are able to combine with the oxygen in the copper(II) oxide, and are known as reducing agents:

$$CuO(s) + H_2(g) \rightarrow Cu(s) + H_2O(g)$$
$$CuO(s) + CO(g) \rightarrow Cu(s) + CO_2(g)$$

This use of hydrogen is explained in Chapter 12.

16.5. Using electricity to obtain metals

We have already learned that metals form salts with other elements, in which the metal becomes a positive ion. We have also seen that solutions of such salts will conduct electricity.

Investigation 16e. Passing electricity through a solution of copper(II) sulphate(VI) (see also Investigations 16f and 16g)

Harmful

For this investigation you will need a 6-volt battery (or other source of a 6-volt d.c. supply), two carbon rods (or plates) and one iron and two copper plates (these should be as large as it is possible to fit into a 250 cm³ beaker). You will also need a 250 cm³ beaker about half full of one molar copper(II) sulphate(VI) solution and some connecting wires. If you have an ammeter reading up to about 3 amperes, or a 6-volt bulb in a holder, it will be useful.

Take the two carbon rods, make sure that they are clean and then attach a wire to each (crocodile clips are best for this). Now, place the two carbon rods in the beaker of solution, making sure that they do *not* touch. Do *not* make any other connections at this stage. Leave the rods in the beaker for about two minutes. After this time, remove the rods and see if they have changed in any way.

The carbon should not show any change.

Now, connect one of the wires from a carbon rod to the positive terminal of the battery. If you have a bulb or ammeter, connect the wire from the other carbon rod to the bulb or to the positive terminal of the ammeter. Finally, connect the negative terminal of the battery to the other terminal of the ammeter or bulb. If you have not got a bulb or ammeter, connect the negative terminal of the battery directly to the second carbon rod. Lower the two carbon rods into the solution of copper(II) sulphate(VI) taking care that they do *not* touch. If you are using a meter, it should now give a reading; if you are using a bulb, it should glow. Figure 16.4 shows the arrangement of the apparatus using an ammeter. The figure also shows the circuit diagram, in electrical symbols.

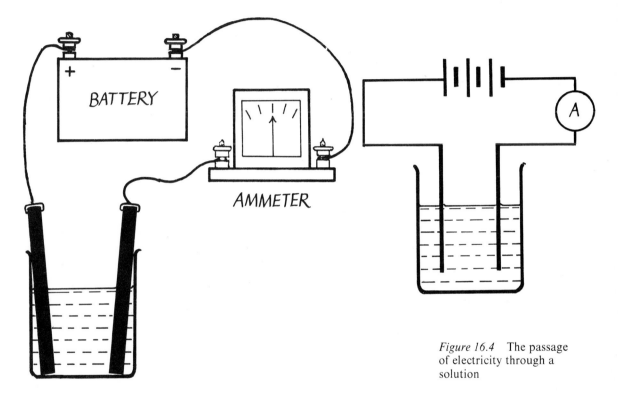

Figure 16.4 The passage of electricity through a solution

After two minutes, disconnect the wire attached to the positive terminal of the battery and lift out the two carbon rods. Examine them to see if there is now any change.

When you are satisfied that you have made a note of any changes that you can see, replace the rods, making sure that they do *not* touch each other. Reconnect the wire to the positive terminal of the battery. Allow the experiment to continue for another ten minutes and then again disconnect the wire from the positive terminal of the battery. Compare the colour of the solution with the colour of the original solution. (If the solution in the beaker appears cloudy, allow the beaker to stand so that the cloudiness clears.) Remove and again examine the two carbon rods, keeping the one that was connected to the negative terminal of the battery. (Note: this connection may have been made through the ammeter or bulb.)

Investigation 16f. Recovering copper

Take the carbon rod that was connected to the negative terminal of the battery in the last investigation. You will have noticed that this became covered with a red-brown coloured substance during the last investigation. What do you think this substance is?

Now, connect this carbon rod to the positive terminal of the battery. Clean the iron plate thoroughly with steel wool and then connect it to the negative terminal of the battery, again using the ammeter or bulb, if available. Make sure that the carbon rod and the iron plate do *not* touch. Empty the beaker and rinse it with water. Then about half fill the beaker with one molar sulphuric(VI) acid solution. Lower the carbon rod and the iron plate into the beaker, taking care that they do not touch. Watch the contents of the beaker carefully. After about twenty minutes, disconnect the battery, lift out the iron plate and the carbon rod and rinse them with water. What has happened?

Harmful

Investigation 16g. Plating with copper

Take the two copper plates and clean them thoroughly with steel wool. Wash them well in distilled water, and then rinse them in ethanol or propanone and allow to dry. (The ethanol or propanone rinses off the water and evaporates quickly.) Mark one plate, in some way, near the top and then weigh both of the plates as accurately as possible. Put the plates into a 250 cm³ beaker and about half fill the beaker with one molar copper(II) sulphate(VI) solution. Connect the circuit as before and allow electricity to pass for about thirty minutes. At the end of this time, disconnect the plates from the battery. Rinse the plates well in distilled water and again dip them into ethanol or propanone. When the plates are dry, reweigh them and compare the new masses with the masses at the beginning of the experiment. Does anything happen to the copper(II) sulphate(VI) solution used during the experiment?

Flammable

Harmful

If you do not have an accurate balance available, the careful cleaning and weighing may be omitted. If no balance is available, the plates could be replaced by wire, and the thickness of the wire before and after the experiment could be compared.

Whilst you are waiting for this experiment to finish, look back to Investigation 5b. Compare the results of that investigation with those you have observed in the last three investigations (16e, 16f and 16g).

In Investigation 16e, copper from the electrolyte, copper(II) sulphate(VI), was deposited on the electrode, carbon, that was connected to the negative terminal of the battery. No change was noticed when the battery was not connected, nor was any change observed on the other carbon rod. The electrode connected to the negative terminal of the battery is called the **cathode** and the one connected to the positive terminal is called the **anode**. The copper(II) sulphate(VI) solution becomes paler in colour because copper ions, which are positively charged, are deposited on the cathode. Ions which move to the cathode are called **cations** and ions that move to the anode are called **anions**.

In Investigation 16f, the copper on the surface of the carbon anode is eventually transferred to the steel cathode. In this case, the copper must be transferred from the anode since there is no copper in the solution.

In Investigation 16g, copper is lost from the anode and gained by the cathode. If you were able to find the mass of the two electrodes, you would have observed that the loss in mass of the anode was nearly equal to the gain in mass of the cathode. The concentration of the copper(II) sulphate(VI) solution remains constant.

16.6. Electroplating

By using the process of electrolysis, a metal surface may be covered with a thin layer of another metal. The process is rather similar to that used in Investigation 16f and is known as **electro-**

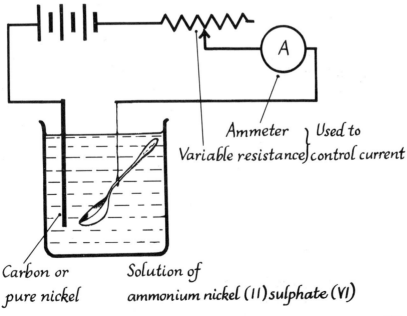

Carbon or pure nickel

Solution of ammonium nickel (II) sulphate (VI)

Ammeter ⎱ Used to
Variable resistance ⎰ control current

Figure 16.5 Nickel-plating a spoon

plating. The object to be plated is made the cathode and either carbon, or a pure sample of the plating metal, is made the anode. The electrolyte is a salt of the metal to be deposited. Figure 16.5 shows how a spoon could be plated with nickel.

Plating is usually used to deposit a thin layer of expensive metal on the surface of a cheaper metal, for example silver-plating. The solution used for this is a poisonous cyanide.

★ *Warning – cyanide should NEVER be used in a school laboratory.*

16.7. Extraction and purification of copper

Copper is one of the few metals that has been found in the free state. When it does occur uncombined, it is known as **boulder copper**. Copper can be removed from boulder copper in the same way that the copper was recovered in Investigation 16f. The boulder copper is made the anode in electrolysis and a piece of pure copper is used as the cathode. When the electricity is switched on, copper is removed from the boulder and is deposited on the cathode.

In a similar way, pure copper can be obtained from an impure sample. The impure sample of copper is used as the anode and the copper is deposited on a pure copper cathode. In both of these cases copper(II) sulphate(VI) is usually used as the electrolyte.

Electrolysis is used in the extraction of several metals (see Chapter 17) and also some non-metals, for example chlorine (see Chapter 29).

16.8. Explanation of electrolysis

a. *Acidulated water, with platinum electrodes.* In Chapter 12 we saw that if an electric current is passed through a dilute solution of sulphuric(VI) acid in water, two volumes of hydrogen are liberated at the cathode and one volume of oxygen is liberated at the anode. Further, the sulphuric(VI) acid remains unchanged. However, the same number of electrons must leave the solution as enter it. Hence we can represent the changes as:

At the cathode	*At the anode*
$4H^+(aq) + 4e^- \rightarrow 4H(g)$	$2H_2O(l) \rightarrow 4e^- + O_2(g)$
$4H(g) \rightarrow 2H_2(g)$	$+ 4H^+(aq)$
Product 2 moles of hydrogen given off	1 mole of oxygen given off
Electrical change 4 moles of electrons enter	4 moles of electrons leave

b. *Copper(II) sulphate(VI) solution.* In this case, there are four ions: $Cu^{2+}(aq)$ and $SO_4^{2-}(aq)$ from the copper(II)

sulphate(VI), and H^+(aq) and OH^-(aq) from the water. The copper ions lose their charge at the cathode in preference to the H^+(aq) ions. As a result, copper is deposited on the cathode:

$$Cu^{2+}(aq) + 2e^- \rightarrow Cu(s)$$

If the anode is made of copper, this dissolves in preference to a discharge of either sulphate(VI) or hydroxyl ions.

$$Cu(s) \rightarrow Cu^{2+}(aq) + 2e^-$$

Hence, we can represent the change as:

At the cathode	*At the anode*
$Cu^{2+}(aq) + 2e^- \rightarrow Cu(s)$	$Cu(s) \rightarrow Cu^{2+}(aq) + 2e^-$
Product 1 mole of copper deposited	1 mole of copper dissolved
Electrical change 2 moles of electrons enter	2 moles of electrons leave

Notice that the number of metal atoms deposited depends upon the number of electrons that pass through the electrolyte, that is upon the current used and the time for which the electricity passes.

To obtain a good firm metal deposit, a low current is used. About 0.02 ampere per 100 mm² of the surface to be plated should give a good result.

Test your understanding

1. Why do you think that copper is a suitable metal for coins?
2. What do you think would happen if you dipped a pen-knife blade into a solution of a silver salt?
3. If you made a simple electrical cell, which pair of metals would you expect to give you the higher voltage: (i) iron and magnesium, (ii) aluminium and zinc?
4. You are given a black powder. When you heat this on a charcoal block it gives a red-brown substance that forms into little balls. If you dissolve some of the original substance in strong acid, it forms a blue solution. What was the substance? Write an ionic equation for the change that occurred on the charcoal block.
5. What type of charge is carried by a cation?
6. If you had a battery that would last forever, can you think of any reason why the electrolysis in Investigation 16g should not go on for ever?
7. A solution contains both magnesium and lead ions. Which would you expect to be deposited first in electrolysis?
8. You have a piece of metal, 15 mm × 30 mm, that you want to plate with silver. What current would you suggest passing through the solution?

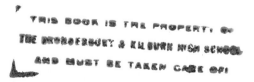

Chapter 17

The Industrial Extraction of Metals

In Chapter 16, we discussed the general methods by which it is possible to obtain samples of metals from their compounds. In this chapter, we shall examine the use of some of the methods on the large scale to obtain metals for industry. You will remember that all the methods used to obtain samples of metals are reduction processes. The method which is used to obtain the metal depends upon the ease or difficulty of this reduction process. The sources, or ores, of some metals are given in Table 17.1; also listed in this table are the methods used for the extraction of the metal.

TABLE 17.1. METAL ORES

Metal	Compound	Extraction Method
Potassium	Chloride	
Sodium	Chloride	Electrolysis
Magnesium	Chloride and	
	Carbonate	Converted to chloride, then electrolysis
Calcium	Carbonate	
Aluminium	Oxide	Electrolysis
Iron	Oxide	Heated with coke
Tin	Oxide	
Zinc	Sulphide	Converted to oxide, then heated with carbon
Lead	Sulphide	
Copper	Sulphide	Heated alone
Mercury	Sulphide	

Gold and silver, and, to a lesser extent, copper and mercury are found as 'free' metals as well as compounds.

Table 17.1 is by no means complete; other compounds are used as ores and other methods are used for extraction.

17.1. The sulphide ores

The ore is first powdered and then agitated with water, which often contains oils. The powdered ore floats on the surface in a

TABLE 17.2. THE SULPHIDE ORES

Metal	Ore	Formula
Zinc	Zinc blende	ZnS
Copper	Copper pyrites	$CuFeS_2$
Lead	Galena	PbS
Mercury	Cinnabar	HgS

froth, whilst impurities settle at the bottom. The purified ore is then skimmed from the surface.

Zinc and lead sulphides

When these sulphides are heated in a current of air, they are converted to oxides:

$$2ZnS(s) + 3O_2(g) \rightarrow 2ZnO(s) + 2SO_2(g)$$

The oxide is then heated with excess coke, either in a fire-clay retort or in a blast furnace, to obtain the metal:

$$ZnO(s) + C(s) \rightarrow Zn(l) + CO(g)$$

Mercury sulphide

This is heated in air; the mercury metal distils as vapour from the hot solid:

$$HgS(s) + O_2(g) \rightarrow Hg(g) + SO_2(g)$$

Copper pyrites

The extraction of copper from the mixed iron and copper sulphide is a rather involved process. The pyrites is roasted in air so that the iron sulphide is converted to iron oxide, whilst most of the copper remains as sulphide:

$$2CuFeS_2(s) + 4O_2(g) \rightarrow Cu_2S(s) + 2FeO(s) + 3SO_2(g)$$

Sand is then added which forms a 'slag' with the iron oxide; this can be removed from the top of the fused copper(I) sulphide. When air is then blown through the copper(I) sulphide, *some* of the sulphide is converted to oxide:

$$2Cu_2S(s) + 3O_2(g) \rightarrow 2Cu_2O(s) + 2SO_2(g)$$

The copper(I) oxide reacts with more of the sulphide, producing sulphur dioxide and copper metal:

$$Cu_2S(s) + 2Cu_2O(s) \rightarrow 6Cu(l) + SO_2(g)$$

Sulphur dioxide is an important by-product of these extraction processes and is often used to make sulphuric(VI) acid.

Iron pyrites

Iron also occurs as a sulphide – iron pyrites, FeS_2 – but it is not economical to extract iron from this ore.

17.2. The extraction of tin

Tin occurs as an oxide, SnO_2, known as **tinstone** or **cassiterite**. When the crushed ore is washed with water, the lighter particles float away and the heavier tinstone sinks. The oxide is then heated with carbon at a high temperature and the liquid tin is tapped off:

$$SnO_2(s) + 2C(s) \rightarrow Sn(l) + 2CO(g)$$

The tin is purified by allowing the liquid tin, which melts at a low temperature to flow away from other solids.

17.3. Purification of metals

The methods used to purify metals depend upon the physical properties of the metals.

Mercury, a liquid, is purified by distillation, in the same way that other liquids are purified. Zinc, which boils at about 1 170 K (about 900 °C), is also distilled from impurities.

Copper is purified by electrolysis, as described in Chapter 16. Electrolysis is also used to obtain very pure samples of lead and zinc. The impure metal is made the anode and the pure metal is collected at the cathode.

17.4. The extraction of sodium

The higher a metal is in the activity (electrochemical) series, the more difficult it is to obtain a sample of the uncombined metal. Sodium is obtained by passing an electric current through a molten compound.

Investigation 17a. To obtain sodium in the laboratory

★ Warning – this investigation should be carried out only by a teacher.

The ore from which sodium is obtained is sodium chloride, which has a melting-point of 1 076 K (803 °C). If, however, an impurity is added, the melting-point is depressed; a mixture of 40 per cent sodium chloride and 60 per cent calcium chloride is used which melts at about 870 K (about 600 °C).

A large porcelain crucible is filled with a sodium chloride/calcium chloride mixture and is placed in a suitable crucible furnace (see Figure 17.1). The construction of such a furnace has been described by Munro in S.S.R. 153 p. 423. After gently heat-

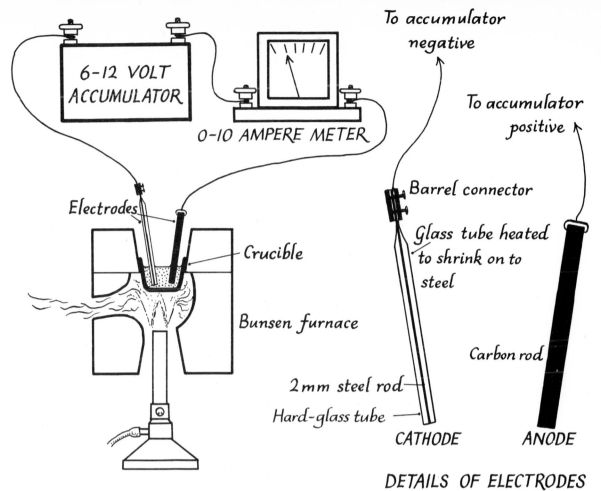

Electrodes

Crucible

Bunsen furnace

6-12 VOLT ACCUMULATOR

0-10 AMPERE METER

To accumulator negative

To accumulator positive

Barrel connector

Glass tube heated to shrink on to steel

Carbon rod

2mm steel rod

Hard-glass tube

CATHODE

ANODE

DETAILS OF ELECTRODES

ing the crucible to drive off any moisture, the crucible is heated strongly to melt the salt mixture. More salt mixture is added until the crucible is about three-quarters full of molten salt mixture.

When the salt has melted, the electrode assembly is lowered into the crucible. This should have been arranged so that both electrodes can be lowered together into the crucible at the same time and should already be connected to the external circuit as shown in Figure 17.1. The cathode, which is made from a length of steel rod about 2 mm in diameter, is protected by a piece of hard-glass tubing.

Whilst the current is passing, hold a piece of moist starch-iodide paper, *in tongs*, over the crucible. What do you observe?

After about ten minutes, remove the electrodes and allow them to cool. At this stage the burner may be turned off. When the cathode is cool, carefully lower it into about 2 cm³ of distilled water in a small beaker. Observe what happens and test the liquid in the tube with a piece of litmus paper.

The ammeter showed that a current was passing through the

Figure 17.1 Extraction of sodium in the laboratory

Toxic

Flammable

Harmful

cell. Since the electrolysis is of fused sodium chloride and there is no water present, the sodium ions are discharged at the cathode and the chloride ions at the anode:

$$2Na^+(l) + 2e^- \rightarrow 2Na(l)$$

$$2Cl^-(l) \rightarrow Cl_2(g) + 2e^-$$

The starch iodide paper is turned blue by the chlorine, and the sodium reacts with the water to give an alkaline solution. The results of the electrolysis of fused sodium chloride should be compared with those obtained by the electrolysis of a solution of sodium chloride (see Chapter 29).

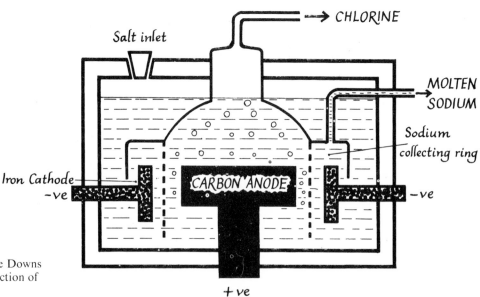

Figure 17.2 The Downs cell for the extraction of sodium

In the **Downs** industrial cell (see Figure 17.2), the molten sodium floats up from the cathode since it is less dense than the molten salt. Any calcium that is formed will be a solid and is filtered off. Sodium melts at 370 K (97 °C) and calcium at 1 118 K (845 °C).

17.5. The extraction of aluminium

Like sodium, aluminium is obtained by passing an electric current through a molten compound. The only satisfactory source of aluminium is the mineral **bauxite**, which is a form of aluminium oxide. The bauxite is purified and dried to produce a form of oxide known as **alumina**. By itself, the alumina will not conduct electricity and so it is dissolved in a molten mineral, **cryolite** (aluminium sodium fluoride), which is a conductor. The casing of the cell is lined with carbon, which acts as the cathode. The anode is made of several large carbon or titanium plates that are lowered into the molten mixture. Like the Downs cell for sodium, no heating is

needed since the passage of electricity through the electrolyte produces sufficient heat to keep it molten. The molten aluminium
settles at the bottom of the cell and becomes the cathode of the
cell. Oxygen is produced at the anode. As the anode is burnt away
it is lowered further into the cell.

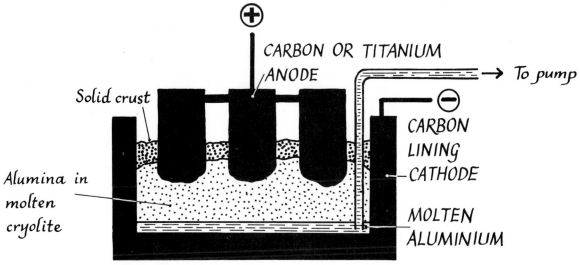

Figure 17.3   The extraction of aluminium

17.6. The extraction of iron

Iron ores, which are fairly widely distributed through the world,
are either iron(III) oxide (**haematite** and **limonite**) or magnetic
oxide of iron (**magnetite**), the most common being iron(III) oxide.
The oxide is reduced to iron in a **blast furnace** by carbon monoxide.

Iron ore, coke and limestone are fed through a double hopper
at the top of the blast furnace, which may be as much as 30 metres
high. The use of a double hopper prevents the loss of hot gases.
A powerful jet of hot air is blown in at the bottom of the furnace
through small-bore pipes.

The coke burns at the bottom of the furnace:

$$C(s) + O_2(g) \rightarrow CO_2(g)$$

This reaction evolves a large amount of heat, and the temperature rises to about 1 700 K (about 1 400 °C). At this temperature the carbon dioxide is reduced to carbon monoxide:

$$CO_2(g) + C(s) \rightarrow 2CO(g)$$

The carbon monoxide reduces the iron(III) oxide to metallic iron:

$$Fe_2O_3(s) + 3CO(g) \rightarrow 2Fe(l) + 3CO_2(g)$$

The iron which is produced gradually melts and runs to the bottom
of the furnace.

The limestone is added to remove the chief impurity, which is in

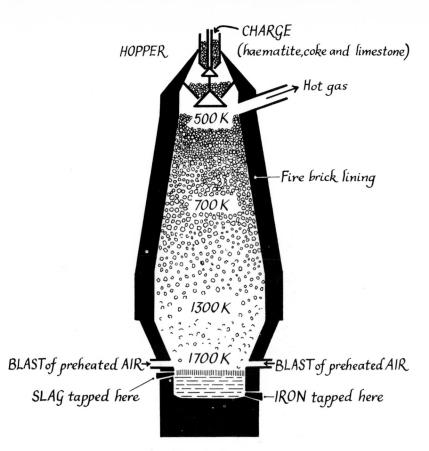

CHARGE
(haematite, coke and limestone)

HOPPER

Hot gas

500 K

Fire brick lining

700 K

1300 K

1700 K

BLAST of preheated AIR →

← BLAST of preheated AIR

SLAG tapped here

— IRON tapped here

Figure 17.4 The blast furnace

the form of sand and is known as silica (SiO_2). At the temperature of the furnace, the limestone decomposes to form quicklime and carbon dioxide:

$$CaCO_3(s) \rightarrow CaO(s) + CO_2(g)$$

(The carbon dioxide is reduced to carbon monoxide and this, in turn, reduces some of the iron oxide.) The quicklime, which is basic, combines with the silica, which is acidic:

$$CaO(s) + SiO_2(s) \rightarrow CaSiO_3(l)$$

The product, which is calcium silicate, is known as **slag**. The molten slag runs down the furnace and floats on top of the molten iron.

The hot waste gas which leaves from the top of the furnace contains a large amount of unburnt carbon monoxide. The gases are mixed with air and burnt; the heat produced is then used to preheat the air blast for the furnace. The molten slag is run out at the bottom of the furnace and is tipped into heaps to cool. This process of running out a molten substance is known as **tapping**. The slag is used in road construction. The molten iron is tapped from the bottom of the furnace and is either run into large moulds to form

186

'**pig iron**' or, more often, is used at once to manufacture steel.

The furnace is operated as a continuous process, raw materials being added at the top as the iron and slag are removed at the bottom. The operation is only stopped when the furnace lining burns out. This is not very frequent, since modern furnace linings last for up to ten years.

17.7. The conversion of iron to steel

Although iron is an element and hence should be pure, the product of the blast furnace, which contains some carbon, is commonly known as 'iron'. Mixtures of metals or metals with other elements are very important to modern industry, because the addition of other elements can be used to improve the properties of a metal. Such mixtures are known as **alloys**.

The so-called 'pig iron' is an alloy of about 95 per cent iron and 5 per cent carbon with traces of other elements, for example silicon and phosphorus. After leaving the blast furnace, the 'iron' is treated to remove most of the carbon content.

Many processes have been used to purify iron after it leaves the blast furnace, but the process that is most successful is known as the LD process. The letters LD are taken from the initial letters of Linz and Donawitz, two Austrian towns where the process was developed.

The process is carried out in a converter arranged as shown in Figure 17.5.

Figure 17.5 The LD converter

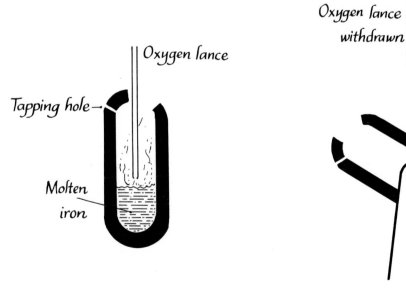

TILTED FOR CHARGING

The converter is tipped forward on its trunnions and molten iron from the blast furnace is poured in from a ladle. The converter then returns to the vertical position and the oxygen lance is lowered into the converter. Oxygen at high pressure is then blown on to the surface of the molten iron. The oxygen causes the carbon that is present to burn and form carbon dioxide, which escapes and so the iron becomes purer.

If the iron from the blast furnace contains a large amount of phosphorus, then lime is blown in with the oxygen. The lime combines with the phosphorus to form calcium phosphate(V).

The product is a low carbon content alloy of iron, usually known as **mild steel**, which is suitable for most purposes. Carefully controlled amounts of carbon and other elements are added to this steel in order to obtain alloys with special properties. Details of these alloys are given in Chapter 18.

Test your understanding

1. Name the process that must always be used to 'win' a metal from its ore.
2. Which metals are extracted from sulphide ores?
3. Why can mercury be purified by distillation?
4. Why is calcium chloride added to sodium chloride when sodium is to be extracted by electrolysis?
5. Why is cryolite used in the manufacture of aluminium?
6. Give the names of two iron(III) oxide ores.
7. Why is a blast furnace so called?
8. Why is limestone added to a blast furnace?
9. Which is denser, molten iron or molten slag?
10. What is the main impurity in iron obtained from a blast furnace?
11. What has to be done to remove this impurity?
12. What is the name given to a mixture of metals?

18.3. The uses of metals

In this section, we shall consider the uses of metals listed in Table 18.1. Iron and steel will be considered in a separate section.

Sodium

Because of its low melting-point and density, sodium is used to transfer heat from the centre of atomic reactors to boilers. It is made to circulate by an electromagnetic method, hence avoiding the use of a pump and the chance of leakage. Its high thermal conductivity and specific heat are other properties that make it suitable for this purpose. It is used in the manufacture of sodium compounds and in the extraction of some metals. Small quantities are used in the manufacture of alloys and sodium street lamps.

Calcium

This metal is used to harden lead for bearings

Magnesium

Because it is very light, magnesium is used in the production of light alloys. Since it burns readily, producing an intense light, it is used in the manufacture of flares.

Aluminium

Since aluminium is a light metal, it is used in the manufacture of aircraft and other large machines that need to be light in weight. It is usual to use an alloy for this purpose since pure aluminium is not very strong. Although a chemically clean aluminium surface reacts fairly rapidly, the aluminium becomes coated with a layer of oxide that protects the metal from further attack. This makes it an ideal metal for packaging food, a use that is well known. Similarly, since the oxide-coated aluminium is still a good reflector of both light and heat, it is being used for reflectors and for thermal insulation.

Since aluminium is high in the activity series, it is used in the extraction of metals by the **thermite process**. It is also a good conductor of heat, so is used in the construction of cooking utensils, and is a good conductor of electricity, so is used for overhead cables. (It is not as good a conductor as copper, by volume, but, weight for weight, it is twice as good. As it would stretch too easily, the cables are made of a central steel core with several strands of aluminium around it.) Aluminium is also used, in powder form, for the preparation of paint.

Zinc

The biggest single use of zinc is the coating of steel to prevent corrosion (see Chapter 10). Several processes are used:

a. *Galvanizing.* The steel object, after completion, is dipped into molten zinc.

b. *Electroplating.* Details of this process, in which electricity is used to produce a thin coating, were given in Chapter 16.

c. *Sherardizing.* The steel object is covered in zinc powder and heated to about 650 K (about 370 °C).

d. *Painting.* Zinc powder suspended in a liquid is painted on to the metal surface.

Combined with copper, zinc produces a well-known alloy, **brass**. Zinc is also used for the negative electrode in the so-called 'dry battery'. Since it is higher in the activity series than gold and silver, it is used to displace these two metals from solutions of their salts.

Tin

The best known use for this metal is the tin-plating of steel containers; it is ideal for this use since it resists attack by most vegetable juices. The process used is similar to that used in galvanizing. Tin is used in the manufacture of alloys, of which the most common are **pewter** and **soft solder**. It is also used to make tubes for food and toothpaste.

Lead

Originally used extensively for roofing and waste water pipes, its high cost is causing these uses to decline. Large amounts of lead are used today in the manufacture of accumulators (electrical storage batteries) and in the manufacture of alloys for engine bearings and solders. Since it is the most effective absorber of atomic radiation, it is used to screen radioactive material. Other uses of lead include the manufacture of lead shot and of white lead, used as a paint and sealing compound.

Copper

Because it is a good conductor of electricity, copper is used for electrical wires and, since it is a good conductor of heat, for steam pipes. It is also used in the manufacture of alloys, for example brass. Copper is used for small-bore water pipes, although improved heat-resistant plastics are beginning to replace it for some purposes.

Mercury

The special property of mercury is that it is a liquid metal. Since it also has a fairly regular rate of expansion, it is used in the manufacture of thermometers. It is also used in the manufacture of some special electrical contacts. Another use for which it is particularly suitable is the manufacture of alloys that are liquids initially but which soon harden; these are known as **amalgams**.

18.4. The properties of iron and steel

As explained in Chapter 17, the blast furnace produces an iron alloy with quite a high carbon content. This product is known commercially as *cast iron*. The metal obtained after treatment, for example in the LD process, contains less carbon and is known as *mild steel*. Table 18.2 gives a list of different steels and their carbon content.

TABLE 18.2. THE CARBON CONTENT OF
IRON AND STEELS

Type of Steel	Percentage of Carbon
Wrought iron	Less than 0.1
Mild steel	0.1 to 0.25
Medium carbon steel	0.25 to 0.5
High carbon steel	0.5 to 1.5
Cast iron	2.5 to 5.0

The carbon content is not the only factor that controls the properties shown by different types of steel; the rate of cooling also has a considerable effect.

Investigation 18a. The heat treatment of steel

For this investigation you will need a few pieces of spring steel (pieces of clock spring or steel knitting needles are ideal).

Keep one of the pieces aside as a standard, for comparison with samples that you experiment with. Take another piece and, holding it in tongs, heat it in a bunsen flame until it is bright red. When it is red-hot, plunge it into a large container of cold water. This process is known as **quenching**. Compare the springiness of the standard and *quenched* samples. Take care not to bend the *quenched* sample too much, since it may snap. Steel treated in this way is said to have been **hardened**.

Take a third sample and heat it in a bunsen flame as before. Keep it at red-heat for about a minute and then remove it from the flame very slowly; this is best done by slowly raising it out of the flame into the hot air above the flame. Allow this sample to cool slowly in the air. This is a process known as **annealing**. When it is quite cold, again compare the properties with the standard sample. Try to bend the annealed piece and see if it straightens when you take away your hands.

Now harden the soft annealed piece of steel in the same way as before. When it is cold, test the end that was heated to see if it has hardened, and then clean the end with emery cloth until it is shiny. Now hold the sample above the bunsen flame and watch the *shiny end* carefully. As soon as the end is *blue*, take the sample of steel away from the flame and allow it to cool. Compare this with the standard sample.

Compare your observations with those given in Table 18.3.

TABLE 18.3   HEAT TREATMENT OF STEEL

Treatment	Properties of Steel
Heated to redness and cooled quickly	Hard and liable to snap (brittle)
Heated to redness and cooled slowly	Soft, can be bent
Hardened, then heated under control and cooled carefully	Springy

The properties of steels can be further modified by the addition of other elements. Details of these alloy steels are given in Table 18.4.

18.5. The uses of iron and steel

Iron and steel are the metals most commonly met in everyday life, and are used for most constructional work. Mild steel, which, unlike cast iron, can be worked with tools, is used for most constructional work. It is used for railway lines, car bodies, angle iron, steel cabinets and many other purposes. Cast iron, which is brittle and cannot be worked with tools, is used, as its name suggests, for **castings**. The molten iron is poured into sand moulds. In a steel foundry, articles such as fire-grates and drain covers are made by this method. Cast iron, that is iron with 4 to 5 per cent carbon, has the advantage that it expands slightly on cooling and so produces a good impression of its mould.

Figure 18.2   Iron and steel in use

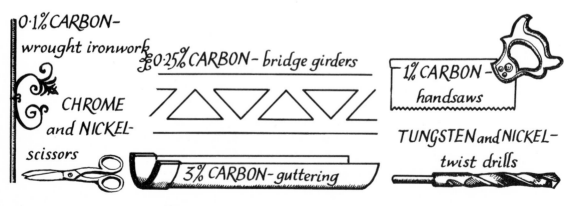

194

TABLE 18.4. ALLOY STEELS

Added Element	Special Property	Uses
Carbon (about 1 per cent)	Hard, can be sharpened	Cutting tools
Manganese	Very hard	Rock crushing tools, steel helmets
Chromium and nickel	Resists rust	Stainless steel, cutlery, surgical instruments
Tungsten, vanadium and chromium	Not softened, even when hot	High speed cutting tools, for example drills
Aluminium, cobalt and nickel	Improved magnetic properties	Permanent magnets (known as Alnico)
Silicon	Improved electrical properties	Electrical switch-gear

This table only gives an outline of the different types of alloy steels. Many other elements are used in the production of steel alloys. The properties of an alloy steel can be varied by using different percentages of the added elements.

18.6. Other alloys

Just as the properties of iron can be modified by alloying with other elements, so can the properties of other metals be changed. A good example is found in soft solder.

Investigation 18b. The melting of solder

Take a circular steel (not tin-plated) or brass sheet about 100 mm in diameter. Mark the centre with a punch and mark a circle, with a radius of about 40 mm, around the centre point. Place the piece of metal on a tripod and arrange a bunsen burner so that it is directly underneath the centre mark. Evenly spaced around the marked circle, place a piece of pure tin, a piece of pure lead and a piece of solder. If you can obtain solders of different composition, use a piece of each. The samples should be, as far as is possible, of the same size. Light the bunsen burner so that the roaring flame just touches the centre of the metal sheet. Watch to see the order in which the metals melt.

By alloying different metals it is possible to obtain a metal that melts at a lower temperature than the pure metals. A most striking example of this is **Wood's alloy**. This alloy, which is made from lead, tin, bismuth and cadmium metals, melts at 343 K (70 °C). It is used in automatic fire extinguishers: as soon as the temperature reaches 343 K a small piece of the alloy melts and unblocks a sprinkler jet. It can also be used to make *joke* teaspoons which melt when stirred in hot water. In Table 18.5 the melting-points of different solders are compared with those of the pure metals. Did your results give the same order?

TABLE 18.5. MELTING-POINTS OF SOLDERS

Percentage of Lead in Solder	Percentage of Tin in Solder	Melting-point in K (°C)	Name Given to the Solder
100	0	600 (327)	
70	30	533 (260)	Plumbers'
50	50	493 (220)	Tinmans'
40	60	458 (185)	Electricians'
0	100	505 (232)	

In addition to changing the melting-point of metals, alloying can cause very marked changes in the strength of metals. Probably the best example of improved strength is the alloy **Duralumin**. To obtain this alloy, aluminium is melted with about 4 per cent copper, 0.5 per cent magnesium and a little manganese. The alloy reaches full strength after standing some time, a process known as **age-hardening**. When fully hardened, Duralumin is about twice as strong as mild steel of the same weight. This increase in strength cannot be obtained without some loss. Duralumin is not as resistant to corrosion by the atmosphere as aluminium. To overcome this disadvantage, the Duralumin sheet is 'sandwiched' between two very thin sheets of aluminium. This process is known as **cladding** and is illustrated in Figure 18.3.

Aluminium

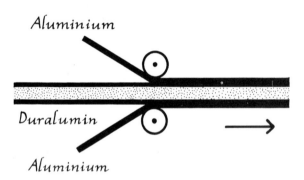

Duralumin

Aluminium

Figure 18.3 The cladding of Duralumin sheet

By mixing metals in different proportions, it is possible to obtain alloys with varying properties. Brass, an alloy of copper and zinc, is a good example of a type of alloy that has properties dependent upon the ratio of the two metals present. Pure copper is a soft easily worked metal, whereas zinc is a hard rather brittle metal. A brass that contains about 30 per cent zinc can be pressed into shape when cold; during the pressing process it gradually becomes harder. Brass of this composition, that is 30 per cent zinc and 70 per cent copper, is used in the manufacture of cartridge cases; hence, it is often called **cartridge brass**. Brass which contains 40 per cent or more of zinc cannot be worked when it is cold, but

196

is usually shaped when hot. Thus, we obtain from copper and zinc a metal that is not as brittle as zinc, but that is much stronger than pure copper. If brass is required for lathe work a little lead is added. This causes the metal to break into small pieces rather than to stay in long spirals.

We handle alloys every day when we use money. *Copper* coins are hardened by alloying and the *silver* coins are not silver at all.

TABLE 18.6 THE COMPOSITION OF
COINAGE METALS IN GREAT BRITAIN

Coin	Composition
Copper	97% copper, 2.5% zinc, 0.5% tin
Silver	75% copper, 25% nickel

Even more surprising properties are found with alloys of mercury. Mercury is a liquid at normal temperatures and freezes at 234 K (− 39 °C). It dissolves gold, silver and tin to produce alloys that rapidly harden to solids. These alloys are known as **amalgams**. They are very useful for 'filling' teeth. When you visit your dentist next, watch to see the amalgam mixed; it will usually be tin and mercury.

Test your understanding

1. Why does aluminium not appear to react with air?
2. Why is sodium used in certain heat exchanges?
3. Aluminium is used to displace metals by the thermite process. Make a list of metals that it cannot displace.
4. Why is aluminium used in 'space' clothing?
5. Why do you think that tin is used to make toothpaste tubes?
6. Why are the accumulators used in motor cars so heavy?
7. If you were making a metal object and found that the metal had become hard, how would you try to soften it?
8. Why should you avoid overheating drills made from carbon steel? How can steel be improved so that drills will work at higher temperatures?
9. What sort of brass would you choose for beating into a fruit bowl?

Chapter 19

Carbon and Fuels

19.1. The unique nature of carbon among the elements

Of all the elements, both natural and artificial, carbon forms a vastly larger number of compounds than all the others put together. This is because carbon atoms have the ability to join to each other in a way that is not possessed by any other element, although **silicon**, which belongs to the same chemical 'family' as carbon, does show the property to a very limited extent. We have seen (Section 6.7) this ability of carbon to join its atoms together in the allotropes, diamond and graphite. Some 1 000 000 carbon compounds have been recognized, and the number is continually being added to.

Organic Chemistry is the study of the compounds of carbon, and much more will be said about this in Chapter 33. In this chapter it is intended to limit our study to those compounds, both natural and synthetic, which are chiefly important because they are used as **fuels**.

19.2. What is a fuel?

A fuel is a chemical substance which reacts with oxygen to liberate energy, usually in the form of heat. The liberation of heat is often accompanied by the evolution of light. We are not here concerned with the so-called 'atomic fuels': these are not strictly fuels within the normal meaning of the term, though they do supply energy.

All normal fuels are compounds containing carbon and hydrogen. The free elements are also used, together with carbon monoxide.

The chemical reactions by which the energy is liberated are:

$$C(s) + O_2(g) \rightarrow CO_2(g)$$

and

$$H_2(s) + \tfrac{1}{2}O_2(g) \rightarrow H_2O(l)$$

In the case of carbon, one mole (12 g), on complete reaction with oxygen, will liberate 393.7 kJ of energy. In the case of hydrogen, one mole (2 g) gives 241.8 kJ of energy if the product is steam and 285.8 kJ if the product is liquid water.

These amounts of energy are available even when the two elements are combined together, or are combined with other elements, although the total quantity of energy liberated will be modified by the energy needed to break up the compound.

The very large quantities of heat energy given by these two reactions, coupled with the availability of carbon and hydrogen compounds (chiefly in the form of **hydrocarbons** – that is compounds containing carbon and hydrogen *only*), makes them ideally suited as energy producers.

19.3. Coal and its origins

All living organisms, plant and animal, are largely made up of compounds containing carbon. It is for this reason that the study of the compounds of carbon was originally called Organic Chemistry, as it was then believed that the compounds found in living things were in some way different from those found in non-living things. It was believed that some 'life-force' – hence the term 'organic' – was needed to prepare them.

Although today we know that this is not true, and that there is no real difference between the carbon compounds found in living things and those mineral substances containing only other elements, we still use the term 'organic' as a useful way to separate the very large part of chemistry that deals with the carbon compounds from the chemistry of the rest of the elements. This latter is called **Inorganic Chemistry**.

Coal, the most important of our solid fuels, has been formed in the course of time from living plants that existed some 200 000 000 years ago. That coal originated from plants is suggested by the fact that countless **fossil plants** are found in the coal seams. Whole fossilized trees have been found.

We can see the beginnings of the long process of transformation happening even today in the formation of **peat**. This is composed of the leaves, stems and roots of many types of plant, particularly the mosses, that grow in marshy areas. These have been only slightly changed from the original plant material, which is largely composed of cellulose, $(C_6H_{10}O_5)_n$, where 'n' is a large number – of the order of several hundred. There has been some loss of the gaseous elements, hydrogen and oxygen, and, as a result, the proportion of carbon in the residue has gone up.

The second stage in the formation of coal is that known as **lignite** or **brown-coal**. This is made up of fossil wood and contains a still higher proportion of carbon.

A typical household coal comes next. Again, the percentage of carbon is increased. The final stage is the formation of 'hard' coal or **anthracite**.

All these changes are brought about by the action of pressure and heat on the original plant remains. In the course of time, layers of other materials have been deposited on top of the decayed material, and it is the weight of these that has been mainly responsible for the chemical changes that have taken place. Table 19.1 shows how the percentages of carbon, hydrogen, oxygen, nitrogen and ash content vary as the series of substances mentioned above is formed from the original plant material.

TABLE 19.1. PERCENTAGE COMPOSITION OF WOOD AND COAL

Fuel	Carbon	Hydrogen	Oxygen and Nitrogen	Ash
Wood (cellulose)	44.4	6.2	49.4	—
Peat	58.0	5.0	28.0	9.0
Lignite	70.0	6.0	21.0	3.0
Household coal	80.0	6.0	12.0	2.0
Anthracite	94.0	2.0	2.0	2.0

As the percentage of carbon rises, the **calorific value** of the fuel increases. This is the heat energy which may be obtained by completely burning a stated mass of the fuel. Table 19.2 gives average figures for the calorific values of the fuels referred to. The units are kilojoules per kilogramme of fuel.

TABLE 19.2. CALORIFIC VALUES OF FUELS

Fuel	Calorific Value in kJ kg^{-1}
Wood	21 760
Peat	23 430
Lignite	26 780
Household coal	34 310
Anthracite	36 820

Investigation 19a. The destructive distillation of coal

Fit up the apparatus shown in Figure 19.1.

Put some small pieces of household coal into the hard-glass test-tube and heat the tube with a bunsen burner – gently at first and then with increasing strength. Note what happens. When a 'smoke' appears at the burner, try to light it. Describe the appearance of the liquid which forms in the cooled boiling-tube. What happens to the glass wool?

The experiment you have just carried out is a small-scale version of what used to be done on the large scale at a gas-works.

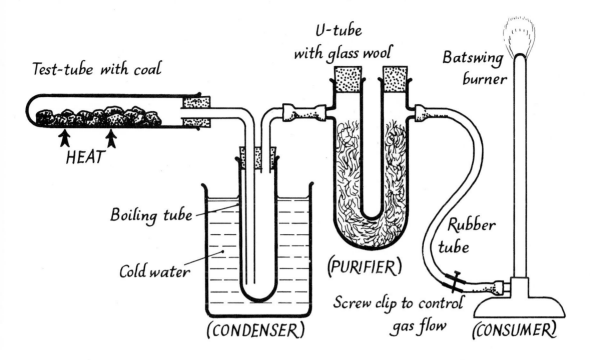

Figure 19.1 Apparatus for the destructive distillation of coal

19.4. Coke

The solid residue left in the retorts when coal has been heated in the absence of air is an impure form of carbon. It has many important industrial uses, not the least of these being as a **smokeless fuel**.

Atmospheric pollution has become an ever-increasing problem, and areas of many industrial and other urban sites have been declared to be 'smokeless zones'. In these areas, it is an offence to burn fuels which produce smoke. Coke, which is coal from which all the tar and other impurities which produce smoke have been removed, cannot produce any more smoke when it is burned.

Coke is also important as the starting material for the synthesis of many organic compounds, as the material from which water-gas and producer-gas (see Section 4.6) are made, and as a reducing agent in the extraction of many metals from their ores.

For this last purpose, it is either used in the solid form as carbon itself, for example in the reduction of zinc oxide:

$$ZnO(s) + C(s) \rightarrow Zn(s) + CO(g)$$

or in the gaseous form of carbon monoxide:

$$Fe_2O_3(s) + 3CO(g) \rightarrow 2Fe(s) + 3CO_2(g)$$

19.5. The combustion of coke

When coke is burned in a suitable grate or brazier, the primary reaction is the formation of carbon dioxide gas:

$$C(s) + O_2(g) \rightarrow CO_2(g)$$

However, if there is insufficient air, the carbon dioxide which is formed in this first reaction is then reduced by more of the coke to carbon monoxide:

$$CO_2(g) + C(s) \rightarrow 2CO(g)$$

For this reason, it is very dangerous to burn coke in any situation where there is liable to be insufficient air for its complete combustion. This may happen if, for example, the chimney over the grate where the coke is being burned gets blocked up, possibly through a bird building its nest in the chimney while the grate is out of use. There have been many recorded incidents of deaths being caused in this way. During the First World War, when trench warfare was the normal practice, many thousands of soldiers lost their lives through carbon monoxide poisoning when they tried to keep warm and dry in their dug-outs by lighting fires without enough ventilation.

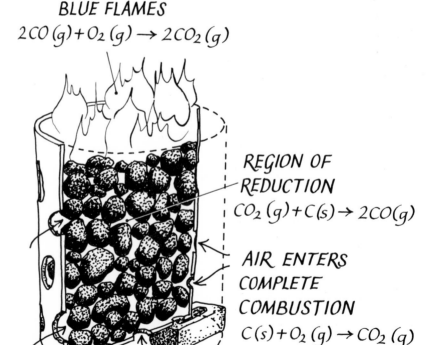

BLUE FLAMES

$$2CO(g) + O_2(g) \rightarrow 2CO_2(g)$$

REGION OF REDUCTION

$$CO_2(g) + C(s) \rightarrow 2CO(g)$$

AIR ENTERS COMPLETE COMBUSTION

$$C(s) + O_2(g) \rightarrow CO_2(g)$$

Figure 19.2 The watchman's brazier

When coke or coal is burned, it is a common sight to see blue-coloured flames on top of the burning material. This is a sign of the presence of carbon monoxide gas, which burns with a blue flame to form carbon dioxide:

$$2CO(g) + O_2(g) \rightarrow 2CO_2(g)$$

Figure 19.2 shows the reactions that take place in a watchman's brazier when coke is being burned.

19.6. Ignition temperature

In order to set any fuel alight, it is necessary to raise its temperature above a certain value. The temperature required depends on the fuel used. This is why, when we set out to light a coal fire, we make up the fire by using a lower layer of paper, then a layer of wood and finally a layer of coal on top. It is possible to bring the paper up to its ignition temperature with a match. The burning paper then raises the wood to its ignition temperature. Finally, the burning wood raises the coal to its ignition temperature. It would be very difficult, if not impossible, to set fire to the coal directly with a match!

19.7. What is a flame?

A flame is simply a visible sign that a chemical change is taking place. Reactions in which light is produced are usually reactions in which heat is produced at the same time, but this is not necessarily the case. It is quite possible to have **cold flames**! Indeed, it has been suggested that the lighting of the future might possibly be a chemical lighting where use is made of a cold flame.

Investigation 19b. A chemical light

Make up the following solutions:

Solution A. Dissolve about 0.1 g of Luminol in 10 cm³ of 10 per cent sodium hydroxide solution and dilute to a total volume of 100 cm³ with water.

Solution B. Mix 20 cm³ of 3% potassium hexacyanoferrate(III) solution with 20 cm³ of 3% hydrogen peroxide solution and dilute with water to a total volume of 200 cm³.

Take 25 cm³ of solution A and dilute with 175 cm³ of water. Add this solution to solution B in a darkened room. A bright blue light will be seen. The brilliance may be intensified by using more sodium hydroxide. (See *School Science Review*, vol. 54, no. 188, March 1973, for details.)

★ Warning – this investigation must only be carried out by an experienced teacher.

Harmful

19.8. Explosions

A mixture of a fuel gas and air (or oxygen), within certain proportions of the gases, will explode on heating. In this case, the reaction between the gas and the oxygen is so rapid that it proceeds through the mixture at a very fast rate. This is what happens whenever a mixture of gases explodes. The sudden change in volume that takes place results in the loud sound that we can hear: the surrounding air is given a sudden 'shock' which is transmitted through the air to our ears. A change of volume also takes place when a solid or a liquid explosive is used.

19.9. The bunsen burner

The bunsen burner is probably the most useful piece of apparatus in the laboratory. It was designed in 1855 by Robert von Bunsen.

Investigation 19c. The bunsen flames

Completely close the air-hole of a bunsen burner. Now, turn on the gas fully and light the burner. Describe the appearance of the

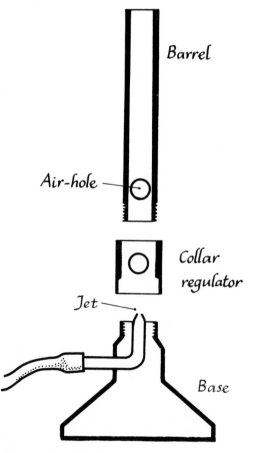

Figure 19.3 The construction of a bunsen burner

flame. Take a wood splint and place it for a short time across the base of the flame. Remove the splint as soon as it starts to catch fire and examine it carefully. Describe its appearance.

Take a small evaporating dish and partly fill it with water. Support the dish on a tripod or on a retort stand ring so that the base is exposed. Allow the yellow part of the flame to play on the outside of the dish for a short time. Remove the flame and examine the dish. What do you think has caused the appearance of the outside of the dish?

Next, gradually open the air-hole. Describe the appearance of the flame (a) when the air-hole is partially open and (b) when it is fully open. Repeat the investigations with the splint and the evaporating dish. What differences, if any, do you notice?

Close the air-hole of the bunsen and turn down the gas supply a little. Adjust the supply of gas and air so that, with the air-hole wide open, the flame is small and is making a roaring noise. Now, slowly reduce the supply of gas to the burner. It is probable that the flame will disappear down the tube of the bunsen and that a 'screeching' sound will be heard. If this happens, turn off the gas supply and take care not to touch the burner in the region of the air-hole as it will probably be very hot. This occurrence is known as **burning-back**.

Investigation 19d. The burning-back of a bunsen burner

For this investigation, a length of glass tubing, about one metre long and about 20 mm internal diameter, is needed. It should be fitted at one end with a cork which is bored so that it may be loosely fitted over the tube of a bunsen burner. The glass tube should be supported vertically in a clamp.

★ Warning – this investigation should be done as a demonstration by the teacher.

Close the air-hole of the bunsen, turn on the gas and light it at the upper, open end of the glass tube. The typical, unsteady, quiet yellow flame will be seen. Now, gradually open the air-hole. The flame will pass through all the phases shown by a bunsen.

When the flame is that of the ordinary bunsen with its air-hole open, gradually reduce the gas supply. The flame will burn back and will be seen to burn down the glass tube at an accelerating rate, while at the same time a noise will be heard. If the gas supply is turned up again before the flame has burned right back, it is sometimes possible to raise it again to the top of the tube.

This investigation shows that the reason for the burning-back is that the gas/air mixture is being supplied to the flame at too slow a rate. Consequently, if the mixture is capable of supporting combustion, the flame will travel through it quicker than the mixture can be supplied. Every inflammable mixture of gas and air has a definite **explosion rate** – that is a definite speed at which a

205

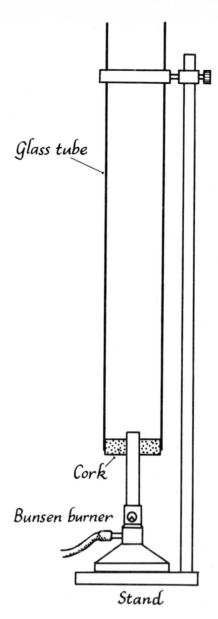

Glass tube

Cork

Bunsen burner

Stand

Figure 19.4 The apparatus arranged for Investigation 19d

flame will run through it. The reason that a bunsen does not burn back when the air-hole is closed is that the gas on its own is incapable of burning. There is no air mixed with it. Air is only available at the top of the bunsen tube and so the flame remains there. For all inflammable gas/air mixtures there are limits of composition between which the mixtures are explosive. Outside these limits, the mixture will not burn.

Investigation 19e. The 'exploding can'

For this investigation a can (of the dried milk type) with a press fitting lid is needed. A *small* hole is made (with a nail) in the base, and a *larger* hole is made in the lid.

Fit the lid to the can – not too tightly – and, with the aid of a length of rubber tubing which is inserted through the larger hole in the lid, displace the air in the can with gas.

Stand the can upside-down (on its lid) on the bench and ignite the gas as it comes out of the small hole. STAND CLEAR! The flame will be seen to get smaller and smaller and will eventually seem to go out. DO NOT APPROACH THE CAN AT THIS POINT! After some time (it can be quite long) there will be a loud explosion and the tin should be propelled upward with some force, separating from the lid.

This investigation illustrates the fact that, in order to explode, a gas/air mixture must have certain proportions. At the start, when the gas issuing from the small hole is ignited, there is insufficient air mixed with it for an explosion to occur. As the gas burns at the small hole, air is drawn in through the larger hole at the bottom and dilutes the remaining gas until eventually an explosive mixture is formed.

★ *Warning – this investigation should only be done by the teacher.*

Explosive

Test your understanding

1. What is a fuel?
2. What two elements are contained in nearly all fuels?
3. How much energy is given out when (a) 24 g of carbon, (b) 48 g of methane, (c) 13 g of ethyne (C_2H_2) are burned in air?
4. What are 'hydrocarbons'? Name FOUR hydrocarbons.
5. From what original source has coal been formed?
6. Through what stages has the original source passed on its way to becoming hard coal (anthracite)?
7. Describe the changes that take place when some pieces of coal are heated strongly in the absence of air. What is the residue left?
8. Explain the meaning of the term 'calorific value' as applied to a fuel.
9. Why is it dangerous to burn carbon-containing fuels without sufficient ventilation?
10. Draw a labelled diagram to show the chemical changes that take place when coke is burned in an open grate.
11. What is meant by the term 'ignition temperature'?
12. What is a flame?
13. Describe, using labelled diagrams, the appearance of a bunsen flame (a) when the air-hole is wide open, (b) when the air-hole is completely closed.
14. Why does an explosive make a 'bang' when it goes off?
15. What is meant by the term 'burning-back' in relation to a bunsen burner?
16. What is meant by the term 'Organic Chemistry'?

Chapter 20

The Oil Industry

20.1. The importance of the oil industry

It is no exaggeration to say that the oil industry is far and away the most important single chemical industry today. We are, more and more, becoming dependent upon oil and the natural gas associated with it, not only for power and heating, but also for the supply of the basic chemicals needed for the synthesis of a great number of organic compounds.

20.2. Mineral oil

Mineral oil, otherwise called **crude oil** or petroleum, is found in many parts of the world, but, for economic production, the main areas from which it is obtained are:

1. The Middle East
2. The Americas, North and South
3. The Far East
4. The U.S.S.R.

In addition, large deposits have been located under the North Sea, and this new source is being developed rapidly. It should be in full production by the middle 1980s and the United Kingdom will then be independent of foreign sources.

The oil brought into this country at present, and that under the North Sea, consists mainly of **hydrocarbons** – i.e. compounds made up of the elements carbon and hydrogen only – belonging to the **alkane** series (Section 20.5).

20.3. Prospecting for oil

The finding of economic sources of oil is a very important, very difficult and *very* expensive process. Many millions of pounds may be spent before a new economically profitable source is located.

Broadly speaking, the search for new sources of oil follows the following pattern:

1. In areas where access is difficult, aerial survey is usually the first step.

2. This is followed by a geologist's report on the region, from the point of view of the rocks occurring on the surface.
3. If this report is favourable, then a much more detailed survey of the area is undertaken. There are only a few geological formations in which oil may be found and every effort is made to work out the structures of the rocks lying under the surface. Among the methods which may be used for this purpose are:
   a. Measurement of the pull of gravity. This alters slightly with the nature of the rocks beneath the surface.
   b. Measurements of the magnetism in the area. Again, this is affected by the rocks.
   c. Seismic surveys. These involve the firing of explosive charges buried in the ground and measurements of the time taken for the shock wave of the explosion to travel to microphones placed at different distances from the point of explosion. Different rock layers reflect the explosion with different intensities. Figure 20.1 illustrates the ideas involved in seismic surveying. The interpretation of the results obtained is very complicated.

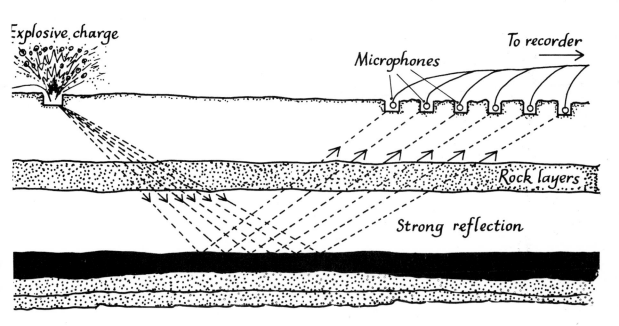

Figure 20.1 Seismic surveying

4. If the results of these tests still suggest the possibility of oil being found, then a trial boring is made. This is an expensive process. Great care is taken to extract a 'core' of the rocks through which the drill-bit passes, and a good picture of the underlying structures is now obtained. Only some 20 per cent of these trial borings 'strike oil'.

When oil is located, it normally has to be pumped to the surface. Occasionally, the pressure of the gas above the oil is sufficient to force the oil to the surface. Sometimes, but very rarely, if the drilling has been done properly, the oil rushes out of the well with great force. This is called a **gusher**.

20.4. Natural gas

In many parts of the world, oil is found in association with large quantities of what is called **natural gas**. Sometimes, the gas is found on its own, or with only small quantities of oil. The gas is largely methane (CH_4), the simplest of all hydrocarbons.

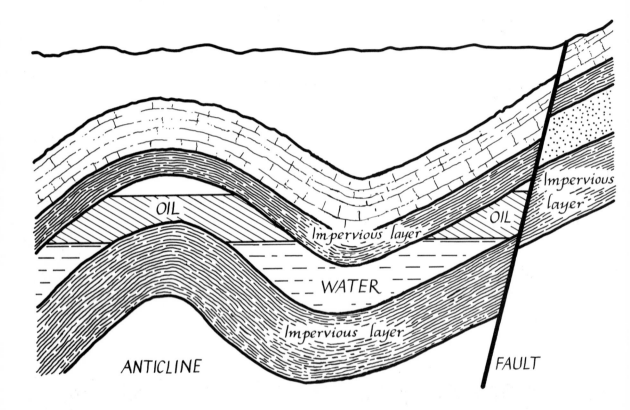

Figure 20.2 Some rock formations in which oil may be found

It is believed that natural gas and oil have been formed in the course of time by the action of heat and pressure on marine organisms, both animal and vegetable, for example, seaweeds.

Natural gas, in very large quantities, is found in North America, in the Sahara Desert and other parts of Africa and, in association with the mineral oil referred to in Section 20.2, under the North Sea. Great Britain is now using this natural-gas supply and it will be sufficient to supply all the gas requirements of the country for a considerable number of years.

210

20.5. The alkanes

The **alkanes** form a 'family' of compounds. They all have a 'general formula' and possess similar chemical properties. Their physical properties vary in a regular manner from member to member. A family of compounds like this is known as a **homologous series**.

The general formula is C_nH_{2n+2}. That is, for every carbon atom in the compound there are twice as many hydrogen atoms plus two more. The simplest compound, where 'n' has the value of one, is thus CH_4 – **methane**. When 'n' is two, the formula is C_2H_6 and the compound is called **ethane**.

Each carbon atom in the alkanes is attached to four other atoms – either of carbon or of hydrogen. Each compound consists of a 'backbone' or skeleton of carbon atoms, the hydrogen atoms 'clothing' the skeleton.

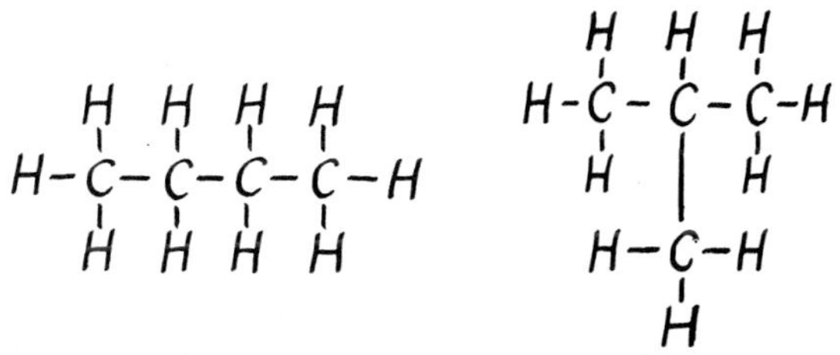

Figure 20.3 The isomers of C_4H_{10}

When 'n' has a value of four or more, then it is possible to have more than one arrangement of the atoms of carbon forming the backbone. Such different compounds, having the same numbers of carbon and hydrogen atoms, but with different arrangements, are known as **isomers**. When the carbon atoms are joined together in a line or chain, without any branches, the compound is named as shown in Table 20.1.

TABLE 20.1   THE NORMAL ALKANES

Formula	Name	Boiling-point
CH_4	Methane	113 K (-160 °C)
C_2H_6	Ethane	180 K (-93 °C)
C_3H_8	Propane	228 K (-45 °C)
C_4H_{10}	Butane	274 K (1 °C)
C_5H_{12}	Pentane	309.3 K (36.3 °C)
C_6H_{14}	Hexane	342.0 K (69.0 °C)

It will be seen from Table 20.1 that, at ordinary temperatures, the first four members of the family of alkanes are gases. When 'n' has a value of five and above, the compounds are liquid, and when 'n' reaches a value of sixteen, solids are found.

20.6. The refining of crude oil

The separation of crude oil into useful materials is known as **refining**. The crude oil is imported in very large ships or **tankers**. Some modern tankers have a deadweight of two hundred thousand tonnes or more.

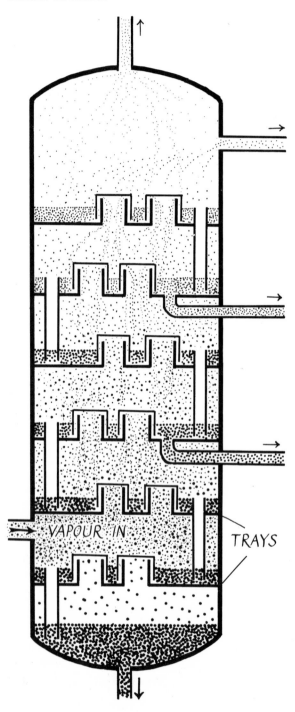

Figure 20.4 Diagram of an oil fractionating column

The main process used to separate the substances in the crude oil is that of **fractional distillation**.

The oil is first changed into vapour by being passed through pipes which are strongly heated from the outside. The vapours are then passed to a **fractionating column**. They enter the column at a point about one-third of the way up from the bottom. The column has a large number of 'trays' fixed across it (see Figure 20.4). The construction of each individual tray is shown in Figure 20.5.

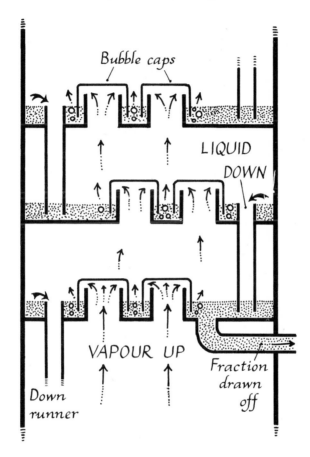

Figure 20.5 Detail of the construction of a tray in a fractionating column

As the hot vapours ascend the column, they become cooled, those vapours which have the higher boiling-points condensing first. The liquids thus formed collect in the trays and the vapours ascending the column are forced to bubble through them. This causes more of the higher boiling-point liquids to condense and to re-vapourize any lower boiling-point liquids that may have condensed. When a tray is full, the liquid overflows to the tray below by means of the **down-runner**. When the whole column is working properly, each tray will contain a mixture of liquids with boiling-points fairly close together. Pipes leading from the side of the column are used to draw off the liquids as required. The control of the column is effected by returning a small portion of the liquids with the lowest boiling-points to the top of the column (see Figure 20.6).

The liquids with very high boiling-points descend the column from the point where the vapours enter and form a residue at the bottom of the column. This residue is taken away and is further treated by distillation under reduced pressure.

The fractions obtained in this first or **primary distillation** are put through many other processes and treatments before they are suitable for commercial use, but the final main uses of these first fractions are given in Table 20.2.

TABLE 20.2. THE MAIN FRACTIONS FROM CRUDE OIL

Fraction	Boiling-point Range	Number of Carbon Atoms
Refinery gases	—	C_1 to C_4
Gasoline (petrol)	Up to about 420 K (about 150 °C)	C_5 to C_9
Kerosene (paraffin)	420 K to 510 K 150 °C to 240 °C)	C_{10} to C_{12}
Diesel oil	510 K to 540 K (240 °C to 270 °C)	C_{13} to C_{15}
Lubricating oils	Above 540 K (Above 270 °C)	e.g. C_{18}
Paraffin waxes (separated with lubricating oils)	Solids	e.g. C_{25}
Heavy fuel oils and bitumens	From residue	Larger numbers

20.7. The refinery gases

The most important of the gases obtained in the distillation process is methane. The gases are used as fuels. Propane and butane are easily liquefied by a small increase in pressure and can then be stored in liquid form. The liquefied gas is sold in small containers under various trade names; for example, 'Camping gaz' and 'Calor Gas'.

20.8. Gasoline (petrol)

When the properties of mineral oil were first investigated, and its commercial importance understood, it was for heating and lighting that it was wanted. Thus the kerosene (paraffin) content was the important factor. The gasoline (petrol) fraction was of little value and was discarded. Today, if crude oil were only

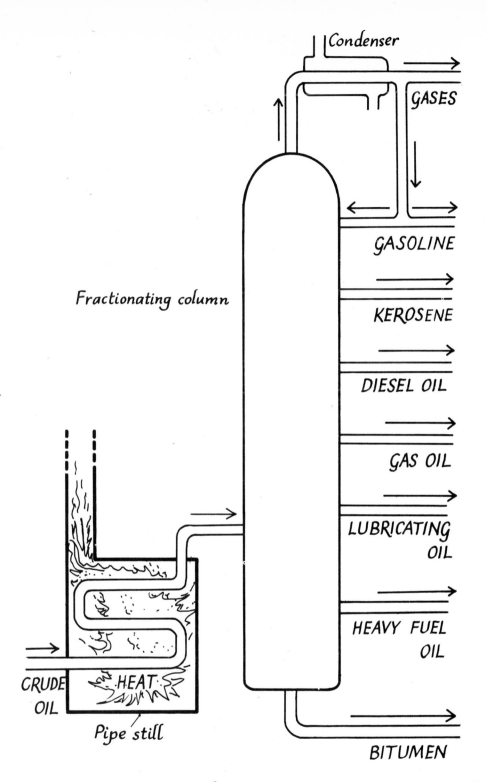

Condenser

GASES

Fractionating column

GASOLINE

KEROSENE

DIESEL OIL

GAS OIL

LUBRICATING OIL

HEAVY FUEL OIL

CRUDE OIL

HEAT

Pipe still

BITUMEN

Figure 20.6 Diagram of a fractionating column and the main fractions obtained

distilled, it would not provide enough of the gasoline fraction for the needs of the internal combustion engine to be satisfied, and vast quantities of the other fractions would have to be wasted. Fortunately, it is possible to increase the proportion of hydrocarbons with numbers of carbon atoms in the gasoline range by the process of **cracking**.

20.9. The cracking of oil

In this process, the larger molecules obtained in the primary distillation of crude oil are broken down into smaller units. This is done by the heating of the vapours containing the larger molecules, in the presence of suitable catalysts which are kept in a fluid state by being blown by a strong blast of the vapours. Changes take place like the one indicated here:

$$C_{12}H_{26}(g) \rightarrow C_8H_{18}(g) + 2C_2H_4(g)$$

One molecule of the C_{12} hydrocarbon gives one molecule of the C_8 (octane) hydrocarbon, together with two molecules of the gas **ethene**.

The gas, ethene, is not a member of the alkane series, but belongs to an entirely different series of hydrocarbons, the **alkenes**, which are **unsaturated**. For further details of the unsaturated hydrocarbons, Chapter 33 should be read.

These unsaturated hydrocarbons may be separated quite easily from the alkanes and are of great importance for the synthesis of many important organic chemicals, in particular, plastics.

Many other useful substances are formed during the cracking process.

20.10. The octane number of a petrol

Not all the hydrocarbons in the gasoline range have the same power and effect when used in an internal combustion engine. One of the most important properties which a satisfactory petrol must have is a resistance to detonation or **knocking**.

During the compression stroke of a petrol engine, a mixture of petrol vapour and air is compressed, before it is sparked to make it burn. Normally, mixtures of petrol vapour and air burn quite quietly, but when detonation takes place, a sharp metallic noise is heard. Some of the mixture ahead of the flame explodes instead of burning quietly. This wastes energy and is damaging to the engine, because it means that the piston receives its downward thrust while it is still not sufficiently near the end of its stroke. The efficiency of an internal combustion engine is improved as the degree of compression of the petrol vapour/air mixture is increased, but increased compression also increases the tendency to knocking.

The hydrocarbon of the alkane series which is most resistant to knocking is one of the isomers of octane. In general, the hydrocarbons with straight unbranched carbon chains are very liable to knock. Normal heptane is one of these (see Figure 20.7).

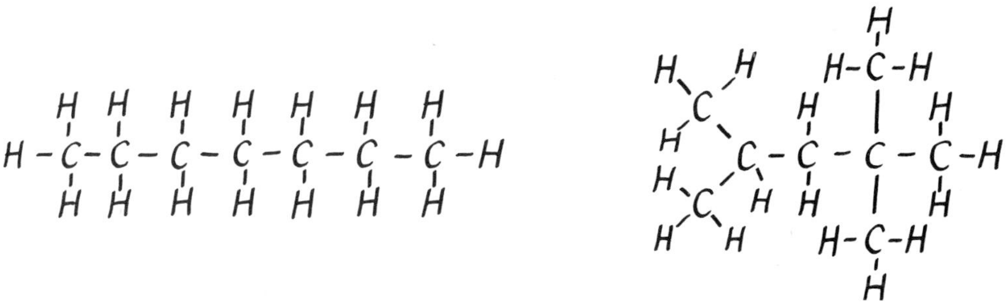

Heptane

2,2,4-trimethylpentane

Figure 20.7

A petrol which has the same resistance to knocking as pure 2,2,4-trimethylpentane is said to have an **octane number** of 100. A petrol which has as little resistance to knocking as pure normal heptane is given an octane number of 0. A petrol which behaves in the same manner as a mixture of 95 parts of pure 2,2,4-trimethylpentane with 5 parts of pure normal heptane is said to have an octane number of 95. Octane numbers of the petrol used in the average 'family' car vary from about 93 for a lower-grade petrol to 101 for a 'super' blend. An engine which is designed for high compression needs a petrol with a higher octane number than an engine with a lower compression.

Certain substances are useful as **anti-knocks**. These substances are added to petrols to increase the resistance to knocking. Among them are benzene (C_6H_6) and lead tetra-ethyl ($Pb(C_2H_5)_4$). It is by using these substances that octane numbers above 100 may be obtained.

Test your understanding

1. What is meant by the term 'hydrocarbon'? Name the first four members of the alkane series of hydrocarbons and give their formulae.
2. What products are formed when a hydrocarbon fuel is burned in air?
3. Where is natural gas found? What is its chief component?
4. Outline the steps that normally have to be undertaken when looking for new sources of mineral oil.
5. Explain the meaning of the following terms in relation to the oil-producing industry: (a) anticline, (b) gusher, (c) fault, (d) crude oil.

217

6. What are 'isomers'? Draw the structural formulae of the three isomers of pentane (C_5H_{12}).

7. What is meant by 'fractional distillation'?

8. Draw a labelled diagram to show the construction and method of working of an industrial fractionating column.

9. Name the chief fractions obtained in the first (primary) distillation of crude oil.

10. What are 'Calor gas' and 'Camping gaz'?

11. What is an 'unsaturated' hydrocarbon? Name, and give the formula of, one unsaturated hydrocarbon.

12. What is meant by 'cracking'? Why is it an important industrial process?

13. What is 'knocking'? Why is it undesirable?

14. A 'Four Star' petrol may have an 'octane number' of 99; a 'Two Star' petrol may have an 'octane number' of 93. What is meant by the term 'octane number'?

15. What is an 'anti-knock'? Name TWO substances used as anti-knocks.

Chapter 21

Other Fuels – Including Food

In this chapter, we will have a look at some other important fuels and consider in detail how food acts as a fuel for all living organisms.

21.1. Water-gas and producer-gas

We saw in Section 4.6 how these two industrial gases are made by passing steam and air alternately through hot coke. The two processes are complementary to each other and it would be impossible to run them separately.

As you will recall, when steam is blown through white-hot coke, water-gas is formed:

$$C(s) + H_2O(g) \rightarrow CO(g) + H_2(g)$$

It is the mixture of carbon monoxide and hydrogen that is called water-gas.

Producer-gas consists essentially of about 30 per cent of carbon monoxide mixed with about 70 per cent of non-combustible gases.

The comparative calorific values of some important fuel gases are given in Table 21.1. The figures given are only average values. Particular samples of the gases referred to may give values considerably different.

TABLE 21.1. THE CALORIFIC VALUES OF SOME GASES

Fuel	Calorific Value in $kJ \, m^{-3}$
Natural gas	37 200
Coal-gas	18 600
Water-gas	11 600
Producer-gas	3 900

21.2. Ethyne

Ethyne is a hydrocarbon in which the proportion of carbon to hydrogen is much higher than in the case of the alkanes. It is the parent member of another 'family' of hydrocarbons. Its formula is C_2H_2. The high proportion of carbon in the compound causes it to burn in air with an extremely yellow and smoky flame.

Ethyne is commonly known as **acetylene** and this name is still used industrially. When the gas is mixed with oxygen and burned, the flame becomes colourless and is extremely hot. This **oxy-ethyne** flame is used for cutting and **welding** metals. Welding is the process of melting two pieces of metal so that they flow and the two pieces form one, the edges fusing together.

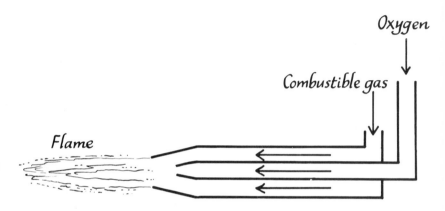

Figure 21.1 A blow-torch for the oxy-ethyne or oxy-hydrogen flames

Explosive

Flammable

Harmful

Investigation 21a. The preparation of ethyne gas in the laboratory

Ethyne is most conveniently prepared by the action of water on calcium dicarbide. Calcium dicarbide itself is made by strongly heating lime and coke together in a furnace. The temperature used is about 2 300 K (about 2 000 °C) and the heating is done electrically.

$$CaO(s) + 3C(s) \rightarrow CaC_2(s) + CO(g)$$

This reaction is strongly endothermic.

When water is added to the dark grey solid, a vigorous reaction takes place and ethyne gas is given off:

$$CaC_2(s) + 2H_2O(l) \rightarrow C_2H_2(g) + Ca(OH)_2(s)$$

Set up the apparatus as shown in Figure 21.2.

Put a thin layer of sand in the bottom of the flask and then place some small pieces of calcium dicarbide on top of the sand. Fill the tap funnel with water. Add the water a little at a time to the dicarbide so that the supply of gas is controlled. Remember that the first gas coming from the flask will be mixed with air from the apparatus, so reject the first two jars of gas collected.

Now, collect some jars of the gas. Observe its simple properties (see Investigation 9e – COWSLIPS). The smell of the gas is due

220

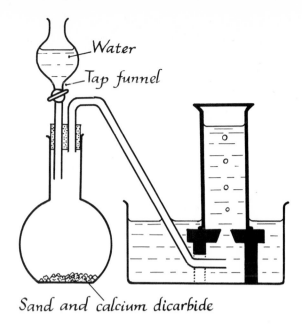

Water

Tap funnel

Sand and calcium dicarbide

Figure 21.2 Apparatus
for preparing ethyne

to impurities; when pure, the gas smells rather like ethoxyethane (ether). In particular, test a jar of the gas with a lighted splint. As mentioned in Section 33.7, the gas burns with a very smoky flame. For this reason, it is advisable to cover up any books, papers, etc., or the particles of carbon formed may dirty them. Shake up a jar of the gas with a little dilute potassium manganate(VII) solution and another jar with a little dilute bromine water. What happens in each case? Further use of these tests with potassium manganate (VII) and bromine water is made in Investigations 33a and 33e. If a compound both decolourizes bromine water *and* changes the colour of potassium manganate(VII) solution, then in all probability it has a double or a treble link between the carbon atoms in its molecule, that is it is **unsaturated**. See Investigation 33f for more details.

21.3. Alcohol

When we speak of 'alcohol' we usually mean that compound correctly known as **ethanol**. Ethanol is dealt with in detail in Chapter 34.

Ethanol is a very efficient fuel with a high calorific value (29 700 kJ kg^{-1}). As a fuel, it is usually obtained as methylated spirits (see Section 34.2). Besides its use as a fuel, ethanol is a very effective solvent.

Investigation 21b. Burning ethanol

Place about 2 cm³ of ethanol or methylated spirit in a small evaporating dish. Stand the dish on a gauze on a tripod and set

Flammable

fire to the liquid. Notice the type of flame and compare it with the flame from burning methane (Investigation 33a), burning hexane (Investigation 33c), burning hexene (Investigation 33f) and burning ethyne (Investigation 21a).

21.4. Benzene

This compound is a hydrocarbon in which the ratio of the numbers of carbon and hydrogen atoms is exactly the same as the ratio in ethyne, that is one atom of carbon to one atom of hydrogen. The molecular formula is, however, C_6H_6, and the compound is a liquid.

In benzene, the carbon atoms are arranged in a ring, the ring taking the shape of a hexagon. One hydrogen atom is attached to each carbon atom.

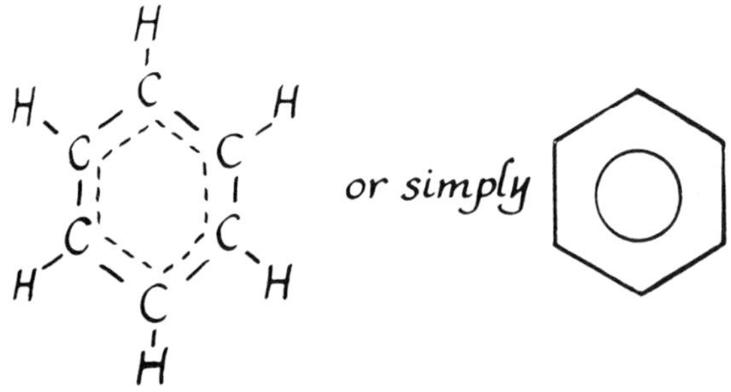

Figure 21.3   The formula of benzene

Benzene is a dangerous substance to handle and it should not be used in a school laboratory. However, for most elementary experimental work, methylbenzene (commonly known as toluene) may be substituted.

Investigation 21c. The burning of methylbenzene

Toxic

Flammable

* *Warning – benzene, methylbenzene and all other organic liquids should not be allowed on the skin, nor should their vapours be inhaled.*

As in the case of ethanol, place about 2 cm³ (not more) of methylbenzene in an evaporating dish on a gauze on a tripod. Ignite the liquid and note the very smoky flame. Compare the flame with the flames from burning methanol, ethanol, methane, hexane, hexene and, of course, ethyne.

Like ethanol, benzene is a very important solvent. It is also used as a starting material for making a large number of **aromatic** organic compounds. The word 'aromatic' is due to the pleasant (though dangerous) smells of many of the compounds that contain a **benzene ring** in their molecules. All compounds that contain at least one of these hexagonal rings of six carbon atoms are known as aromatics.

222

21.5. Food as a fuel

All living organisms need fuel in order to obtain the energy required for their life processes – for moving, growing, etc. Part of the food that an organism takes in is used for this purpose.

In living organisms, these food fuels are burned just as truly as a piece of coal is burned in a grate. The main difference is that the burning takes place at a low temperature and is far more efficient. The general situation is as follows:

food + oxygen → carbon dioxide + water + energy

This process is called **tissue respiration**.

Plants and animals both need foodstuffs, but, whereas the latter must obtain their foods ready-made, the former have the ability to manufacture their own foods if they are of the 'green' kind – that is if they have **chlorophyll** in their cells.

Green plants take in water through their roots, carbon dioxide through their leaves and, in the presence of light (normally from the sun) and using the chlorophyll, convert the water and carbon dioxide into complex organic substances, for example sugars and starch, as well as into the cellulose which forms the greatest part of the cell walls of plants. This process of building up complex organic substances is known as **photosynthesis**.

In the process of photosynthesis, oxygen gas is returned to the air. It is through this process that the balance of oxygen to carbon dioxide in the atmosphere is maintained.

All burning, whether of fuels in fires or of foods in both animals and plants, results in oxygen being used up and in an increase of carbon dioxide in the air. The process of photosynthesis is the reverse of the burning process and may be represented as:

carbon dioxide + water + energy → food + oxygen

21.6. The carbon cycle

This circulation of carbon, into and out of the air, is known as the **carbon cycle**. The carbon cycle is illustrated diagrammatically in Figure 21.4.

It is important to understand that the processes of respiration go on in plants, as well as animals, throughout the twenty-four hours. This means that plants, as well as animals, are putting carbon dioxide into the air. However, during the hours of daylight, green plants carry out photosynthesis at such a rate that, on balance over the twenty-four hours, they put much more oxygen back into the air than they take from it.

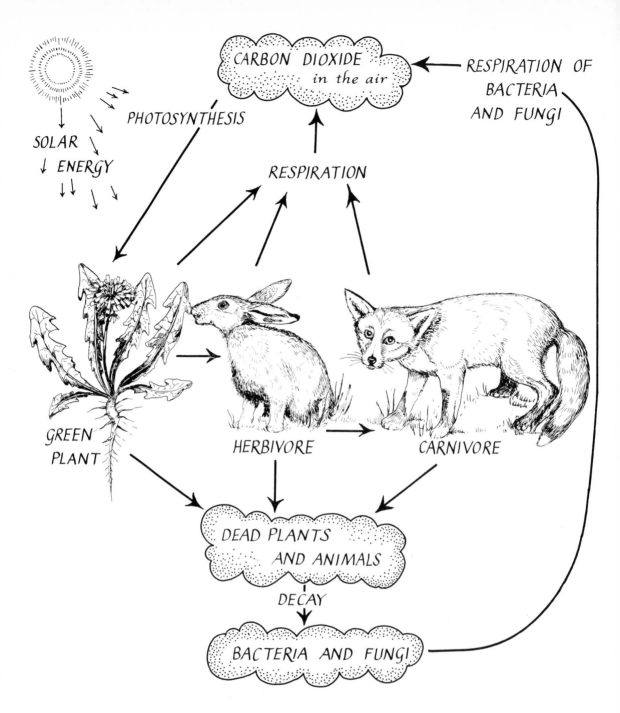

CARBON DIOXIDE *in the air*

PHOTOSYNTHESIS

RESPIRATION OF BACTERIA AND FUNGI

SOLAR ↓ ENERGY

RESPIRATION

GREEN PLANT

HERBIVORE

CARNIVORE

DEAD PLANTS AND ANIMALS

DECAY

BACTERIA AND FUNGI

Figure 21.4 The carbon cycle

21.7. Foodstuffs

The 'burning' of the foodstuffs in the animal body takes place in every living tissue.

Foodstuffs are divided into three main groups. These are **carbohydrates**, **fats** and **proteins**. Carbohydrates and fats are both energy-producing foodstuffs, but proteins are required mainly to build up the animal body. Table 21.2 groups common foodstuffs according to their composition.

TABLE 21.2. CLASSIFICATION OF FOODSTUFFS

Foods Rich in Carbohydrates	Foods Rich in Fats	Foods Rich in Proteins
Bread	Fat meat	Lean meat
Rice	Kippers	Fish
Cornflakes	Egg yolk	Egg white
Potatoes	Butter	Peas
Cakes	Margarine	Beans
Biscuits	Nuts	Nuts
Sugar	Chocolate	Chocolate
Jam	Cheese	Cheese
Milk	Milk	Milk

Carbohydrates

These are compounds which contain carbon, hydrogen and oxygen. The hydrogen and oxygen in the molecules are in the proportions of two atoms of hydrogen to one atom of oxygen – that is in the proportions in which they occur in water. Important carbohydrates are the sugars and starches. Typical sugars are **glucose** ($C_6H_{12}O_6$), **sucrose** and **maltose** (both $C_{12}H_{22}O_{11}$). Starch has a formula $(C_6H_{10}O_5)_n$, where 'n' is not known exactly, but is about 200. The starch molecule is a long chain-like structure.

Fats

These also are compounds of carbon, hydrogen and oxygen. They are insoluble in water, but form an **emulsion** with water, milk being an example. In the body, the fats are stored under the skin as a food 'reserve'.

Proteins

In addition to carbon, hydrogen and oxygen, these substances contain nitrogen and sometimes sulphur and/or phosphorus. They are extremely complex substances, but are made from a relatively small number (about 20 in all) of **amino acids**. Several different amino acids are involved in each protein molecule and the whole molecule may contain fifty or sixty acid units. The properties and characteristics of a particular protein depend on the exact order in which the amino acid units are joined together. It is easy to see that almost countless combinations are possible.

21.8. The digestion of foodstuffs

In order to be taken into the blood, that is to be **assimilated**, the complex materials of the foodstuffs have to be broken down

225

into simpler materials. This change is brought about by the action of water in the presence of catalysts. The action of water is known as **hydrolysis**.

The catalysts involved are known as **enzymes**. The enzymes are produced by the body. For example, the saliva, produced by the salivary glands, contains an enzyme called **ptyalin**. Ptyalin brings about the hydrolysis of starches, breaking them down into sugar.

Investigation 21d. The hydrolysis of starch

Take some plain boiled potato and chew it thoroughly in the mouth for some time. Notice the sweet flavour produced.

The digestion of carbohydrates results in the formation of glucose. The glucose is absorbed into the blood-stream through the walls of the small intestine. It is known as **blood-sugar** and is the chief source of energy.

Proteins are broken down into the amino acids from which they are made, and the body makes use of suitable amino acid units to build up its own structure.

Fats are stored under the skin and around certain organs, such as the heart and kidneys. When extra energy is required, they break down into suitable energy-giving substances.

21.9. Tissue respiration

When air is taken into the lungs, some of the oxygen is absorbed into the blood-stream. Here, it combines with the protein, **haemoglobin**, to form a compound, **oxy-haemoglobin**. In this way, oxygen is carried around the body in the blood. Where energy is needed, the oxygen is released from its compound with haemoglobin and reacts with the blood-sugar (glucose), as outlined in Section 21.5.

The waste-products of the energy-producing process are carried away by the blood and are disposed of. Water is removed in the urine, in sweat and, to a certain extent, through the lungs. Carbon dioxide is disposed of through the lungs; the air breathed out is richer in carbon dioxide than the air breathed in. As mentioned in Section 21.5, the process of burning the foodstuffs is known as tissue respiration.

Test your understanding

1. What is 'water-gas'? How is it made? Give the chemical equation for the reaction.
2. What is 'producer-gas'? Why are the water-gas and producer-gas processes necessarily carried out together?

226

3. How is ethyne gas prepared in the laboratory? Give a labelled diagram of the apparatus you would use in this preparation and give the chemical equation for the reaction.
4. A gas is found to decolourize both bromine water and an alkaline solution of potassium manganate(VII). What deduction about the nature of the gas may be made from these observations?
5. Why does ethyne burn with a very smoky flame?
6. What is the formula of calcium dicarbide? How is this substance made?
7. What are 'aromatic' compounds and why are they so called?
8. What is the molecular formula of benzene? Describe, using a labelled diagram, how the atoms making up the benzene molecule are arranged.
9. Draw a simple diagram to show the circulation of carbon in nature.
10. Explain the meaning of the term 'tissue respiration'.
11. Explain clearly what happens in the process of photosynthesis.
12. Plants brought by visitors to cheer up a sick person are often taken out of the sick-room at night. Can you think of a 'scientific' explanation of why this might be a good idea? Explain your suggestion.
13. Name the three main groups of foodstuffs. To which group or groups does each of the following foodstuffs belong: (a) meat, (b) bread, (c) cheese, (d) butter, (e) lard, (f) chocolate, (g) cabbage, (h) milk, (i) glucose, (j) fish, (k) marmalade?
14. What happens in the process of digestion?
15. What are 'enzymes'? What part does the enzyme ptyalin play in digestion?
16. What is the function of haemoglobin in blood?
17. How are the waste-products of tissue respiration disposed of?
18. What do fuels and foodstuffs have in common?

Chapter 22

Carbon Dioxide and Carbonates

22.1. The preparation of carbon dioxide and some of its properties

Investigation 22a. The preparation and some properties of carbon dioxide gas

Set up the apparatus shown in Figure 22.1. Place some marble chips in the flask and cover them with water so that the end of the thistle funnel is well covered. See that the gas-jar is covered by a piece of card or paper, so that the collection of the gas is not disturbed by draughts. This method of collection is known as **downward delivery** and may be used for any gas with a density greater than the density of air. The gas is led to the bottom of the jar and, as it collects, the air is displaced by the denser gas which gradually fills up the gas-jar from the bottom.

When all is ready, pour some hydrochloric acid down the thistle

Harmful

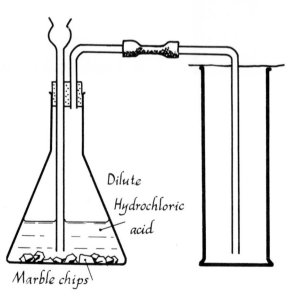

Figure 22.1 The preparation and collection of carbon dioxide gas

Dilute Hydrochloric acid

Marble chips

funnel. It is best to use the concentrated acid but, for safety, dilute (3M or 2M) acid may be used. The acid should be added a little at a time so that the evolution of the gas is controlled at a convenient rate.

As we shall see later in this chapter, marble is a form of calcium carbonate. The equation for the reaction is:

$$CaCO_3(s) + 2HCl(aq) \rightarrow CaCl_2(aq) + CO_2(g) + H_2O(l)$$

The products, in addition to the carbon dioxide, are calcium chloride (a salt) and water. This type of chemical change takes place whenever an acid is added to a compound containing the carbonate ion, CO_3^{2-}. You will remember that one of the general properties of *all* acids is that they effervesce when added to chalk. Like marble, chalk is a form of calcium carbonate.

The general equation for the action of an acid on a carbonate is:

carbonate + acid → a salt + carbon dioxide + water

Collect several gas-jars of the gas and examine its properties, using the idea of COWSLIPS (see Investigation 9e). Pay particular attention to the action of the gas on a lighted splint. Also test its solubility by immersing a jar of the gas upside-down in a bowl of water and then removing the gas-jar cover. Rock the jar gently so that the water washes the inside of the jar, taking care that none of the gas escapes.

Take another jar, add a little water and shake well. Test the resulting liquid with blue litmus paper (see also Investigation 9f). The product formed in this reaction is **carbonic acid**:

$$CO_2(g) + H_2O(l) \rightarrow H_2CO_3(aq)$$

The gas is slightly poisonous, and exposure to an atmosphere containing too high a proportion causes headache and a general feeling of unease. A still higher concentration causes death by suffocation.

Investigation 22b. The action of carbon dioxide on lime-water

Lime-water is made by shaking up a quantity of calcium hydroxide (slaked lime) with water and allowing the resulting suspension to settle. The clear liquid is then decanted.

Take some *fresh* lime-water and arrange for a supply of carbon dioxide gas to bubble through it. This is best done by immersing the end of the delivery tube from the preparation apparatus under the surface of the lime-water in a test-tube and by adding acid to the flask through the thistle funnel so that a steady supply of the gas is maintained.

Allow the gas to pass for a long time until no further change is visible in the test-tube. Then remove the test-tube and boil the liquid in it carefully over a small flame. Observe and describe all the changes that take place.

When the gas is first passed into the lime-water, the precipitate formed is calcium carbonate:

$$Ca(OH)_2(aq) + CO_2(g) \rightarrow CaCO_3(s) + H_2O(l)$$

When more carbon dioxide is passed into the resulting suspension, a further change takes place:

$$CaCO_3(s) + CO_2(g) + H_2O(l) \rightarrow Ca(HCO_3)_2(aq)$$

resulting in the formation of calcium hydrogencarbonate, which is soluble.

Calcium hydrogencarbonate is not very stable and, when the solution containing it is boiled, it is decomposed:

$$Ca(HCO_3)_2(aq) \rightarrow CaCO_3(s) + CO_2(g) + H_2O(l)$$

It is this last reaction that is responsible for the 'furring-up' of kettles and hot-water pipes (see Section 23.10).

22.2. The large-scale uses of carbon dioxide

The slight solubility and very weak acidic properties of the gas give rise to properties which make **soda-water** pleasant to drink, either on its own, or in the large range of 'fizzy' drinks. The solubility of any sparingly soluble gas increases with increased pressure. In the manufacture of 'fizzy' drinks, carbon dioxide gas is forced into the liquid under pressure and the stopper (or some other form of cap) is then sealed over the top of the bottle. When the stopper is removed, the gas is no longer held under pressure and, as a result, is less soluble. It then gradually comes out of the solution as the carbonic acid decomposes:

$$H_2CO_3(aq) \rightarrow CO_2(g) + H_2O(l)$$

You will have found out when testing the properties of the gas that carbon dioxide does not support combustion. Therefore another large-scale use is in fire extinguishers.

There are a number of different types of fire extinguisher in which carbon dioxide plays different roles. The simplest type is that which contains the gas under pressure in a cylinder. When the release lever is squeezed, the pressure on the gas is released; it expands rapidly and comes out of the nozzle. This type of gas extinguisher (Figure 22.2) is particularly valuable for chemical and electrical fires in the laboratory where water would be dangerous. Another advantage is the lack of 'mess' when the extinguisher is used.

230

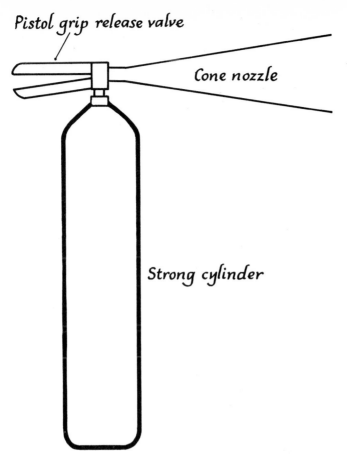

Cone nozzle

Strong cylinder

Figure 22.2 A carbon dioxide gas fire extinguisher.

In another type of extinguisher, the purpose of the carbon dioxide is to force a jet of water on to the seat of the fire from a distance. A bulb of compressed carbon dioxide gas is used to set up a pressure in the container and this forces out the liquid. In the foam type of extinguisher, a suitable foaming agent is added to the liquid in the container. In this case, the liquid coming out forms a stiff foam which 'blankets' the burning material. Foam-type extinguishers are particularly useful for dealing with large fires in the open air, such as may occur when an aircraft crashes, and for petrol fires.

The same principle is seen in the 'Sparklet' type soda syphon (see Figure 22.3).

Yet another important large-scale use of carbon dioxide is in the solid form. In this form it is used for refrigeration and is particularly useful for this purpose because it does not melt and turn into a liquid, but sublimes directly into the gas. The temperature of sublimation is 195 K ($-78\,°C$). In use, lumps of solid carbon dioxide are packed with the material to be kept cold – ice-cream, for example – in a heat-insulated container. Solid carbon dioxide is sometimes called *dry ice*. It is sold under various trade names, 'Drikold' being one.

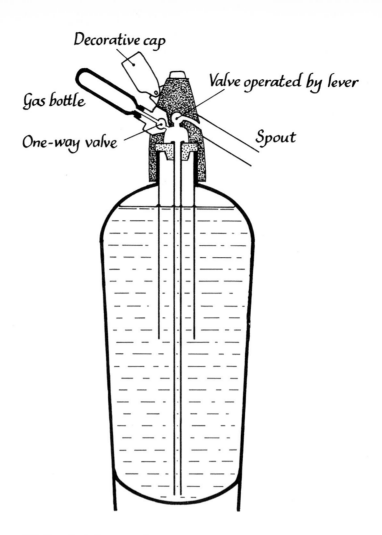

Decorative cap

Valve operated by lever

Gas bottle

One-way valve

Spout

Figure 22.3 A 'Sparklet' syphon

22.3. Calcium carbonate

Calcium carbonate is widely distributed in nature, where it occurs in many different forms. Marble and chalk have already been met; other forms include **limestone, calcite** and **coral**. In addition, the shells of many marine animals are largely calcium carbonate.

Indeed, much of the naturally occurring calcium carbonate in its different forms has come originally from the shells of dead organisms. This can be seen clearly in the cases of chalk and limestone which are frequently found to contain **fossils**. Fossils are traces of animals or plants which lived many millions of years ago. We have seen in Section 19.3 that coal was formed in the same way from plants, and that plant fossils are found in the coal measures.

The differences in the forms of calcium carbonate are due to the action of heat and pressure, caused when the layers of calcium carbonate become covered by great weights of overlying rock. In the coral reefs of today we can see the continuing process of the formation of calcium carbonate deposits.

22.4. Limestone and its uses

Limestone is one of the principal raw materials of industrial chemistry.

Lime is made by roasting the limestone (see Section 22.5).

Cement (see Section 32.4) is made by roasting limestone with sand and aluminium oxide.

Glass (see Section 32.5) is made by strongly heating limestone with sand and sodium carbonate.

Limestone is used in the extraction of iron (Section 17.6), as the source of carbon dioxide for the manufacture of sodium hydrogencarbonate and sodium carbonate by the Solvay process (Section 22.8) and in the manufacture of calcium dicarbide (Investigation 21a).

22.5. The manufacture of lime from limestone or chalk

The process of making lime by heating limestone or chalk to a very high temperature has been known since ancient times.

Investigation 22c. Making lime in the laboratory

For convenience, in the laboratory, the preparation of lime (calcium oxide) can best be done from marble. Take a small lump of marble (a piece about a centimetre cube is convenient) and wrap a length of nichrome wire around the lump. Support the lump so that it may be heated directly in the bunsen flame. Alternatively, the marble may be supported in the corner of a tripod. The arrangement is shown in Figure 22.4.

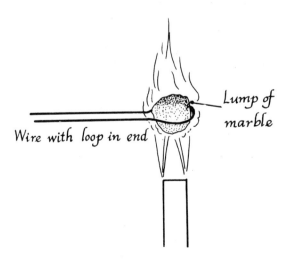

Lump of marble

Wire with loop in end

Figure 22.4 Making lime from marble

Now, roast the piece of marble for about fifteen minutes as strongly as you can, using the hottest part of the flame. When the heating is completed, allow the lump to cool thoroughly.

The equation for this reaction is:

$$CaCO_3(s) \rightarrow CaO(s) + CO_2(g)$$

The lime made in this way should be tested as described in Investigation 22d.

On the large scale, the heating of the limestone or chalk is carried out in a **lime-kiln**. The limestone may be mixed with coke, or, in the more modern type of kiln, the heating may be done by a fuel gas. Figure 22.5 is a diagram of a lime-kiln.

LIMESTONE AND COKE

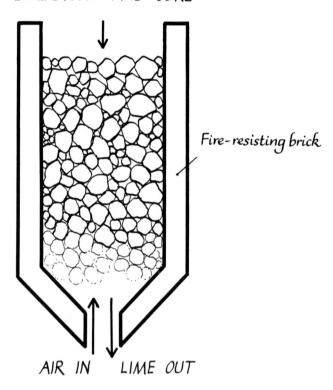

Fire-resisting brick

AIR IN   LIME OUT

Figure 22.5 A lime-kiln

It is important to make sure that the carbon dioxide formed in the burning is swept away by a current of air. This is because the break-up of the calcium carbonate is a reversible process and, if the carbon dioxide is not removed, it will react with the lime and prevent the reaction from taking place completely.

22.6. The properties of lime

Investigation 22d. The reaction of lime with water

When lime is freshly made, it is often called **quicklime**. The word 'quick' means 'living'.

Take the piece of lime that you made in Investigation 22c and place it in an evaporating dish. Stand the dish on a tripod. Now, cautiously add one or two drops of water to the lime. What do you see happen? Continue adding more water, a few drops at a time, until there is no further change and the mixture is a pasty solid. Why do you think the term 'quicklime' is used?

Gently heat the dish with the moist lime until the product is dry. It is now calcium hydroxide, or **slaked lime**. The process of adding water to lime is known as **slaking**. Perhaps you have heard the expression 'to slake one's thirst'? Do you think the term 'slaking' is a good one when used with reference to the adding of water to quicklime?

$$CaO(s) + H_2O(l) \rightarrow Ca(OH)_2(s)$$

Investigation 22e. Making lime-water

Take some of the slaked lime you have made and put it into a test-tube. Now, add some pure water (distilled or de-ionized) to it, cork the tube, and shake it thoroughly for a minute or so. Allow the solid to settle in the tube (this will take some time) or filter some of the suspension into a clean test-tube. If you filter the suspension, do you notice anything about the surface of the filtrate?

When you have collected a sufficient volume of the **lime-water**, breathe out gently through it for some time, using a straw or a glass tube. What happens? Have you seen anything like this before? What is the reason for what you observe to happen?

Did you notice, when you were making quicklime from marble in Investigation 22c, how the product appeared to shine brightly in the heat of the bunsen flame? This property of giving out a strong white light when heated was made use of in bygone times for illuminating stage performances. Hence we get our expression, 'being in the **limelight**'!

22.7. The uses of lime and slaked lime

Lime has an extremely high melting-point – about 2 800 K (about 2 530 °C) – and is used for lining furnaces, that is it is used as a **refractory**.

Another use of lime, usually in the slaked form, is for making **lime mortar**.

Mix some powdered slaked lime with about three times its bulk of sand, making sure that the mixing is done thoroughly. Add water, a little at a time, to your mix, until a thick paste is formed. Mould some of the paste into shapes and put the shapes aside to stand for a few days. Examine the product. You have made some lime mortar.

Mortar is used for joining bricks together. It hardens first by drying out, but later on the action of carbon dioxide from the air on the slaked lime forms some calcium carbonate. The action of the carbon dioxide is very slow, even though the sand in the mixture helps to make it more porous. It has been found possible to 'date' old buildings by measuring the depth below the surface of the mortar to which the carbon dioxide has penetrated. You can test this for yourself by adding some dilute hydrochloric acid to samples of fresh mortar (made by yourself) and mortar obtained from old brickwork. Only the old mortar effervesces when the acid is added.

The continued action of carbon dioxide and water on the calcium carbonate formed causes soluble calcium hydrogen-carbonate to be formed (see Investigation 22b) and the mortar crumbles away at the surface. Figure 22.6 shows this effect diagrammatically.

Brickwork where the mortar has been so affected must be **pointed**; that is fresh mortar must be put in to gaps between the bricks so that the effect will not spread too deeply.

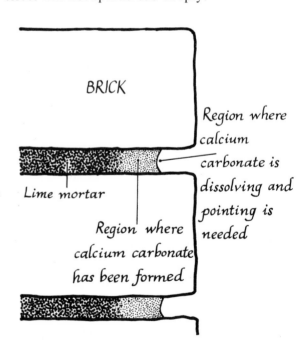

BRICK

Region where calcium carbonate is dissolving and pointing is needed

Lime mortar

Region where calcium carbonate has been formed

Figure 22.6 The action of carbon dioxide on lime mortar

Modern bricklaying mortar contains some cement as well as lime. A common mixture is one part lime, one part cement and six parts sand. This mortar has more strength than the simple lime mortar.

Another use of lime is in the Gossage process for making sodium hydroxide (caustic soda). When a suspension of slaked lime in water, **milk of lime** as it is known, is shaken up with a solution of sodium carbonate, a precipitate of calcium carbonate is formed and a solution containing sodium hydroxide is left:

$$Ca(OH)_2(aq) + Na_2CO_3(aq) \rightarrow CaCO_3(s) + 2NaOH(aq)$$

The precipitated calcium carbonate is filtered off and the solution is evaporated to obtain the sodium hydroxide.

Still another use for lime is in the extraction of the metal, magnesium, from sea-water. When lime is added to sea-water, a precipitate of magnesium hydroxide is thrown down:

$$Mg^{2+}(aq) + Ca(OH)_2(aq) \rightarrow Mg(OH)_2(s) + Ca^{2+}(aq)$$

Slaked lime is familiar to us all as **whitewash**, the white marking used for the lines of football pitches, cricket pitches and tennis courts. It is also used to coat the walls of buildings, particularly buildings on farms. It is cheap, and it has the effect of killing small insects.

One of the most important uses of slaked lime is in agriculture, where it is used as an alkali to reduce acidity in the soil.

22.8. Sodium carbonate and sodium hydrogencarbonate

Sodium carbonate and sodium hydrogencarbonate are both familiar household chemicals. Sodium carbonate is used as **washing soda** ($Na_2CO_3 . 10H_2O$), while sodium hydrogencarbonate ($NaHCO_3$) is used as **baking soda** or for the relief of indigestion, though its continued use for this latter purpose may well be dangerous.

Both compounds are manufactured by the one process – the **Solvay process**. The starting materials are brine (sodium chloride solution), limestone or chalk (to provide both carbon dioxide and lime) and a supply of ammonia gas (NH_3) which is recovered and used again, so that only replacement of unavoidable losses is necessary.

The brine is trickled down a tower. This tower is packed with a material which breaks up the flow and ensures that the liquid descending the tower comes into very close contact with the ammonia gas which is passed up the tower from the bottom. The result is a solution which contains both sodium chloride and ammonium hydroxide:

$$NH_3(g) + H_2O(l) \rightarrow NH_4OH(aq)$$

This **ammoniated brine** is now pumped to the top of a second tower and trickles down this tower while carbon dioxide gas is blown up from the bottom. This stage is known as **carbonation** and results in the formation of ammonium carbonate:

$$2NH_4OH(aq) + CO_2(g) \rightarrow (NH_4)_2CO_3(aq) + H_2O(l)$$

The solution is now transferred to a third tower, down which it trickles while more carbon dioxide is blown up. This stage is known as **bicarbonation**. It produces ammonium hydrogencarbonate, which immediately reacts with the sodium chloride to form a precipitate of the sparingly soluble sodium hydrogencarbonate:

$$(NH_4)_2CO_3(aq) + CO_2(g) + H_2O(l) \rightarrow 2NH_4HCO_3(aq)$$
$$NH_4HCO_3(aq) + NaCl(aq) \rightarrow NaHCO_3(s) + NH_4Cl(aq)$$

The sludge of sodium hydrogencarbonate is now filtered off from the solution of ammonium chloride. The solid sodium hydrogencarbonate is washed well with water.

To make sodium carbonate from the hydrogencarbonate, the hydrogencarbonate is heated strongly:

$$2NaHCO_3(s) \rightarrow Na_2CO_3(s) + CO_2(g) + H_2O(g)$$

The anhydrous sodium carbonate that is formed is dissolved in water and is then crystallized from the solution to give washing soda crystals.

The carbon dioxide formed in this stage is returned to the bottom of the carbonation or bicarbonation towers. This provides one-half of the carbon dioxide needed in the process. The rest of the carbon dioxide comes from limestone or chalk. This is heated to give lime and carbon dioxide:

$$CaCO_3(s) \rightarrow CaO(s) + CO_2(g)$$

The lime is slaked; it is added to the ammonium chloride solution left from the filtration of the sodium hydrogencarbonate. On heating, the ammonium chloride is decomposed by the slaked lime and ammonia gas is regenerated:

$$2NH_4Cl(aq) + Ca(OH)_2(s) \rightarrow CaCl_2(aq) + 2NH_3(g) + 2H_2O(l)$$

The only waste-product in the whole process is thus calcium chloride.

Figure 22.7 illustrates the Solvay process by means of a 'flow-sheet'.

Washing soda, as its name implies, is used in washing, where its function is to **soften** hard water (see Section 23.15).

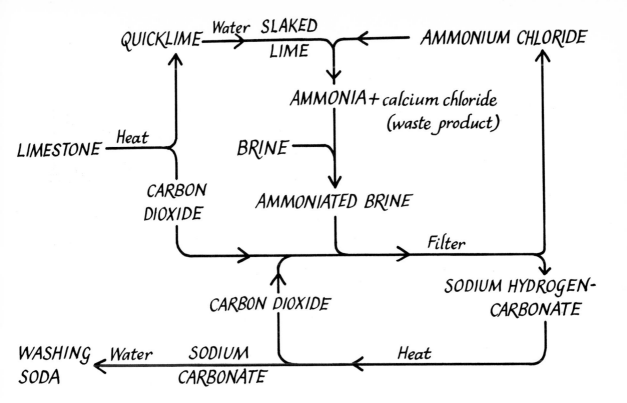

Figure 22.7 Flow-sheet for the Solvay process

When left standing in air, crystals of washing soda lose part of their water of crystallization to the air, that is, they effloresce:

$$Na_2CO_3.10H_2O(s) \rightarrow Na_2CO_3.H_2O(s) + 9H_2O(g)$$

Washing soda is soluble in water; most other carbonates are insoluble in water.

On heating, baking soda (as we have seen) liberates carbon dioxide gas. If baking soda is mixed with dough which is then heated in the oven, the carbon dioxide gas 'aerates' the dough and makes it rise. The product, however, contains sodium carbonate, and this gives a rather bitter flavour.

Sodium hydrogencarbonate will react with excess acid in the stomach, producing carbon dioxide and a salt.

Investigation 22g. Making some carbonates

Prepare a 3M solution of sodium carbonate. Also prepare solutions of copper(II) sulphate(VI), zinc sulphate(VI) and lead(II) nitrate(V), each also about 3M. Put some of these solutions into separate test-tubes and add some of the sodium carbonate solution to each. Observe what happens. Filter off the precipitated carbonates, wash them on the filter papers and dry them, taking care not to heat the copper(II) carbonate. The equations are as follows:

Toxic

$$CuSO_4(aq) + Na_2CO_3(aq) \rightarrow CuCO_3(s) + Na_2SO_4(aq)$$
$$ZnSO_4(aq) + Na_2CO_3(aq) \rightarrow ZnCO_3(s) + Na_2SO_4(aq)$$
$$Pb(NO_3)_2(aq) + Na_2CO_3(aq) \rightarrow PbCO_3(s) + 2NaNO_3(aq)$$

Corrosive

To prepare sodium carbonate itself, take some 3M sodium hydroxide solution and divide it into two exactly equal volumes. Now, from a carbon dioxide generator, bubble carbon dioxide gas through one part until there is no further change. Sodium hydrogencarbonate will be precipitated:

$$NaOH(aq) + CO_2(g) \rightarrow NaHCO_3(s)$$

When the reaction is completed, add the second portion of the original solution, warm and evaporate to crystallization or to dryness:

$$NaHCO_3(s) + NaOH(aq) \rightarrow Na_2CO_3(s) + H_2O(g)$$

The formulae given for the precipitated 'carbonates' of copper, zinc and lead are not strictly correct. The composition of these carbonates is variable, depending on the conditions under which the precipitation occurs. Thus, copper 'carbonate' is really a combination of the carbonate with some copper(II) hydroxide.

Lead 'carbonate', in the form of **white lead**, is used as the basis of most lead paints. White lead is not pure lead 'carbonate', but corresponds to the formula $2PbCO_3.Pb(OH)_2$. Lead paints are being used less and less today because of the poisonous nature of lead compounds. They are being replaced by paints containing polymers (see Chapter 36), which leave a plastic film over the surface being treated. Titanium(IV) oxide is also used.

Ammonium 'carbonate', which is really largely the hydrogencarbonate, NH_4HCO_3, decomposes very easily, liberating ammonia and carbon dioxide gases. It is the substance used as **smelling-salts**.

22.9. The decomposition of carbonates by heat

The carbonates of most metals are decomposed by heat, sodium and potassium carbonates being the exceptions to this rule. In practice, samples of sodium carbonate and potassium carbonate often do give carbon dioxide gas when they are heated. This is due in all probability to the fact that the substances often contain some of the hydrogencarbonate.

Different metal carbonates show considerable differences in the ease with which they may be decomposed by heat.

Investigation 22h. Comparing the ease of decomposition of some carbonates

To compare the effect of heat on different carbonates, set up the apparatus shown in Figure 22.8.

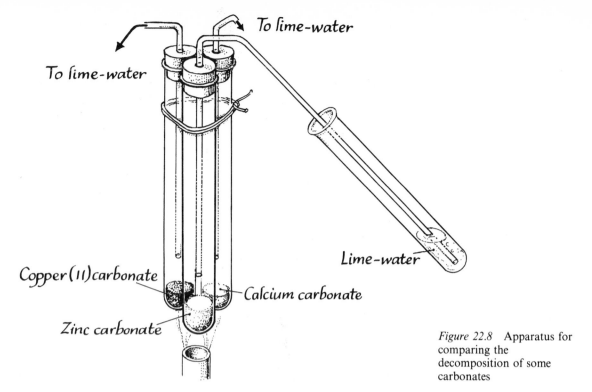

To lime-water

To lime-water

Lime-water

Copper(II) carbonate

Zinc carbonate

Calcium carbonate

Figure 22.8 Apparatus for comparing the decomposition of some carbonates

A convenient size for the test-tubes, which should be made of Pyrex or other 'hard' glass, is 100 mm × 16 mm. Three tubes are fitted with delivery tubes and arranged in a triangle as shown. They are heated by a bunsen burner arranged so as to heat the three tubes as evenly as possible. Each tube contains a different metal carbonate. Try the following carbonates: sodium, zinc, copper(II), lead(II) and calcium.

Place *fresh* lime-water in each of the collecting tubes. Arrange to have the same volume of *powdered* carbonate in each of the heating tubes and make sure that it is the anhydrous form of the carbonate that is used. It is a good plan to have a small plug of 'Rocksil' fibre above the surface of the powder and just below the delivery tube so that no powder can be blown over into the lime-water.

When all is ready, start the heating and at the same time start a stop-clock or note the time on a watch or clock with a seconds hand. Note the time at which each sample of lime-water starts to turn milky.

From your results, arrange the carbonates according to the order in which they start to turn the lime-water milky most quickly. Is there any connection between your order and the order in which the metals react with water (see Section 11.3) or with dilute acids (see Investigation 13g)?

Harmful

1. Draw a clearly labelled diagram of the apparatus you would use to prepare and collect some carbon dioxide gas in the laboratory. Name the substances you would use in the preparation and give the equation for the reaction.
2. Name the acid that is formed when carbon dioxide gas is dissolved in water and give the equation for the reaction.
3. Describe all the changes you would see if some carbon dioxide gas were bubbled for a long time through lime-water.
4. Name the products formed, and write the equations for the reactions involved, when carbon dioxide gas is bubbled through lime-water, (a) for a short time, (b) over a long period.
5. Describe what happens when a solution of calcium hydrogen-carbonate is boiled. What is the significance of this reaction to people living in hard-water areas?
6. Give THREE large-scale uses of carbon dioxide gas.
7. Describe, giving a labelled diagram, the 'Sparklet' type of soda-water syphon.
8. What advantages does solid carbon dioxide have over ice for the purposes of refrigeration?
9. Name FIVE different naturally occurring forms of calcium carbonate.
10. Calcium carbonate is an important industrial chemical. Outline THREE large-scale uses of this substance.
11. What do you see happen when some water is added to freshly prepared quicklime? Give the equation for the change that takes place.
12. How is lime manufactured?
13. Give, and explain, THREE large-scale uses of lime or slaked lime.
14. What is meant by 'pointing' brickwork? Why is this process necessary?
15. Give the chemical formulae of (a) washing soda, (b) baking soda. What happens each of these substances is heated?
16. What is the 'Gossage' process?
17. How would you prepare some lead(II) carbonate in the laboratory? For what purpose was this substance used? Why is its use for this purpose now being discontinued? What has taken its place?
18. If you were given some strontium carbonate and some rubidium carbonate, explain how you would set about finding out which of the two was more easily decomposed by heat.

Chapter 23

Water in the Home

23.1. The importance of water

There is one substance, other than air, which we take for granted, but which is of the utmost importance to us; it is **water**.

Water is an unique substance. Its properties, perfectly suited to so many uses, are not at all what we would expect from its formula, H_2O. Its freezing-point and boiling-point are both far higher than such a small molecule suggests; it is an excellent solvent; it occurs abundantly in nature; and it has the very unusual property in the liquid state of having its greatest density at $277 \, K \, (4 \, °C)$, a temperature above its freezing-point of $273 \, K \, (0 \, °C)$. It is this last property that is responsible for the fact that the seas, rivers and lakes do not freeze solid, but only form a solid layer on the surface which then prevents the liquid underneath from becoming solid.

The provision of an adequate water supply is one of the great problems of an industrial society. In this chapter we are going to consider the supply of water and some of the problems associated with the use of water in the home.

23.2. The water cycle in nature

The processes of evaporation and condensation that we studied (see Investigation 1h) go on in nature on the large scale.

The heat from the sun is continually evaporating water from the seas, rivers, lakes and other sources of liquid water and even from solid snow and ice. In this process, only the water is changed into vapour; the substances dissolved in the water are left behind.

The amount of water vapour that the air can hold depends largely on the temperature. More vapour can be held at higher temperatures and less vapour at lower temperatures. If a mass of air containing a large amount of water vapour is cooled, the water which it can no longer hold in the form of vapour will be deposited as liquid water.

The temperature at which the air will start to deposit the moisture it holds as the temperature falls, is known as the **dew-point**.

To determine this temperature, take a metal container (a copper calorimeter does very well) and polish a strip down one side. Fill the container about two-thirds full of cold water and stand it on a wooden surface. Place a sheet of glass or transparent plastic between the container and your face, in order to prevent the moisture in your breath from affecting the result of the experiment, and put a thermometer into the water.

Carefully add *small* pieces of ice to the water, taking care to stir thoroughly after each addition. Watch the polished strip carefully and note the temperature at which it clouds over. This is the temperature at which the air outside the container starts to deposit the moisture it holds. The outside of the container should be dried after the contents have warmed above the dew-point, and the investigation repeated. Figure 23.1 shows the arrangement for the investigation.

As the warmer air near the earth's surface rises, being less dense than the colder air, it becomes cooled. When the temperature falls below the dew-point, condensation takes place and **clouds** are

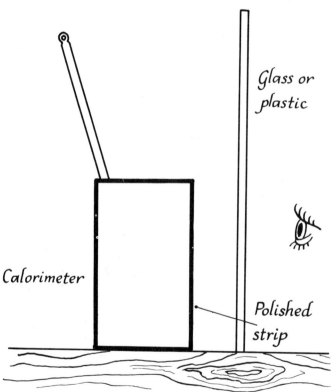

Glass or plastic

Calorimeter

Polished strip

Figure 23.1 Apparatus for finding the dew-point

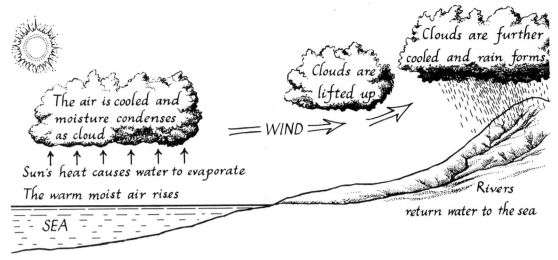

Figure 23.2 The water cycle in nature

formed. Eventually, when conditions are right, the moisture returns to the surface of the earth as rain, snow, hail or sleet. Figure 23.2 illustrates the cycle of water.

When the temperature of the air near the surface of the earth becomes cooled at night, **dew** may be deposited or, if the temperature of the earth's surface is very cold, **frost** may be formed.

The water which falls on the surface of the land eventually appears as streams and rivers, which return the water to the sea and complete the cycle.

23.3. The storage of water

The very large quantities of water that are needed for domestic and industrial use have necessitated the making of large **reservoirs** in which the water may be stored until it is required. These reservoirs may be natural (for example lakes) or artificial.

Large artificial lakes have been formed by building great concrete walls, or dams, across suitable valleys, so that the rivers running into the valleys are held back and the water piles up. The region from which water is taken, that is the ground area on which the rain has fallen, is called a **catchment area**.

For distribution to towns and other places where it is wanted, the water is pumped into local reservoirs or water-towers, from where it is supplied by the force of gravity. A small hill, or other rise, however, between the final reservoir and the supply tap is not an obstacle, provided that the total height of the obstacle is not more than about 10 m above the level of the water in the final reservoir, and the tap itself is lower than the water level.

23.4. The purification of water

Water that is to be used for drinking must be very carefully purified before it is used. This is because the water, as it runs over the ground in rivers and streams or while it is in the reservoirs, picks up solid matter and may also pick up many of the bacteria which can cause diseases in man.

One of the most important processes used in the purification of water is filtration. This is done by allowing the water to seep through layers of sand and fine gravel. A typical filter-bed is shown in Figure 23.3.

Water from deep wells is usually sufficiently filtered by its passage through the ground, but water from shallow wells is often dangerous and needs further treatment before it is safe to drink.

On the small scale, water may be purified by boiling. On the large scale, chemical treatment is often necessary. For this purpose, chlorine gas is often used. Only a very small proportion of chlorine is needed. In normal times, about three parts of chlorine per million parts of water are used and we are not able to taste this small amount of chlorine. In times of epidemics, ten parts of chlorine per million parts of water are used and this concentration gives a distinct taste to the water. At still higher concentrations, chlorine is used to purify the water in swimming baths so that the chances of disease being passed from one swimmer to another are lessened. Ozone (O_3) is also used in the purification of water.

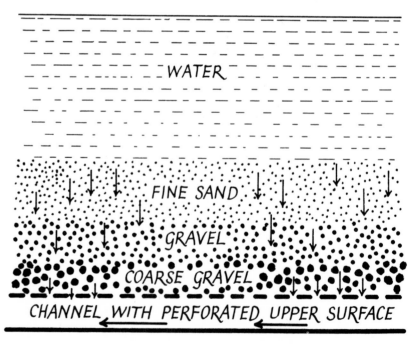

Figure 23.3 The filtration of water

23.5. Washing and soap

To remove grease and dirt from fabrics and surfaces in general, by the use of water, it is necessary also to use substances which help the transfer of the dirt from the material into the water. Such substances are known as **detergents**. Soap is one form of detergent, but it is common today to talk of 'soap *and* detergents', meaning by the word 'detergent' those washing substances which are not soaps, but which act in a similar manner.

Soap is made by boiling an oil or a fat with sodium hydroxide (caustic soda) solution, or with potassium hydroxide (caustic potash) solution.

Investigation 23b. Making soap

Put about 3 cm³ of castor oil into a 100 cm³ beaker and add about 15 cm³ of a 20 per cent solution of sodium hydroxide. Gradually heat the beaker until the solution comes to the boil, stirring gently all the time. Boil very gently for a few minutes, continuing with the stirring to prevent any frothing over. Dilute by adding about 20 cm³ of pure water and about 10 g of salt. Re-heat to boiling and keep at the boiling-point, with stirring, for about two minutes. Allow the mixture to cool thoroughly. Filter off the solid which has formed, breaking up any large pieces, and wash with a little pure water.

Corrosive

When the solid has dried, a little should be shaken with some pure water in a test-tube. The production of a lather will confirm that soap has been made.

castor oil + sodium hydroxide → soap + alkanol

In many cases, the alkanol left after the reaction is **propane-1,2,3-triol**.

On the large scale, soap is made by boiling together animal fats or vegetable oils with caustic soda or caustic potash in large steel pans, using super-heated steam. When the reaction is completed, salt is added to precipitate the soap. This part of the process is known as **salting-out.** The crude soap is filtered from the liquid and is washed. It is then heated with water and colouring materials and perfumes are added. The final product is then left to solidify.

One of the commonest soaps is sodium octadecanoate and it is made by boiling mutton fat with caustic-soda solution. Soft soaps and toilet soaps are usually made from oils, rather than fats, and potassium hydroxide (caustic potash) is used in place of the caustic soda. Much of the propane-1,2,3-triol is retained in the soap.

23.6. How soap works

The atoms in sodium octadecanoate and other soaps are arranged in a long chain – rather like a snake – with a 'head' and a 'tail'. The 'tail' is made up of a long chain of carbon atoms, seventeen of them in the case of sodium octadecanoate, with hydrogen atoms attached. The 'head' is made up of a group of atoms, one carbon, two oxygen and one sodium (see Figure 23.4). The sodium atom is joined ionically to the rest of the atoms.

Figure 23.4 The arrangement of the atoms in sodium octadecanoate

When sodium octadecanoate is put into water, the sodium ion separates from the rest of the atoms. The remaining atoms are left with a negative electric charge on the atoms of oxygen at the end. This end of the group of atoms is attracted to the water molecules. The other end of the chain is, however, repelled by the water molecules, but is attracted towards grease and dirt. The result is that, when an article is washed in water which contains soap, the particles of grease and dirt are 'rolled up' and are drawn into the water (see Figure 23.5).

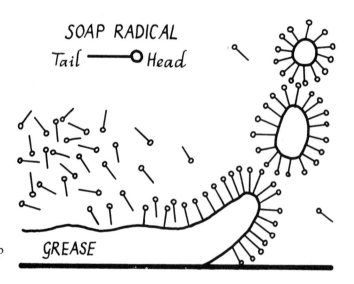

Figure 23.5 How soap works

23.7. Synthetic detergents

Although soap itself is a synthetic detergent, that is a detergent artificially made, the term is used to describe a large number of substances which behave in the same way that soap does, though they have special properties which make their behaviour different from that of soap.

There are two main differences in the structure of a synthetic detergent from that of soap:

a. The 'head' consists of a sulphur atom, three oxygen atoms and an ionically bound sodium atom.
b. The 'tail', instead of being in the form of an unbranched chain, may be branched.

Figure 23.6 shows a typical synthetic detergent structure.

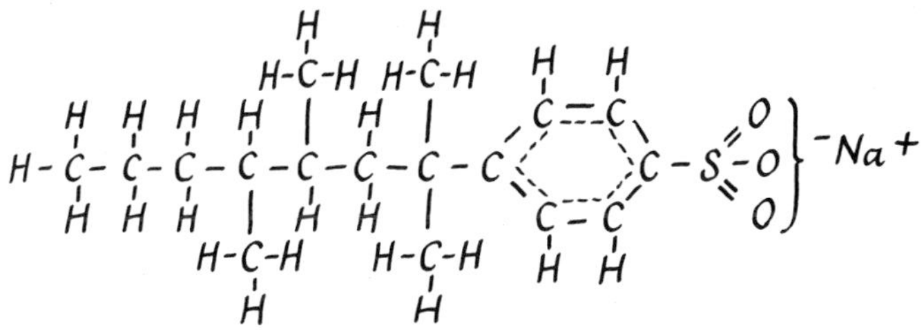

Figure 23.6 A synthetic detergent structure

23.8. Soap and hard water

Investigation 23c. Producing a lather with soap

Take four small beakers (100 cm³), each with a glass rod for stirring. Into the first, put 50 cm³ of distilled water; into the second, put 50 cm³ of rain water; into the third, put 50 cm³ of tap water; and into the fourth, put 50 cm³ of a solution of lime-water that has been diluted with nine times its own volume of distilled water.

Take a cake of ordinary household soap and shave off some small fragments. Alternatively, soap-flakes may be used. Add a small quantity of the shavings to each of the four beakers and stir well. What happens? Add more soap to each beaker until there is no further change. Compare the appearance of the liquids in the beakers. What happens in the beaker with the tap water will depend on the region of the country in which you live, and on where the catchment area (see Section 23.3) for your water is.

The soap forms a **lather** with the water. The amount of soap needed to form a *permanent* lather varies with the type of water used. Water which needs a lot of soap in order to form a permanent lather, and which produces a dirty grey-white 'scum' or precipitate when soap is added to it, is said to be **hard**.

Repeat Investigation 23c, using a small quantity of any available synthetic detergent in place of the soap. What differences do you notice?

Investigation 23d. A more accurate investigation into the action of soap on different types of water

For this investigation, a *solution* of soap is needed. A suitable stock solution may be prepared by dissolving 20 g of sodium octadec-9-enoate in a mixture of 1 dm³ of distilled water with 1 dm³ of industrial methylated spirits. When required for use, one part of this solution is diluted with four parts of distilled water.

Using a pipette, measure 50 cm³ of the water to be tested into a conical flask. Add the soap solution from a burette, 1 cm³ at a time, shaking thoroughly after each addition, until a permanent lather is obtained. A lather is usually taken to be 'permanent' when it lasts for at least half a minute.

Record your results in the form of a table as illustrated in Table 23.1.

TABLE 23.1.   TABLE FOR THE RESULTS OF INVESTIGATION 23d

Type of Water	Volume of Soap Solution Needed	Comment on Hardness
Distilled		
Local tap		
Boiled tap		
Washing-soda treated		
Permutit treated		
De-ionized		
Other area		

The types of water suggested in Table 23.1 can, of course, be altered at choice. It is particularly interesting to compare waters from other parts of the country and from abroad, if specimens are available.

23.9. Hardness in water

As was noted in Section 23.8, water which does not lather readily when soap is added is said to be 'hard' water – that is it is hard to make it lather. Two main types of hardness are recognized: temporary and permanent hardness.

Temporary hardness

This type of hardness is removed by boiling. It is due to the presence in the water of calcium and magnesium ions, together with the hydrogencarbonate ion (HCO_3^-).

Permanent hardness

This type of hardness is *not* removed by boiling. It can, however, be removed in other ways. It is due to the presence in the water of calcium and magnesium ions, together with any other acid radical ions, excluding the hydrogencarbonate ion.

23.10. How water becomes temporarily hard

When rain falls through the atmosphere, it dissolves some of the carbon dioxide gas to form a very weak solution of carbonic acid:

$$CO_2(g) + H_2O(l) \rightarrow H_2CO_3(aq)$$

If the rain water containing this carbonic acid flows over, or through, ground containing calcium carbonate, limestone or chalk, for example, then the carbonic acid will act on the calcium carbonate and change some of it into soluble calcium hydrogen-carbonate:

$$CaCO_3(s) + H_2CO_3(aq) \rightarrow Ca(HCO_3)_2(aq)$$

We saw this reaction taking place in Investigation 22b and also saw the effect of boiling water which contains calcium hydrogen-carbonate:

$$Ca(HCO_3)_2(aq) \rightarrow CaCO_3(s) + CO_2(g) + H_2O(l)$$

As was mentioned there, the deposit of calcium carbonate formed when temporarily hard water is boiled, is known as **fur**. A deposit of calcium carbonate is also formed when water containing calcium hydrogencarbonate evaporates. In caves, this results in the formation of **stalactites** and **stalagmites**. Pot holes and caves in limestone, together with the stalactites and stalagmites they contain, have been formed by this action of carbonic acid and the subsequent decomposition of the calcium hydrogencarbonate. Figure 23.7 illustrates this action.

23.11. How water becomes permanently hard

Permanent hardness in water is caused by the substances dissolved in the water; the water dissolves these substances as it flows over the ground. Calcium sulphate(VI), which is found naturally as **gypsum** ($CaSO_4.2H_2O$), while being only slightly soluble in water, does dissolve to a slight extent. Calcium chloride can dissolve much more readily.

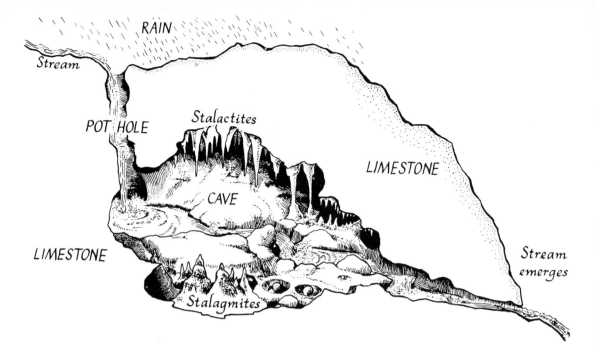

Figure 23.7 The action of
rain water on limestone

It is important to remember that although, in this section and
the last, reference has been made only to *calcium* compounds, the
corresponding magnesium salts produce exactly the same effect.

23.12. The action of soap on hard water

We saw in Investigation 23c that hard water takes a lot of soap
to make it lather and that, at the same time, a dirty greyish-white
scum is formed. This scum is composed of calcium (or magnesium)
octadecanoate if sodium octadecanoate is the soap being used. It
is only when all the calcium and magnesium ions have been removed from the water in the form of these octadecanoates that the
water is softened and a lather can form. It is clear then that hard
water is extremely wasteful of soap.

The modern synthetic detergents, as you will have seen in the
same investigation, are not affected by the calcium or magnesium
ions in the water. They produce a lather at once, and do not form
the scum that soap produces.

23.13. The advantages and disadvantages of hard water

Hard water has two very important disadvantages: it wastes
soap and it deposits fur when it is boiled. This latter effect may
cause water pipes in domestic hot water systems to become
gradually blocked up, their effective diameter being greatly
reduced. Eventually, they may become completely blocked up and

252

this can result in a dangerous situation. The water in the system cannot circulate and the boiler will overheat the water in it. If the safety devices in the system are not working properly, a steam pressure may be set up and an explosion may take place.

In the same manner, kettles become coated with a deposit of calcium carbonate. This results in a wastage of heat, for calcium carbonate is a bad conductor of heat and it is necessary to heat the kettle for a longer time before the water in it will boil.

On the other hand, the calcium salts in hard water are of value in providing the calcium necessary for the proper formation of bones and teeth. It has also been proved that people living in hard-water regions are less liable to coronary thrombosis (heart attacks). The salts also give flavour to the water; pure water tastes very 'flat'. Although we may not realize it, different samples of drinking water from different regions vary in their flavour.

23.14. Methods of softening hard water

If the water in the region in which you live is temporarily hard, you will have noticed that less soap solution was needed in Investigation 23d to produce a permanent lather in the *boiled* tap water than in the tap water itself. This is, of course, because the calcium in the water, which was in the form of calcium hydrogen-carbonate, was deposited as the insoluble calcium carbonate. Boiling is, then, one way to remove *temporary* hardness in water.

Another method used to remove temporary hardness, but *not* permanent hardness, is known as **Clark's method**.

Investigation 23e. The action of lime-water on temporary hardness

Take twelve corked conical flasks (250 cm^3) and pipette into each 50 cm^3 of tap water. Label the flasks from 1 to 12. Now, into flask 1, put 1 cm^3 of lime-water from a burette. Into flask 2, put 2 cm^3 of lime-water. Continue this process until 12 cm^3 of lime-water have been added to flask 12. These flasks must be prepared at least forty-eight hours before the actual experiment is to be carried out, and they should be shaken at intervals.

Soap solution, prepared as described for Investigation 23d, is put into a burette. The soap solution is added to each flask in turn until a *permanent* lather is formed on shaking. The volume of soap solution needed for each of the flask is noted. It is advisable to prepare at least two complete sets of flasks so that, if an error is made by accident in any of the titrations, a replacement is at hand. If the work is spread among the members of a class or group, it need not take long.

Plot a graph of the volumes of lime-water in the flasks against the volume of soap solution needed for each flask. What do you notice about the volumes of soap solution needed?

In actual practice, exactly the required quantity of slaked lime is added to measured volumes of water at the water supply centre. The lime brings about the removal of the calcium which is causing the temporary hardness, causing a precipitate of calcium carbonate to be formed:

$$Ca(HCO_3)_2(aq) + Ca(OH)_2(s) \rightarrow 2CaCO_3(s) + 2H_2O(l)$$

23.15. Methods of removing *all* the hardness from water

Boiling, or adding the correct quantity of slaked lime, only removes temporary hardness. Other methods are available for removing *all* hardness from water.

Adding washing soda

Washing soda, $Na_2CO_3.10H_2O$, brings about the precipitation of all the calcium and magnesium ions in the water, irrespective of the acid radical ions present:

$$Ca(HCO_3)_2(aq) + Na_2CO_3(aq) \rightarrow CaCO_3(s) + 2NaHCO_3(aq)$$
$$CaSO_4(aq) + Na_2CO_3(aq) \rightarrow CaCO_3(s) + Na_2SO_4(aq)$$
$$CaCl_2(aq) + Na_2CO_3(aq) \rightarrow CaCO_3(s) + 2NaCl(aq)$$

These reactions may be summed up ionically as:

$$Ca^{2+}(aq) + CO_3^{2-}(aq) \rightarrow CaCO_3(s)$$

It will be seen from the above equations that the calcium is removed as a precipitate of calcium carbonate.

The 'Calgon' process

The name 'Calgon' (calcium gone!) is a trade name for an important industrial water-softening process which uses a substance that has the power of 'capturing' calcium ions and 'locking them up' so that, although they are not precipitated, and remain in the water, they cannot get at the soap. Addition of Calgon to the water can also bring about the gradual removal of calcium that has already been deposited as fur.

The Permutit process

This was the original 'ion-exchange' process. Although the material used today is prepared synthetically, the original material was a natural type of clay. Although in reality it has a very complicated structure, it may be thought of as being **aluminium sodium silicate**. It is insoluble in water, but when water containing calcium or magnesium ions is passed through a column packed with it, the calcium (or magnesium) ions are removed from the water and are held in the lattice of the clay, an equivalent number of sodium ions being put into the water in their place:

$$Ca^{2+} \quad + \text{ aluminium sodium silicate} \rightarrow 2Na^+ + \text{aluminium calcium silicate}$$
(in water)   (in water)

Sooner or later, all the sodium ions in the Permutit material will have been used up. It is then possible to **regenerate** the material. This is done by passing a strong solution of sodium chloride (common salt) through it:

$$2Na^+ \quad + \quad Ca^{2+} \quad \rightarrow \quad Ca^{2+} \quad + \quad 2Na^+$$
(in solution) (in Permutit) (in water) (in Permutit)

The calcium ions are washed out and the Permutit is again ready for use.

Investigation 23f. A model water-softener

A demonstration Permutit water-softener can be made from a 250 mm length of wide-bore (50 mm) glass tubing. A suggested arrangement is shown in Figure 23.8.

Materials such as Permutit are called **zeolites** or **ion-exchange resins**. (Permutit (natural zeolite) base-exchange material is obtainable from Messrs. B.D.H. (Chemicals), Poole, Dorset.)

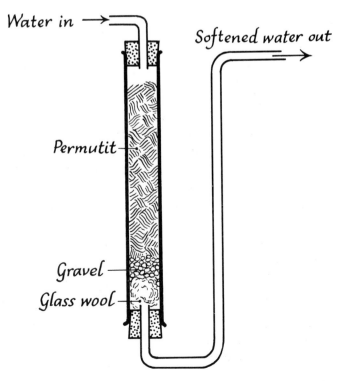

Figure 23.8 A model water-softener

23.16. Pure water

The processes discussed so far for removing all the hardness from water do not make the water pure, though they do soften it.

To produce pure water, it is necessary to remove all ions other than the hydrogen ion ($H^+(aq)$) and the hydroxyl ion ($OH^-(aq)$). One way to do this is by distillation (see Investigation 1h). Another method is that of de-ionization by ion-exchange.

In this latter process, the water is first passed through an ion-exchange resin that exchanges all the metal ions in the water for hydrogen ions. The resulting liquid, which is really a mixture of very dilute acids, is passed through a second ion-exchange resin which exchanges all the acid radical ions for hydroxyl ions. The final result is that the liquid emerging from the second resin is pure water, all but a small fraction of the hydrogen and hydroxyl ions joining together to form water:

$$H^+(aq) + OH^-(aq) \rightarrow H_2O(l)$$

The resins may be regenerated. The first resin is regenerated by passing through it some dilute (1 M) hydrochloric acid. The second resin is regenerated by passing through it a 1 M solution of sodium hydroxide.

The resins may be used separately, as described above, or they may be mixed together. This **mixed-bed resin** produces the best purification, though regeneration is more difficult.

Investigation 23g. Pure water by de-ionization

For this experiment, the resins required are Zeokarb 225 and de-Acidite FF, both of which can be obtained from Messrs. B. D. H. (Chemicals).

Two burettes are taken. A little glass wool is pushed down into the bottom of each. The ion-exchange resins are stirred up thoroughly with pure water and the resulting slurries are poured each into one of the burettes. Care must be taken not to let any air bubbles get trapped in the resins and to keep the water level above the surface of the resins in the burettes at all times. When not in use, the burette tops should be kept tightly closed with rubber bungs.

Harmful

Prepare a solution containing both copper(II) sulphate(VI) and potassium chromate(VI) by dissolving 8 g of the former and 5 g of the latter in a total volume of 1 dm³ of water. The solution will be of a deep-green colour. This green colour is due to the blue colour of the $Cu^{2+}(aq)$ ion and the yellow colour of the $CrO_4^{2-}(aq)$ acid radical ion.

Some of this solution is allowed to trickle slowly through the Zeo-karb 225 column, the liquid running out of the column being collected in a small beaker. Note the colour of this liquid. Notice also the coloured band (blue) that is formed near the top of the column.

Now, pour the liquid obtained from the first column slowly through the de-Acidite FF in the second column. Note the colour of the issuing liquid. Again, note the coloured band (yellow) that is formed near the top of the column.

The first column has removed the blue copper(II) metal ions (and, of course, the colourless potassium ions). The second column has removed the yellow acid radical ions of the chromate(VI) (and, of course, the colourless sulphate(VI) ions). Pure water is left.

To regenerate the first resin, wash out any remaining chromate(VI) with pure water until the emerging liquid is colourless. Now, put a clean beaker beneath the column and slowly pour some dilute, 1 M, sulphuric(VI) acid through the column. After a time, the liquid coming out of the column will be blue. The copper(II)

Harmful

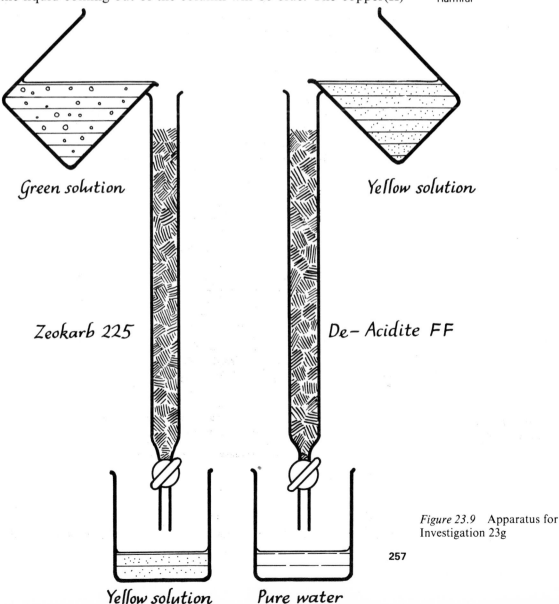

Green solution

Zeokarb 225

Yellow solution

Yellow solution

De-Acidite FF

Pure water

Figure 23.9 Apparatus for Investigation 23g

ions, which had been held by the column, have been replaced by hydrogen ions. Continue with the acid until the liquid emerging is colourless and then wash out the excess acid with pure water.

The second resin is regenerated by passing through it some 1 M sodium hydroxide solution. If the emerging liquid is collected, it will be yellow. The hydroxyl ions from the sodium hydroxide have replaced the chromate(VI) ions. Continue with the sodium hydroxide solution until the liquid emerging is colourless and then wash out the excess sodium hydroxide with pure water. The apparatus for this investigation is shown in Figure 23.9.

The survival of shipwrecked sailors is made easier by the use of this de-ionization process. A double-walled canvas bag has the space between the walls filled with a mixture of the two de-ionizing resins. To obtain fresh water from sea water, the bag is dipped into the sea and is drawn out full of sea water. The bag is then squeezed so that the water is forced through the resins. The issuing water, now pure, is then collected in any suitable container. Figure 23.10 shows such a bag.

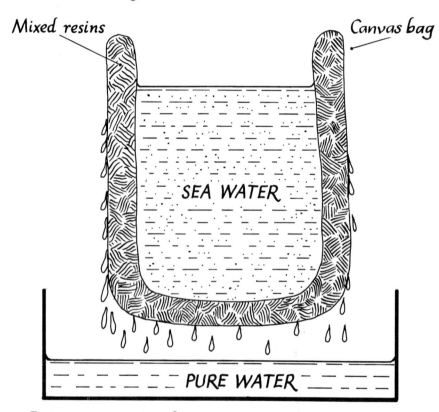

Figure 23.10 A de-ionizing bag

Test your understanding

1. Water is a unique substance: why is this so?
2. What is the meaning of the term 'dew-point'?
3. How is each of the following formed: (a) frost, (b) dew, (c) clouds?

4. Why is it that water from deep wells is usually fit to drink, but water from shallow wells may be a danger to health?
5. Describe, in outline, the methods available to make water fit for drinking.
6. What is the 'water cycle'? Give a labelled diagram to show the various processes taking place during the cycle.
7. What is the difference between a 'soap' and a 'soapless detergent'?
8. Explain, using labelled diagrams, how a soap works in helping to remove dirt from clothes.
9. What is 'hard' water?
10. Explain the difference between (a) temporary, (b) permanent, hardness in water.
11. How would you compare in the laboratory two samples of water taken from different sources with regard to (a) their degree of temporary hardness, (b) their degree of permanent hardness?
12. How does water become temporarily hard?
13. How are stalactites and stalagmites formed? Which grows upward from the ground?
14. Outline (a) the advantages, (b) the disadvantages, of hard water.
15. Explain TWO ways in which temporary hardness may be removed from water.
16. Explain TWO ways in which all the hardness, both temporary and permanet, may be removed from hard water.
17. Explain TWO ways in which pure water (that is, chemically pure water) may be obtained.
18. Explain the difference between 'softening' and 'de-ionization'.
19. $100 \, cm^3$ of water needed $24 \, cm^3$ of soap solution before a permanent lather was obtained on shaking. After boiling, the same volume of water needed $16 \, cm^3$ of the same soap solution for a permanent lather. What proportion of the hardness in the water was permanent?

Chapter 24

Nitrogen

24.1. A very important element

Nitrogen is one of the most important elements to man. As we shall see in this chapter, it is essential for the growth of all plants. Unfortunately, while there is an abundance of nitrogen in the earth's atmosphere, the element itself is not easily available to plant life.

24.2. Nitrogen in the air

In Table 9.1 we saw that the percentage of nitrogen by volume in dry air is almost 79 per cent. Although there is so much nitrogen, the gas itself is very unreactive. Making nitrogen combine with other elements is known as **fixing**. It can be made to combine directly with hydrogen, oxygen and magnesium. The reactions with hydrogen and oxygen can be used in industrial processes for making nitrogen compounds from the gas in the air. These processes will be referred to later in this chapter.

Investigation 24a. Getting nitrogen gas from the air

Corrosive

Set up the apparatus shown in Figure 24.1. The aspirator may be a Winchester bottle ($2.5\,dm^3$) or a larger container.

Air is displaced from the aspirator and driven through the apparatus, by the pressure of water from the mains. The air is passed first through a solution of sodium hydroxide in order to remove carbon dioxide:

$$2NaOH(aq) + CO_2(g) \rightarrow Na_2CO_3(aq) + H_2O(l)$$

The remaining gases are then passed through a heated silica tube containing copper turnings or copper gauze. The copper reacts with oxygen:

$$2Cu(s) + O_2(g) \rightarrow 2CuO(s)$$

The remaining gases, that is **nitrogen** and the so-called **noble gases**, are collected over water.

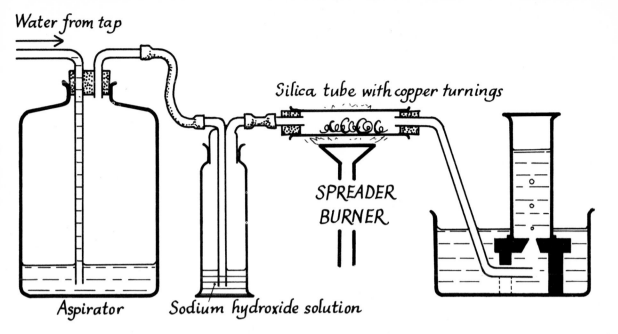

Water from tap

Silica tube with copper turnings

SPREADER
BURNER

Aspirator   Sodium hydroxide solution

It is also possible to separate nitrogen and the noble gases from air by shaking a volume of air with a solution of sodium hydroxide and benzene-1,2,3-triol, as in Investigation 9b.

Figure 24.1   Apparatus for obtaining nitrogen gas from air

24.3. The large-scale separation of nitrogen from air

Two methods are available for separating nitrogen from air on the large scale:
a. by the distillation of liquid air (see Section 9.3); and
b. by a combination of the water-gas and producer-gas processes (see Section 4.6).
This latter method is particularly useful as, if it is carried out under the right conditions, it also results in the formation of hydrogen gas; the nitrogen and hydrogen are in exactly the correct proportions, that is one to three by volume, for the synthesis of ammonia by the Haber process (see Section 24.9).

24.4. The circulation of nitrogen in nature

It was stated in Section 24.1 that nitrogen is essential for the proper growth of all plants. However, in general, plants cannot make use of the uncombined element; they require soluble compounds of nitrogen.

Just as there is a circulation of carbon in nature (see Section 21.6), so there is a circulation of nitrogen. The living plants take in nitrogen, in the form of nitrates, through their roots. This combined nitrogen is used for the proper growth of the plants. In the natural order of events, the nitrogen is eventually returned to the ground and is available again for use. This happens when the plant dies or, if the plant is eaten by an animal, when the animal excretes or eventually dies.

The nitrogen in the plant or animal body is combined in the very complex **protein** molecules (see Section 35.3). When the plant or animal dies, very small organisms, **decay bacteria**, in the soil act upon the proteins and convert them into ammonium compounds (see Section 25.1). Other bacteria, the **nitrifying bacteria**, convert the ammonium compounds into **nitrates(III)** (otherwise known as nitrites) and eventually back into **nitrates(V)**. The nitrogen excreted by animals is in the form of ammonium salts and urea.

Atmospheric nitrogen is converted in nature into useful nitrogen compounds – that is it is fixed – in two ways:

a. by the action of lightning on the nitrogen and oxygen in the air. This results in the formation of some nitrogen oxide gas which, by the further action of air and water, is finally converted into nitric(V) acid and hence into nitrates(V).

b. by the action of the **nitrogen-fixing bacteria**, which are found in the soil and also in the **root nodules** of **legumes**, for example clover, lucerne, peas and beans.

Unfortunately, there are other bacteria in the soil, the **de-nitrifying bacteria**, which can decompose nitrogen compounds, liberating nitrogen gas.

The whole series of processes referred to above may be represented in diagrammatic form (see Figure 24.2).

24.5. Man breaks the nitrogen cycle

Although, in the natural state, the nitrogen compounds needed by plants for their growth are reformed in the soil, when man grows crops for his own use, the chain of events is broken.

The crops are harvested and are removed from the place where they are grown. They are eaten, and the excretory products are not returned to the soil, but are disposed of in other ways, frequently by being discharged into the sea. In this way, the quantity of nitrogen in the soil gradually decreases and, if nothing is done to replace it, the quality and quantity of the crops being grown become less.

24.6. The rotation of crops

Though used relatively little today, some relief for the nitrogen problem can be obtained by **rotating** the crops that are grown. Cereal crops, which remove a considerable quantity of nitrogen from the soil, are followed by crops which require less nitrogen

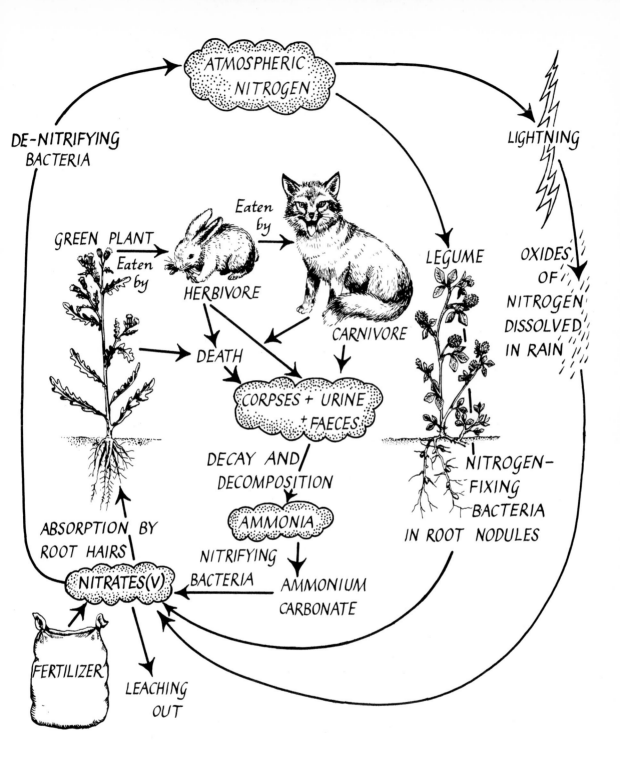

Figure 24.2 The nitrogen cycle

and then by clover or some other crop that can, by the action of the nitrogen-fixing bacteria attached to the roots, actually replace nitrogen in the soil. The rotation of crops also helps in the control of weed growth.

24.7. The use of fertilizers

In past times, the rotation of crops was sufficient for the fertility of the soil to be maintained, particularly when the rotation was accompanied by the addition of **compost** (the decayed remains of plants) and animal manure (the broken-down excrement of animals).

Today, the need for food in sufficient quantities to feed the ever-increasing numbers of human beings makes it impossible to allow the land to rest. Also, the crop yield of the land would not be great enough to support even the present population if traditional methods only were used. Thus, the use of fertilizers, both natural and artificial, is essential.

The natural manures are normally added to the land in the autumn and have a long-term effect. Artificial fertilizers, such as **ammonium sulphate(VI), ammonium nitrate(V), nitro-chalk** and many others, are usually added in the spring and have a short-term effect. Artifical fertilizers are never used at the same time as natural manures; the bacteria in the manures will decompose the nitrogen compounds in the artificial fertilizers and the object will be defeated.

The only large-scale natural source of 'artificials' – the word 'artificial' is used as opposed to the natural animal manures – is **Chile saltpetre**, that is sodium nitrate(V). Most artificial fertilizers are made from atmospheric nitrogen.

24.8. The fixing of atmospheric nitrogen

On the large scale, nitrogen is usually fixed by the **Haber process**. This process was discovered in 1909 by Fritz Haber. It involves the combining of nitrogen from the air with hydrogen to form ammonia gas; the process is dealt with in detail in Section 24.9.

An earlier process, the **Birkeland and Eyde process**, sometimes known as the **electric arc process**, was never very successful and is now obsolete. It involved the passing of air through an electric arc; a small quantity of nitrogen oxide gas is formed. On cooling, this gas gives another gas, nitrogen dioxide, which, when dissolved in water with more air, gives nitric(V) acid.

24.9. The Haber process

In this process, one volume of nitrogen gas is mixed with three volumes of hydrogen gas. The mixture is compressed to a pressure of some two hundred times the normal atmospheric pressure and is heated to a temperature of about 820 K (about 550 °C). The hot compressed mixture of gases is then passed over a catalyst. The

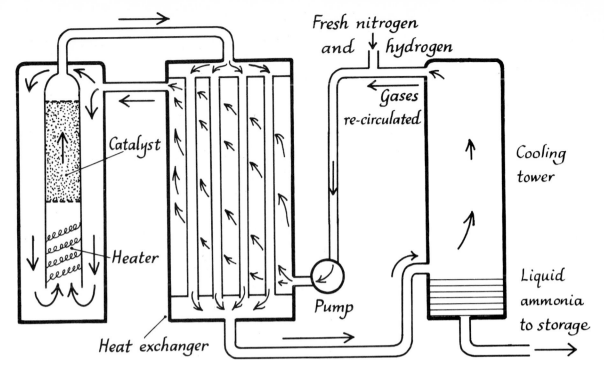

Figure 24.3 The Haber process

catalyst is composed of metallic iron to which are added metallic oxides, such as aluminium and calcium oxides. These 'activate' the iron and make it function more efficiently.

Under these conditions, about 12 per cent of the gases passing through the plant are converted into ammonia:

$$N_2(g) + 3H_2(g) \rightleftharpoons 2NH_3(g)$$

Because of the pressure, on cooling, the ammonia becomes a liquid. It is run off to storage, still under pressure, and the unchanged nitrogen and hydrogen are re-circulated, additional gases being added to restore the original volumes used. In this way, all the hydrogen and nitrogen are eventually converted into ammonia. Figure 24.3 illustrates the Haber process diagrammatically.

The nitrogen needed for the process is obtained from air by the fractional-distillation process (Sections 9.3). The hydrogen is obtained from methane (natural gas). One method of doing this is to pass a mixture of methane and steam over a nickel catalyst at a temperature of about 1 100 K (about 800 °C).

$$CH_4(g) + H_2O(g) \rightarrow CO(g) + 3H_2(g)$$

The carbon monoxide is oxidized to carbon dioxide and is then removed from the mixture by 'scrubbing' – that is, washing the gases with water under pressure:

$$2CO(g) + O_2(g) \rightarrow 2CO_2(g)$$
$$CO_2(g) + H_2O(l) \rightarrow H_2CO_3(aq)$$

24.10. The electric arc process

Although this process is obsolete on the industrial scale, nevertheless it is of interest, because, in thunderstorms, the lightning flashes produce the same effect as the electric arc.

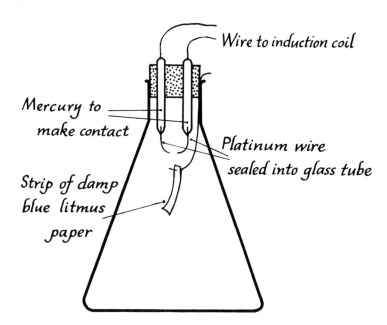

Figure 24.4 Apparatus for Investigation 24b

Investigation 24b. Nitric(V) acid from air

Set up the apparatus shown in Figure 24.4.

When a succession of electric sparks is passed between the pieces of platinum wire, the damp blue litmus paper will be seen to turn red. If, after sparking has been continued for some time, a small quantity of strong ammonia solution is added to the flask, thick white fumes of ammonium nitrate(V) will be seen.

Harmful

When the sparks pass through the air, a small quantity of nitrogen oxide gas is formed:

$$N_2(g) + O_2(g) \rightarrow 2NO(g)$$

On cooling, this gas reacts with more oxygen from the air to form the brown gas, nitrogen dioxide:

$$2NO(g) + O_2(g) \rightarrow 2NO_2(g)$$

Finally, in the presence of more air, this gas dissolves in water to form nitric(V) acid:

$$4NO_2(g) + 2H_2O(l) + O_2(g) \rightarrow 4HNO_3(aq)$$

266

24.11. The more important nitrogenous fertilizers

By far the most important of the artificial nitrogenous fertilizers is ammonium sulphate(VI) – commonly called 'sulphate of ammonia'. This may be made by the action of sulphuric(VI) acid on the ammonia obtained from the Haber process or from other sources:

$$2NH_3(g) + H_2SO_4(aq) \rightarrow (NH_4)_2SO_4(aq)$$

Ammonium nitrate(V) is also used as a fertilizer. It is made by the action of nitric(V) acid on ammonia:

$$NH_3(g) + HNO_3(aq) \rightarrow NH_4NO_3(aq)$$

Ammonium nitrate(V) is a dangerous substance. If not handled carefully, it may explode – indeed, under the right conditions, it may be used as an explosive. To make it safer for agricultural use, it is frequently mixed with calcium carbonate. The mixture is known as **nitro-chalk**.

Calcium cyanamide is another useful artificial fertilizer. It is made by heating calcium dicarbide (see Investigation 21a) at about 1 300 K (about 1 000 °C) in an atmosphere of nitrogen gas:

$$CaC_2(s) + N_2(g) \rightarrow CaCN_2(s) + C(s)$$

The mixture of the cyanamide and carbon is known as **nitrolim**. In the soil, the action of water on the nitrolim slowly liberates ammonia gas:

$$CaCN_2(s) + 3H_2O(l) \rightarrow CaCO_3(s) + 2NH_3(g)$$

24.12. Other elements needed by plants for their growth

In addition to *nitrogen* and, of course, *carbon* (obtained from carbon dioxide in the air), *hydrogen* (obtained from the water in the soil) and *oxygen*, plants need *magnesium, iron, potassium, phosphorus, sulphur* and *calcium*.

These elements are needed in moderate quantities. Other elements are needed in much smaller quantities. These are known as **trace elements**. They include copper, manganese, zinc, boron and molybdenum.

Fertilizers supplying some of these elements include **super-phosphate**, made by treating calcium phosphate(V) with sulphuric(VI) acid, and **bone-meal**, made by the fine grinding of animal bones, which contain a high proportion of calcium phosphate(V) and potassium sulphate(VI).

1. What is meant by the term 'fixing' or 'fixation' with regard to nitrogen?
2. With what other elements can nitrogen be made to combine directly?
3. Describe, giving a labelled diagram of the apparatus you would use, how you would obtain a sample of nitrogen gas, free from carbon dioxide and oxygen, in the laboratory, using the air as source.
4. How, in nature, is nitrogen in the air converted into compounds of nitrogen?
5. Explain the meaning of the expression 'nitrogen cycle'. Draw a labelled diagram to illustrate your answer.
6. Crop yield can be improved by 'rotation'. Explain the meaning of this statement.
7. What methods are normally used for the extraction of nitrogen from air on the large scale?
8. 'Artificial fertilizers are essential for the survival of the human race.' Comment on this assertion.
9. Name THREE artificial fertilizers.
10. Explain the Haber process for the fixation of nitrogen. (You are NOT required to explain the sources of the elements used in the process.)
11. When an electric discharge is passed through air, an acid is formed. Explain, giving a labelled diagram of the apparatus you would use, how you would demonstrate this fact in the laboratory.
12. Besides nitrogen, other elements are essential for the healthy growth of plants. Make a list of these elements.
13. What is a 'trace' element? Name FOUR trace elements.
14. Outline the manufacture of calcium cyanamide.
15. Ammonium nitrate(V) is used as an artificial fertilizer. Why does particular care have to be exercised in its storage and use? How may it be made safer?
16. What is 'compost'? Is it a fertilizer? Find out what it does to help increase crop yields.
17. What is the difference between 'nitro-lime' (nitrolim) and nitro-chalk?

Chapter 25

Ammonia and the Ammonium Salts

25.1. Ammonia and ammonium

It is important that the difference between **ammonia** and **ammonium** should be clearly understood. Ammon*ia* is a gaseous compound, made up of small molecules with the formula NH_3. Ammon*ium* is the name given to a positive ion with the formula NH_4^+. This ion is found combined with various negative ions, for example the chloride ion. The salts formed in this way are the **ammonium salts**. If the negative ion is chloride, then the salt is called ammonium chloride and its empirical formula (see Section 8.5) is $NH_4^+Cl^-$.

Like all salts, the ammonium salts are ionic compounds. Their properties, however differ in a number of ways from the properties of metallic salts such as the sodium or potassium salts.

Investigation 25a. The preparation and collection of ammonia gas

Ammonia is best prepared in the laboratory by the action of a base on an ammonium salt. The most convenient substances to use are ammonium sulphate(VI) and calcium hydroxide (slaked lime). When these substances are heated together, ammonia gas is liberated. The reaction takes place easily; in fact, the gas is liberated slowly even at room temperature:

$$(NH_4^+)_2SO_4^{2-}(s) + Ca^{2+}(OH^-)_2(s) \rightarrow Ca^{2+}SO_4^{2-}(s) + 2NH_3(g) + 2H_2O(g)$$

Harmful

The apparatus used to prepare the gas, dry it and collect it, is shown in Figure 25.1

To dry the gas, calcium oxide (quicklime) is used. The usual drying agents for gases – that is concentrated sulphuric(VI) acid and calcium chloride – both react with ammonia.

The gas has a lower density than that of air, so it may be collected by the method of upward delivery. Collect several small gas jars or boiling tubes of the gas.

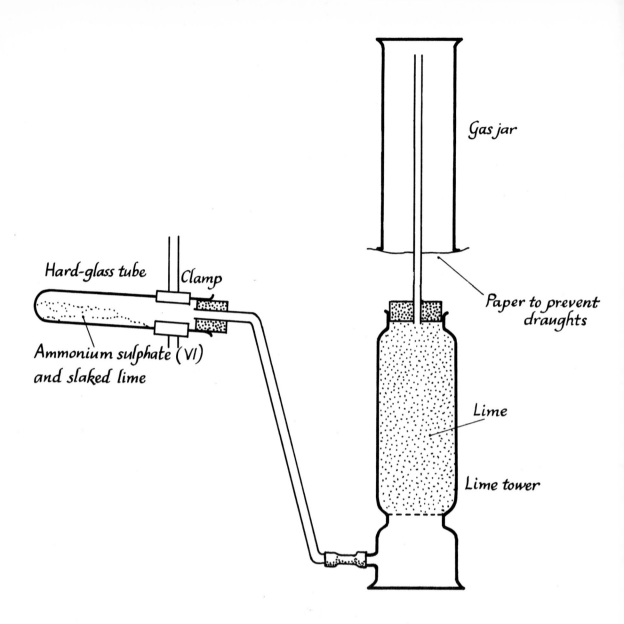

Gas jar

Paper to prevent
draughts

Hard-glass tube

Clamp

Ammonium sulphate (VI)
and slaked lime

Lime

Lime tower

Figure 25.1 The
preparation and collection
of dry ammonia

If the gas is not required to be dry, then it may be prepared
and collected using the simpler apparatus shown in Figure 25.2.

Investigation 25b. The properties of ammonia gas

As usual, when studying the properties of a gas, it is suggested
that the idea of using the word COWSLIPS (see Investigation 9e)
should be followed as far as possible.

The gas is colourless, but has a very distinctive smell. Great
care should be taken when smelling the gas, as it can cause acute
discomfort if taken in too large a dose; it is, in fact, poisonous.
Test the solubility in water by filling a gas jar or a boiling tube

Harmful

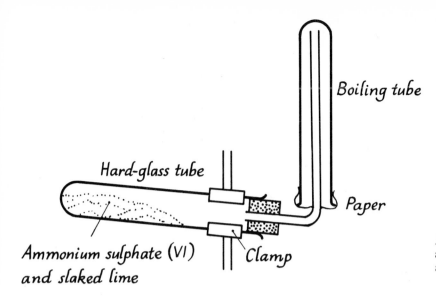

Boiling tube

Hard-glass tube

Paper

Ammonium sulphate (VI) and slaked lime

Clamp

Figure 25.2 Simpler apparatus for preparing and collecting ammonia

with the gas, turning it upside down in a bowl of water and removing the stopper or lid. A more spectacular way of demonstrating the solubility is given in Investigation 25c.

Test the gas with damp litmus paper, both blue and red. What happens? When a gas is very soluble in water, it is because it is reacting with the water. In the case of ammonia, a solution of **ammonium hydroxide** is formed:

$$NH_3(g) + H_2O(l) \rightarrow NH_4^+OH^-(aq)$$

Ammonium hydroxide corresponds in many ways to sodium hydroxide, though it is not caustic.

Try to burn the gas, either by applying a light to a jar or tube of the gas, or by applying the light to the delivery tube from which the gas is emerging. Does it burn at all? If it does burn, does it burn easily? A further account of the burning of ammonia will be found in Investigation 25d.

One of the properties of ammonia gas, and one by which it may be detected, is its action on **turmeric paper**. This is absorbent paper which has been soaked in extract of turmeric. The paper is yellow in colour, but, when it is made damp and exposed to ammonia gas, it turns a deep brown shade.

Try also to test the gas with some hydrogen chloride gas. This may be done most easily by taking the stopper of a concentrated hydrochloric acid bottle and holding the stopper in the ammonia gas. What do you observe? The product formed is a salt – ammonium chloride. This is an example of the way in which ammonia can behave in a basic fashion.

$$NH_3(g) + HCl(g) \rightarrow NH_4^+Cl^-(s)$$

This is a very exciting way of demonstrating the great solubility of ammonia gas in water, and the fact that when it dissolves, the solution formed is alkaline.

Set up the apparatus shown in Figure 25.3.

The water in the bottle should be coloured by adding to it one or two drops of dilute hydrochloric acid and some screened methyl orange indicator (or litmus solution).

Harmful

Remove the large flask (2 dm³) from the apparatus and fill it with ammonia gas, prepared as described in Investigation 25a. Make sure that the delivery tube reaches right to the top of the inverted flask, as shown in Figure 25.4, and lead the delivery tube into the flask through a piece of paper which is twisted around the neck of the flask, in order to minimize draughts.

Replace the flask in position, taking care not to let any air get into it. Open the clip and blow into the tube fixed into the bottle

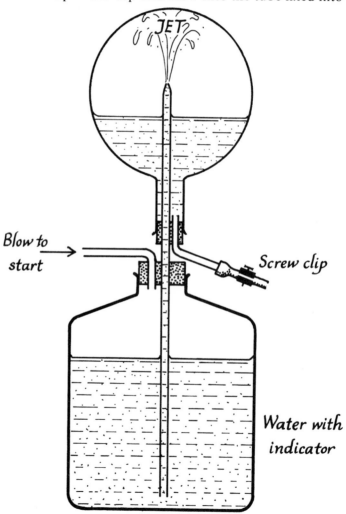

Figure 25.3 Apparatus for the ammonia fountain experiment

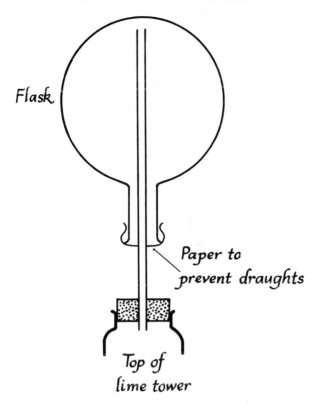

Flask

Paper to
prevent draughts

Top of
lime tower

Figure 25.4 Filling the
ammonia fountain flask
with ammonia

below, so forcing water up the long tube into the flask. As soon
as the water reaches the flask, close the clip tightly and stop
blowing. Note what happens.

Investigation 25d.   The 'burning' of ammonia gas

You will have discovered, while testing the properties of
ammonia gas, that it is very difficult, though not impossible, to
make the gas burn in air. It will burn quite easily in oxygen gas
and it can be made to burn readily in air, *providing that a suitable
catalyst is used.*

The burning of ammonia gas in air in the presence of platinum as
a catalyst – the so-called **catalytic oxidation of ammonia** – can be
shown by setting up the apparatus shown in Figure 25.5.

Harmful

Place a small quantity of strong ammonia solution (880 ammonia – see Section 25.10) in the bottom of the Buchner flask.
Draw air slowly through the flask by means of the water pump.
Remove the rubber bung, together with the air entry tube and the
platinum spiral, heat the spiral red-hot in a bunsen flame, and
quickly replace the bung in the flask before the spiral has had time
to cool down.

As the stream of air is drawn over the surface of the ammonia
solution, ammonia gas is mixed with the air and the mixture is

273

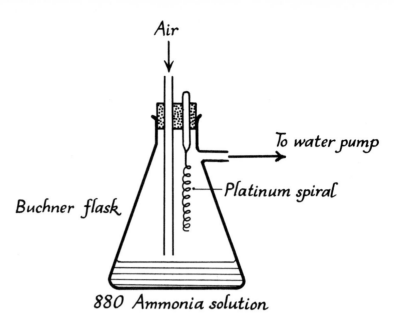

Air

To water pump

Buchner flask

Platinum spiral

Figure 25.5 Apparatus for showing the catalytic oxidation of ammonia

880 Ammonia solution

drawn over the hot platinum spiral. It should be observed that the spiral glows brightly, and continues to glow brightly, as long as a suitable ammonia/air mixture is drawn over it. The reaction is exothermic, considerable heat being evolved:

$$4NH_3(g) + 5O_2(g) \rightarrow 4NO(g) + 6H_2O(g)$$

The formation of water in this reaction is easily seen, as the moisture condenses on the inside of the flask near the top. Also, if the water pump is turned off and the flask is allowed to cool, thick white fumes will be seen. These are due to the formation of ammonium nitrate(V).

Harmful

As the flask cools down, the nitrogen oxide formed in the catalytic oxidation of the ammonia is further oxidized to the gas, nitrogen dioxide:

$$2NO(g) + O_2(g) \rightarrow 2NO_2(g)$$

The nitrogen dioxide, in the presence of more oxygen in the air and water vapour, is changed into nitric(V) acid:

$$4NO_2(g) + 2H_2O(l) + O_2(g) \rightarrow 4HNO_3(aq)$$

The nitric(V) acid then reacts with the ammonia gas in the flask:

$$HNO_3(aq) + NH_3(g) \rightarrow NH_4{}^+NO_3{}^-(s)$$

We shall see, in Chapter 26, how this series of reactions is used commercially for the manufacture of nitric(V) acid from ammonia.

25.2. The large-scale uses of ammonia

We saw, in Chapter 24, that ammonia is manufactured from nitrogen and hydrogen by the Haber process. We saw also that considerable quantities of the compound are converted into **ammonium sulphate(VI)** for use as an artificial fertilizer, by reaction with sulphuric(VI) acid. This is, indeed, the chief use of ammonia gas, but by no means the only use.

Today, large quantities of the gas are used in the manufacture of the plastic, **nylon** (see Section 36.11). The gas is used also for **preserving latex**, the liquid material from rubber trees, from which rubber is prepared.

Quite large quantities were used in refrigeration, although, for this purpose, it is now being replaced by other gases which, while having the same useful properties, have not got the disadvantages associated with ammonia – for example its poisonous nature.

The use of ammonia in refrigeration depends on the fact that the gas may be liquified easily by the application of only a moderate pressure. The liquefaction of the gas causes it to evolve heat. The liquid ammonia is then cooled back to air temperature, and the pressure is released. The liquid evaporates rapidly, absorbing heat from the surroundings. If the change from gas to liquid takes place *outside* the refrigerator, and the change from liquid to gas takes place *inside* the refrigerator, then the temperature inside the refrigerator is lowered (see Figure 25.6).

Another large-scale use of ammonia is found in the treatment of metals. If metals are heated under the right conditions in an

Figure 25.6 The arrangement of a compression-type refrigerator

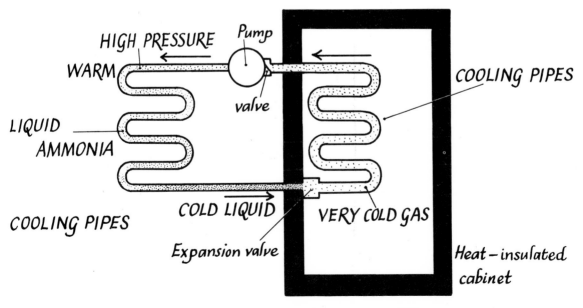

HIGH PRESSURE | Pump

WARM

COOLING PIPES

LIQUID AMMONIA

valve

COOLING PIPES | COLD LIQUID | VERY COLD GAS

Expansion valve

Heat – insulated cabinet

atmosphere of ammonia gas, the nitrogen from the ammonia reacts with the surface of the metal, forming an extremely hard and wear-resisting coat of the nitride of the metal.

25.3. The ammonium salts

These consist, as we saw in Section 25.1, of the ammonium ion, NH_4^+, combined with acid radical ions.

Investigation 25e. The preparation of some ammonium salts

Harmful

Take three 100 cm³ beakers. Into the first, put about 30 cm³ of 3 M hydrochloric acid; into the second, put about 30 cm³ of 3 M nitric(V) acid; and into the third, put about 30 cm³ of 3M/2 sulphuric(VI) acid.

Now, add ammonia solution (ammonium hydroxide) to each beaker, mixing thoroughly, until the contents are strongly alkaline to litmus and each beaker smells strongly of ammonia. Evaporate the contents of the beakers, boiling quite strongly at first, until crystals of the salts are formed on cooling, or solid is left when all the water has been removed. The boiling ensures that any excess ammonia is driven off – it is not necessary to neutralize the acids before evaporating.

The equations for these reactions are:

$$NH_4OH(aq) + HCl(aq) \rightarrow NH_4^+Cl^-(aq) + H_2O(l)$$
$$NH_4OH(aq) + HNO_3(aq) \rightarrow NH_4^+NO_3^-(aq) + H_2O(l)$$
$$2NH_4OH(aq) + H_2SO_4(aq) \rightarrow (NH_4^+)_2SO_4^{2-}(aq) + 2H_2O(l)$$

25.4. Ammonium chloride

This is the substance formed when ammonia gas reacts with hydrogen chloride. It is also known as **sal-ammoniac**. It was known to the alchemists, on account of the ease with which it forms clouds of white fumes (see Investigation 25f), as the **white eagle.**

Investigation 25f. Some properties of ammonium chloride

Take a hard-glass (Pyrex) test-tube (150 mm × 19 mm is a convenient size) and put into it a small quantity of ammonium chloride. Now, heat the tube, gently at first, and then more strongly. What do you observe? Do you see why the alchemists called ammonium chloride the 'white eagle'? Continue heating the tube *very strongly* until there is no further change. What is left in the tube?

Repeat this investigation, but, this time, arrange the test-tube

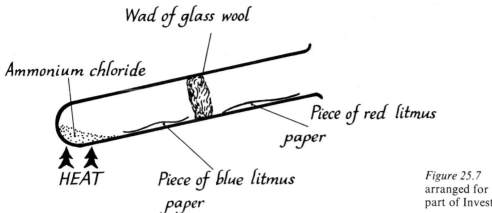

Wad of glass wool

Ammonium chloride

Piece of red litmus
paper

HEAT   Piece of blue litmus
paper

Figure 25.7 The test-tube
arranged for the second
part of Investigation 25f

as shown in Figure 25.7. What happens to the pieces of damp
litmus paper?

These investigations show that when ammonium chloride is
heated it **dissociates**. When we say that a substance 'dissociates',
we mean that it breaks up, but that, on reversing the conditions,
the products join together again to form the original substance.
In other words, dissociation is a **reversible** effect:

$$NH_4^+Cl^-(s) \rightleftharpoons NH_3(g) + HCl(g)$$

Do you remember the word used to describe the change of a
substance directly from the solid state to the gaseous state without
first becoming a liquid?

25.5. The uses of ammonium chloride

By far the most important use of ammonium chloride is in the
dry-cell, or **torch battery**. In the dry-cell (which is not really dry
at all), electrical energy is obtained when metallic zinc dissolves
in a damp paste of ammonium chloride:

$$2NH_4^+Cl^-(aq) + Zn(s) \rightarrow Zn^{2+}Cl_2^-(aq) + 2NH_3(g) + H_2(g)$$

The hydrogen, which would otherwise upset the efficient working
of the cell, is removed as water by reacting with manganese(IV)
oxide (see Figure 25.8).

Another use of ammonium chloride is as a **flux** for soldering
metals. The action of a flux in soldering is to clean the surfaces
of the metals being soldered so that the solder can attach itself
firmly to the metal. It also prevents the formation of metal oxide.

25.6. Ammonium nitrate(V)

This is the salt formed when ammonia reacts with nitric(V) acid.

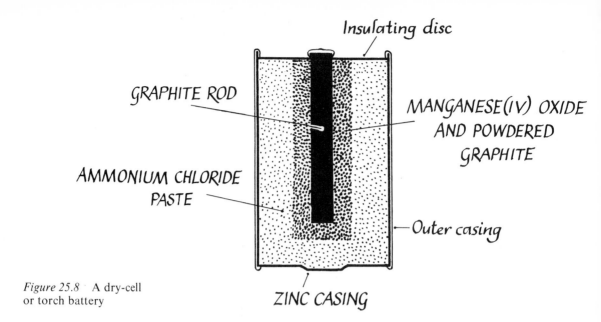

GRAPHITE ROD

AMMONIUM CHLORIDE PASTE

Insulating disc

MANGANESE(IV) OXIDE AND POWDERED GRAPHITE

Outer casing

ZINC CASING

Figure 25.8 A dry-cell or torch battery

Investigation 25g. The action of heat on ammonium nitrate(V)

Place a *very small* quantity of solid ammonium nitrate(V) in the bottom of a hard-glass (Pyrex) test-tube. Heat the tube gently at first, then more strongly and, finally, very strongly indeed. During the heating, make sure that the open end of the tube is pointed away from any person. Try the effect of a glowing splint on the gases evolved in the earlier part of the heating.

In the first part of the heating, the solid ammonium nitrate(V) dissociates, just as ammonium chloride does:

$$NH_4^+NO_3^-(s) \rightleftharpoons NH_3(g) + HNO_3(g)$$

The white fumes of the recombining ammonia and nitric(V) acid are clearly seen.

On further heating, decomposition takes place and a gas, **dinitrogen oxide** – otherwise known as **laughing-gas** – is formed.

$$NH_4^+NO_3^-(s) \rightarrow N_2O(g) + 2H_2O(g)$$

Like oxygen, this gas can relight a glowing splint, though, in practice, it is difficult to make it do so. Dinitrogen oxide is the gas which, when mixed with oxygen, is used by dentists and in operations as a very safe anaesthetic.

On very strong heating, the ammonium nitrate(V) decomposes explosively. Brown fumes of nitrogen dioxide are seen. The reaction is very complex.

25.7. The uses of ammonium nitrate(V)

Ammonium nitrate(V) is used for the manufacture of dinitrogen

278

oxide gas. However, the largest use is as a fertilizer. On account of its explosive character, ammonium nitrate(V) is often mixed with calcium carbonate for agricultural use: the product is sold as nitro-chalk. Ammonium nitrate(V) is particularly good as a fertilizer, for the combined nitrogen is present both in the ammonium group and in the nitrate(V) radical.

For use as an explosive, ammonium nitrate(V) is mixed with aluminium powder to give **ammonal**. It is also mixed with diesel oil.

It must be stressed that such explosive mixtures must **NEVER** be made. Their unstable nature makes them potentially dangerous for even the most careful worker.

25.8. Ammonium sulphate(VI)

Ammonium sulphate(VI) is the most important of the ammonium salts. As we have already seen (see Section 24.11), it may be made by reacting ammonia gas with sulphuric(VI) acid. However, there is another method; in this method, ammonia gas is blown into a suspension of calcium sulphate(VI) (gypsum) and the ammonia is followed by carbon dioxide gas. The reactions may be represented by the equation:

$$CaSO_4(s) + 2NH_3(g) + CO_2(g) + H_2O(l) \rightarrow CaCO_3(s) + (NH_4)_2SO_4(aq)$$

The calcium carbonate is filtered off and the solution of ammonium sulphate(VI) is crystallized.

Investigation 25h. The action of heat on ammonium sulphate(VI)

Put some ammonium sulphate(VI) in the bottom of a hard-glass (Pyrex) test-tube, and heat it, gently at first and then more strongly. Compare the effects produced with those produced when ammonium chloride and ammonium nitrate(V) were heated.

25.9. Ammonium carbonate

On account of its properties, this salt is known as **sal-volatile**. It is very unstable and decomposes slowly at room temperature, liberating ammonia and carbon dioxide gases. For this reason, it is used in **smelling-salts**. Smelling-salts are contained in tightly stoppered bottles which prevent the decomposition of the salt going too far. When the stopper is removed from the bottle, the ammonia gas which has formed can be smelled; the 'shock' produced by the gas can have a beneficial effect.

25.10. Ammonium hydroxide solution

When ammonia is dissolved in water, as we have seen,

ammonium hydroxide is formed. The strongest solution obtainable at room temperature and pressure has a density of 0.880 g cm⁻³. This solution is known as **880 ammonia**.

Test your understanding

1. What is the difference between 'ammonia' and 'ammonium'?
2. Describe, giving a clearly labelled diagram of the apparatus you would use, naming the substances you would need and giving the chemical equation for the reaction taking place, how you would prepare and collect a dry sample of ammonia gas in the laboratory.
3. How would you demonstrate that ammonia gas is very soluble in water? Why is it so soluble?
4. Under what conditions can ammonia be burned in air?
5. A laboratory test used for the detection of ammonia gas is to hold the stopper of a concentrated hydrochloric acid bottle in the gas. If the gas is ammonia, what will be observed? What is happening? Give the equation for the reaction taking place.
6. Explain, using a labelled diagram of the apparatus you would use, how you would demonstrate that, in the presence of platinum metal, ammonia will burn in air. What are the products of this reaction?
7. Outline the chief large-scale uses of ammonia.
8. Explain, giving examples, the difference between dissociation and decomposition.
9. How, starting from ammonia gas, would you prepare some solid ammonium sulphate in the laboratory? Give the equation for the reaction.
10. Make a labelled diagram of a 'dry' cell (torch battery).
11. Describe, giving the chemical equations for the reactions, what happens when (a) ammonium nitrate(V), (b) ammonium chloride, are heated.
12. How is 'laughing-gas' made in the laboratory? What is its use?
13. What are (a) smelling-salts, (b) '880' ammonia, (c) the white eagle?
14. What is a 'flux'? What does it do?
15. What are (a) sal-volatile, (b) ammonal, (c) turmeric paper?

Chapter 26

Nitric(V) Acid and Its Salts

26.1. The manufacture of nitric(V) acid

We have seen (Investigation 25d) that, when ammonia gas is oxidized by air in the presence of platinum as a catalyst, nitrogen oxide gas is formed. When the gas is cooled and dissolved in water in the presence of more air, nitric(V) acid is formed. These reactions

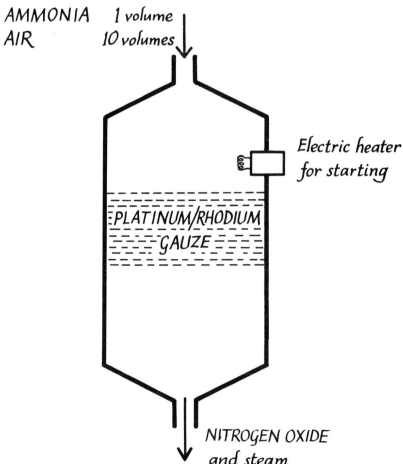

AMMONIA   1 volume
AIR   10 volumes

Electric heater
for starting

PLATINUM/RHODIUM
GAUZE

NITROGEN OXIDE
and steam

Figure 26.1   An ammonia converter

form the basis of the **Ostwald process** by which nitric(V) acid is manufactured on the large scale.

One part by volume of ammonia gas is mixed with ten parts by volume of air and the mixture is passed through a **converter**. This consists of a cylinder, across which are stretched several layers of metal gauze. The metal is an alloy of the metals **platinum** and **rhodium**. Figure 26.1 shows the arrangements.

A temperature of about 1 170 K (about 900 °C) is needed for the reaction to take place, but the reaction is strongly exothermic and it is only necessary to heat the gases in order to start the reaction. Once the reaction is under way, the heat liberated keeps it going (see Investigation 25d). The equation for this catalytic oxidation of ammonia is:

$$4NH_3(g) + 5O_2(g) \rightarrow 4NO(g) + 6H_2O(g)$$

The issuing gases are passed through a series of cooling pipes so that the temperature falls below 420 K (about 150 °C). When this happens, the nitrogen oxide reacts with more oxygen in the remaining air to form nitrogen dioxide:

$$2NO(g) + O_2(g) \rightarrow 2NO_2(g)$$

The ten volumes of air originally present ensure that enough oxygen is left for this reaction, and for the reaction by which, on dissolving in water, the nitrogen dioxide gives nitric(V) acid:

$$4NO_2(g) + 2H_2O(l) + O_2(g) \rightarrow 4HNO_3(aq)$$

26.2. The laboratory preparation of nitric(V) acid

In the laboratory, nitric(V) acid may be prepared in a very concentrated form by the action of hot concentrated sulphuric(VI) acid on one of the salts of nitric(V) acid – that is on a nitrate(V).

Investigation 26a. Making some nitric(V) acid

★ Warning – this investigation should only be carried out by the teacher.

Corrosive

Explosive

Set up the apparatus shown in Figure 26.2. This is one of the few occasions where the use of the **retort** – a piece of apparatus traditionally taken as symbolic of chemical apparatus – is still convenient today.

Put some crystals of sodium nitrate(V), or, better still, potassium nitrate (V) into the retort. Add enough concentrated sulphuric(VI) acid to cover them completely, and re-assemble the apparatus. Use a *new cork* to stopper the retort.

Heat the retort, gently at first, and note what happens. When the contents start to froth, remove the bunsen burner. Heat intermittently, so that the reaction continues smoothly, but does not get out of hand. When sufficient of the liquid has collected

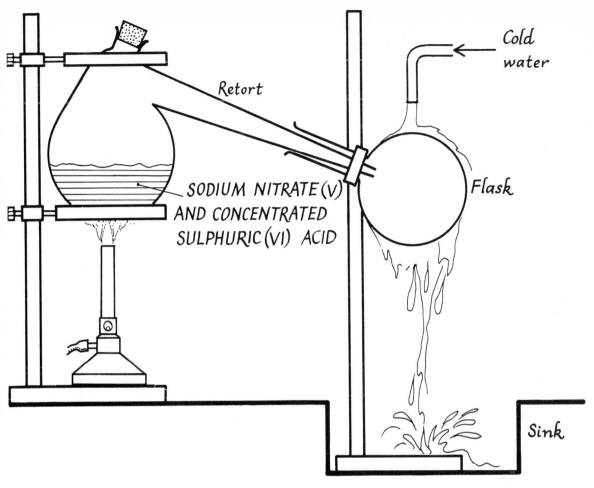

Retort

SODIUM NITRATE (V)
AND CONCENTRATED
SULPHURIC (VI) ACID

Cold
water

Flask

Sink

in the cooled receiver, stop heating and allow the retort to cool. As it cools, note the changes of colour that take place.

When the apparatus is cold, remove the receiver, taking care not to allow any of the liquid to get on the skin or clothes. Examination of the cork used to stopper the retort will show how advisable this is! It is one of the properties of concentrated nitric(V) acid that it has a corrosive action on proteins, forming a yellow compound. This fact is used as a test for protein material. Cork and skin both contain protein and so are stained yellow by the nitric(V) acid.

Care should also be taken when disposing of the contents of the retort. When it is quite cold, it is generally safe to add cold water quickly through the filling-hole, but there is some risk that not all of the sulphuric(VI) acid has been used up, if too much was taken in the first instance.

If sodium nitrate (V) is used for the preparation, the equation for the reaction is:

$$NaNO_3(s) + H_2SO_4(l) \rightarrow NaHSO_4(s) + HNO_3(l)$$

Figure 26.2 The laboratory preparation of nitric(V) acid

It will be noticed that sodium *hydrogensulphate*(VI) – that is, the acid salt – is formed, and not the normal sulphate(VI). It is only possible to get the normal salt by mixing *solid* sodium hydrogensulphate(VI) with more *solid* sodium nitrate(V) and then strongly heating the mixture:

$$NaHSO_4(s) + NaNO_3(s) \rightarrow Na_2SO_4(s) + HNO_3(l)$$

The changes of colour that were seen to take place in the apparatus during the preparation of the acid are due to dissociation of the acid:

$$4HNO_3(l) \rightleftharpoons 4NO_2(g) + 2H_2O(g) + O_2(g)$$

If the temperature of the retort is moderately high, but not too high, the colour is a deep brown, due to the presence of the nitrogen dioxide. At a higher temperature, the nitrogen dioxide itself dissociates:

$$2NO_2(g) \rightleftharpoons 2NO(g) + O_2(g)$$

Thus, at higher temperatures, the colour will not be so intense, as both nitrogen oxide and oxygen are colourless.

Investigation 26b. Some properties of nitric(V) acid

The tests to be described are best done on the fuming yellow liquid obtained in Investigation 26a, but most of them may be done satisfactorily using the ordinary concentrated laboratory acid.

★ *Warning – this first test should only be done by the teacher.*

Corrosive

Toxic

Nitric(V) acid is a vigorous oxidizing agent. It can supply oxygen to other substances. This fact may be shown as follows. Take a sand-tray (òr a small evaporating dish if a sand-tray is not available) and place on it a small pile of sawdust. Warm the sawdust *very carefully* from below, using a small bunsen flame, to make sure that it is quite dry and warm.

Now, with the aid of a glass rod, or by means of a teat pipette, add a few drops of the *freshly prepared acid* to the sawdust. Note the immediate evolution of copious fumes of the poisonous nitrogen dioxide gas and the brightness of the flame formed when the sawdust ignites. The brightness is a sign that oxygen gas is being given off.

Another reaction which shows that nitric(V) acid is a good oxidant (oxidizing agent) is that it will change iron(II) compounds to iron(III) compounds. To show this, make a solution of about 0.5 g of iron(II) sulphate(VI) in 30 cm³ of water, containing a few drops of dilute sulphuric(VI) acid. Divide this solution into two equal parts.

To the first part, add some sodium hydroxide solution. What do you see? To the second part, add 2 cm³ of concentrated nitric(V) acid and warm the solution until it just reaches the boiling-point. Now, cool it thoroughly and then add sodium hydroxide solution until the nitric(V) acid is neutralized and excess is present. What do you see?

The difference in the appearance of the two precipitates formed when the sodium hydroxide solution is added, shows that there has been a change in the valency of the iron. This has been brought about by the action of the nitric(V) acid:

$$Fe^{2+}(aq) + HNO_3(aq) \rightarrow Fe^{3+}(aq) + NO_2(g) + OH^-(aq)$$

You will have noticed the brown fumes of the nitrogen dioxide gas that is formed in the course of the reaction.

26.3. Nitric(V) acid as an acid

When water is mixed with the acid, the acid properties are shown:

$$HNO_3(l) + water \rightarrow H^+(aq) + NO_3^-(aq)$$

The solution will turn blue litmus red and will react with carbonates to give carbon dioxide gas. It is neutralized by bases to form salts and water only. These salts, the **nitrates(V),** are dealt with in the next section.

However, nitric(V) acid does not normally give hydrogen gas when it is added to metals. This is because it is a very good oxidizing agent. With metals, a mixture of the gaseous oxides of nitrogen – nitrogen oxide and nitrogen dioxide in particular – is usually given. There is one case where hydrogen gas *is* formed. This is when *very* dilute nitric(V) acid is added *cold* to magnesium.

26.4. The salts of nitric(V) acid – the nitrates(V)

These contain the acid radical NO_3^-. They are prepared by the usual methods of neutralization.

Investigation 26c. The action of heat on some nitrates(V)

In this investigation, small quantities of the different nitrates(V) are heated in small hard-glass test-tubes.

a. **Sodium nitrate(V).** Heat this substance, gently at first and then more strongly. Test the gas which comes off with a glowing splint. Note the changes of colour which occur. When the

Oxidising

residue is thoroughly cool, add a little dilute hydrochloric acid to it. What happens?

These tests show that, when sodium nitrate(V) is heated, it decomposes into sodium **nitrate(III)** and oxygen gas:

$$2NaNO_3(s) \rightarrow 2NaNO_2(s) + O_2(g)$$

Exactly the same type of change takes place if **potassium nitrate(V)** is heated.

Explosive

'Oxidising'

b. **Ammonium nitrate(V).** Use only a very small quantity of the solid, and make sure that the open end of the test-tube is pointed well away from yourself or from others. Heat gently at first, then more strongly, and finally, very strongly indeed. Be prepared for a surprise! Note all the changes that you see happening. These changes are due to the compound both dissociating and decomposing:

$$NH_4NO_3(s) \rightleftharpoons NH_3(g) + HNO_3(g)$$

and

$$NH_4NO_3(s) \rightarrow N_2O(g) + 2H_2O(g)$$

The gas, N_2O, is dinitrogen oxide – otherwise known as 'laughing-gas'. Mixed with oxygen, it is a good anaesthetic.

The final explosive decomposition of ammonium nitrate(V) leads to a very complicated series of changes, nitrogen dioxide clearly being one of the products.

Toxic

c. **Lead(II) nitrate(V).** Again, take care to use only a very small quantity of the solid and take care not to smell the products too closely or too deeply. Nitrogen dioxide, which is one of the products, is a very poisonous gas.

Heat gently at first, and then more strongly. Note the colour of the gaseous products. Test with a glowing splint. After heating strongly, allow the tube to cool and note the changes in colour of the residue as it cools down. All these observations are in agreement with the following change:

$$2Pb(NO_3)_2(s) \rightarrow 2PbO(s) + 4NO_2(g) + O_2(g)$$

It is often possible to recognize metals in their oxides from the colour of the oxide.

Investigation 26d. Testing for nitrates(V)

★ Warning – great care must be taken in adding the concentrated sulphuric(VI) acid.

There is a very good way to show the presence of a nitrate(V) in solution; it is known as the **brown-ring test**.

To about 5 cm³ of water in a test-tube, add a few drops of dilute nitric(V) acid (to act as the nitrate(V) to be tested for). In another test-tube, prepare a *cold* solution of iron(II) sulphate(VI). Mix the

two solutions in one of the test-tubes. Now, trickle a little concentrated sulphuric(VI) acid, slowly and carefully, down the side
of the inclined test-tube, as shown in Figure 26.3. A teat pipette
may be used with advantage to add the acid.

Corrosive   Harmful

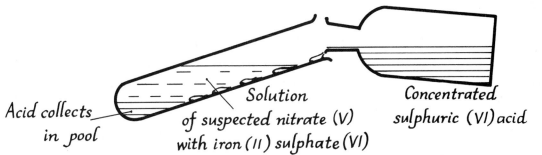

Acid collects in pool

Solution of suspected nitrate (V) with iron (II) sulphate (VI)

Concentrated sulphuric (VI) acid

If the acid has been added carefully, it will form a layer at
the bottom of the test-tube. A clearly visible brown-coloured ring
should be seen at the junction of the sulphuric(VI) acid with the rest
of the solution above it (see Figure 26.4).

The brown-coloured ring is due to a compound of nitrogen oxide
with iron(II) sulphate(VI). The reason why the acid is trickled down
the tube is to ensure that nitrogen oxide may be liberated from the
nitrate(V). This only happens when the sulphuric(VI) acid has a
particular concentration – about 70 per cent. Trickling the acid to
the bottom of the tube makes certain that the correct concentration
will occur somewhere between the bottom of the test-tube and the
aqueous solution above.

Figure 26.3 The brown-ring test

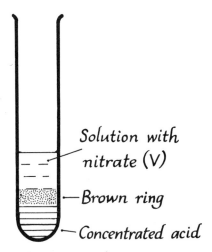

Solution with nitrate (V)

Brown ring

Concentrated acid

Figure 26.4 Appearance
of the test-tube with the
brown ring.

26.5. The large-scale uses of nitric(V) acid

Nitric(V) acid has many uses, but among the more important
uses are:

1. The manufacture of explosives
2. The manufacture of dyestuffs
3. The manufacture of drugs
4. The manufacture of fertilizers

26.6. Explosives

Almost all the normally used explosives are what are known as **nitro compounds**. These contain the nitro group of atoms, $-NO_2$. **Nitroglycerol**, **T.N.T.** and **picric acid** are examples. Another example is **nitrocellulose** or **gun-cotton**.

★ Warning – on no account should any attempt be made to prepare any explosives.

All these explosives are made by the process of **nitration**. This involves the treatment of an organic compound with a mixture of concentrated nitric(V) and sulphuric(VI) acids. On the large scale, the nitration is usually carried out by remote control, in buildings well separated from each other by banks of earth, so that, in the event of an explosion, the damage is limited to the particular building concerned.

Figure 26.5. gives the formulae of some of the nitro-explosives.

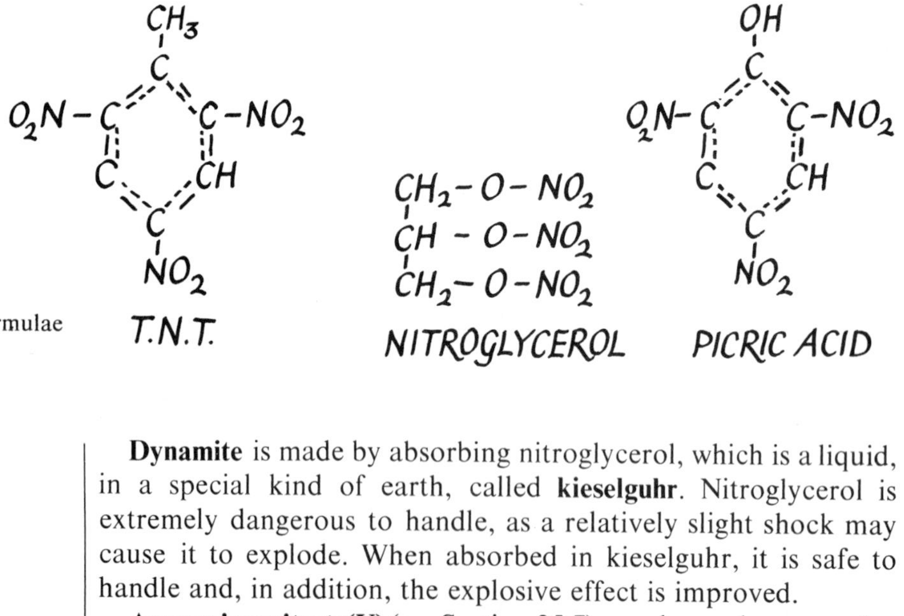

Figure 26.5 The formulae of some explosives

Dynamite is made by absorbing nitroglycerol, which is a liquid, in a special kind of earth, called **kieselguhr**. Nitroglycerol is extremely dangerous to handle, as a relatively slight shock may cause it to explode. When absorbed in kieselguhr, it is safe to handle and, in addition, the explosive effect is improved.

Ammonium nitrate(V) (see Section 25.7) may be used as an explosive, as well as a fertilizer. For explosive use, it is mixed with diesel oil just before it is wanted. It can also be used mixed with aluminium powder as **ammonal**.

Gelignite is made by absorbing nitroglycerol in what is known as **collodion cotton**. This latter is made by treating cellulose with nitric(V) acid under carefully controlled conditions.

The oldest explosive is **gunpowder**. This is a mixture of potassium nitrate(V) (saltpetre) with charcoal and sulphur.

26.7. Dyestuffs

Certain groups of atoms are capable of giving colour to organic substances. Among these groups are the nitro group, referred to above, and the **azo group** (–N:N–). These groups give colours in the orange, yellow and brown range. If, in addition to one of these colour-giving groups, or **chromophores**, there is in the molecule an acid group or a basic group, then the substance may make a good dyestuff. Examples of such dyestuffs are shown in Figure 26.6.

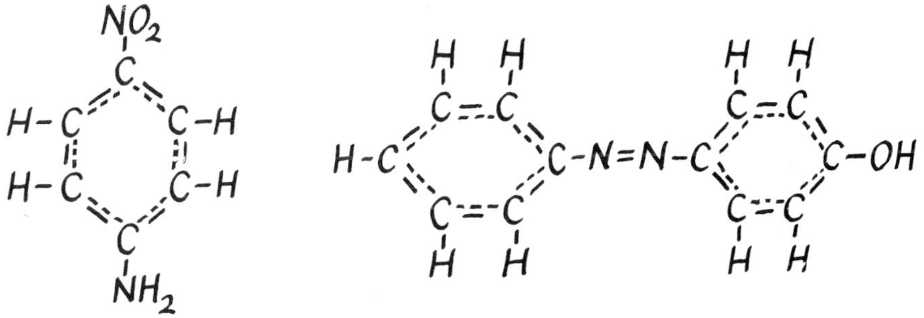

4-NITROPHENYLAMINE   (4-HYDROXYPHENYL) AZOBENZENE

Figure 26.6 Two dyestuff molecules

Many other dyestuffs are based on the compound, **phenylamine** ($C_6H_5.NH_2$). This compound is made from benzene by a series of processes, the first of which involves nitrating benzene to form **nitrobenzene** ($C_6H_5.NO_2$). The phenylamine dyestuffs were discovered by Sir William Perkin in 1856. He prepared the now famous mauve dyestuff, **mauveine**.

26.8. Drugs

There is a remarkable similarity between those molecules which act as dyestuffs, and those molecules which are able to kill bacteria which may produce diseases. The effect produced on the bacteria is, in many cases, due to the drug molecule attaching itself to some part of the bacterial cell – just as a dyestuff molecule will attach itself to a fibre.

Among the many drugs manufactured by the use of nitric(V) acid are the **sulphonamides**. The formula of benzene sulphonamide is given in Figure 26.7. The sulphonamide drugs are particularly effective against pneumonia. It was the drug known as **M and B**

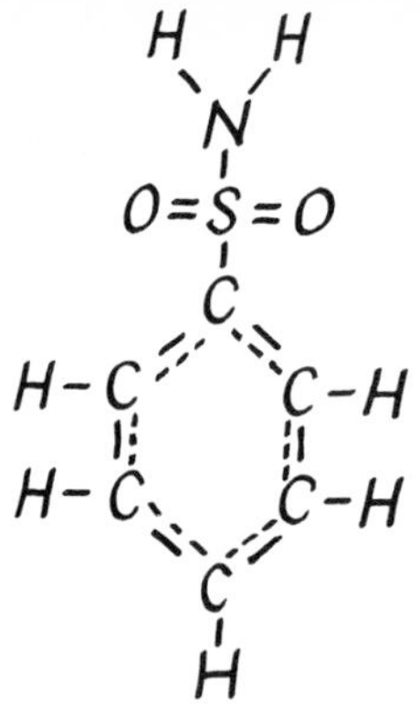

Figure 26.7 The formula of benzene sulphonamide

BENZENE SULPHONAMIDE

693 – the six hundred and ninety-third compound tested by the firm of May and Baker – that was the first really successful anti-pneumonia drug.

Test your understanding

1. Under what conditions can ammonia gas be oxidized by air?
2. What happens when nitrogen oxide gas is cooled in air from 250 °C to below 100 °C? Would you *SEE* any change that took place and, if so, what exactly *WOULD* you see?
3. What happens when nitrogen dioxide is dissolved in water in the presence of air?
4. Describe, giving a fully labelled diagram of the apparatus you would use, how you would prepare and collect some nitric(V) acid in the laboratory. Name the chemicals you would use and give the equation for the reaction.
5. During the preparation of nitric(V) acid in the laboratory, deep-brown fumes are usually seen. Why is this?
6. What happens when a solution containing some iron(II) sulphate(VI) is warmed with a little concentrated nitric(V) acid? How would you show that the change you believe has happened has actually taken place?
7. How would you demonstrate in the laboratory that nitric(V) acid is a good oxidant?
8. What reaction usually occurs when nitric(V) acid is added to metals?
9. What happens when samples of (a) ammonium nitrate(V), (b) lead(II) nitrate(V) and (c) potassium nitrate(V) are heated? Give the equations for the changes that take place.
10. Explain, giving all necessary practical details, how you would test an unknown substance to see if it was a nitrate(V), using the brown-ring test.
11. Outline the chief industrial uses of nitric(V) acid.
12. Name FOUR nitrogen-containing explosives.
13. What is the meaning of the term 'nitration'?

14. Explain the meanings of the following terms: (a) chromophore, (b) azo group, (c) sulphonamide.
15. Why do many drugs and dyestuffs have similar types of molecular structure?

Chapter 27

Sulphur

27.1. The occurrence of sulphur

Sulphur occurs in the free state, and also in combination with many other elements.

Free sulphur

The element is found in volcanic regions, Sicily being one of the main sources. There are also large underground deposits. The American states of Texas and Louisiana are particularly rich in this respect, although the deposits are now gradually being worked out. The extraction of the element from these sources is described in Section 27.2.

Combined sulphur

Sulphur is a very reactive element and is found in combination with many metals as **sulphides** and **sulphates(VI)**. These are more often important as sources of the metals they contain, rather than as sources of sulphur. Some of the more important ores are given in Table 27.1.

TABLE 27.1. SOME IMPORTANT SULPHUR ORES

Ore	Common Name of Ore	Formula
Iron sulphide	Iron pyrites	FeS_2
Copper/iron sulphide	Copper pyrites	$CuFeS_2$
Zinc sulphide	Zinc blende	ZnS
Lead(II) sulphide	Galena	PbS
Mercury(II) sulphide	Cinnabar	HgS
Calcium sulphate(VI)	Gypsum	$CaSO_4.2H_2O$
Calcium sulphate(VI)	Anhydrite	$CaSO_4$

27.2. The extraction of sulphur

The extraction from surface deposits, as in Sicily

The lumps of impure material are piled up in brick kilns. The tops of the heaps are then set on fire. Some of the sulphur on the top burns, and the heat melts a large part of the remaining sulphur which runs down to the bottom of the pile where it is collected in wooden boxes.

The sulphur produced in this way contains some 3 per cent of rocky matter. It is further purified by distillation. Figures 27.1 and 27.2 illustrate this extraction process.

The extraction from underground deposits

In the United States, this is carried out by the **Frasch process**. A drill hole is made and three concentric pipes are put down through the layers of rock until the sulphur deposits are reached. Superheated water at a temperature of about 430 K (about 160 °C) and a pressure of some fifteen times the normal atmospheric pressure is forced down the outer tube. This melts the sulphur at the bottom of the well. Air, also under pressure, is forced down the centre tube. This causes a froth of sulphur, air and water to be forced up to the surface through the tube sandwiched between the other two tubes (see Figure 27.3).

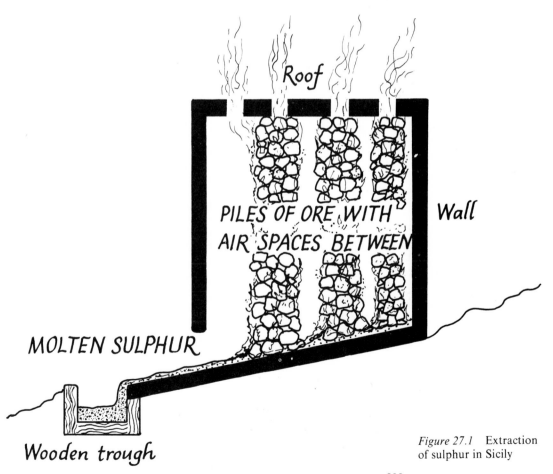

Figure 27.1 Extraction of sulphur in Sicily

293

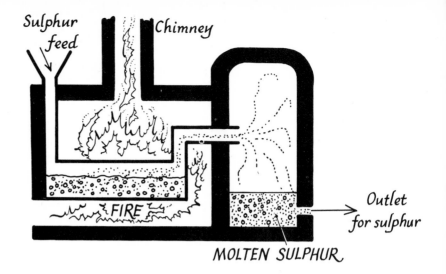

Figure 27.2 Distillation of the impure Sicilian sulphur

The froth of sulphur, air and water is allowed to flow into large 'boxes' constructed on the ground with wooden walls, some two metres high. The sulphur solidifies, and the water drains away. When the box, which covers a large area of ground, is full, the froth from the well is switched to another box. The walls of the first box are dismantled and rebuilt on top of the solid block of sulphur, so that a second layer of sulphur may be deposited on top of the first. In this way, enormous blocks of sulphur, containing thousands of cubic metres, can be built.

The sulphur obtained by the Frasch process is about 99.6 per cent pure and is ready for use. When required, the block is broken up by means of explosives. Figure 27.4 shows how the block of sulphur is built.

27.3. The large-scale uses of sulphur

The larger part of the raw sulphur is used for conversion into sulphur dioxide gas, from which sulphuric(VI) acid can be made by the **Contact process**, as described in Chapter 28. Another large use is for the **vulcanization of rubber**. Rubber is a very soft and pliable material, and it easily perishes, that is it is easily oxidized by the oxygen in the air. If the rubber latex is heated under pressure

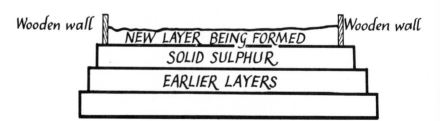

Figure 27.4 The building of the sulphur block

294

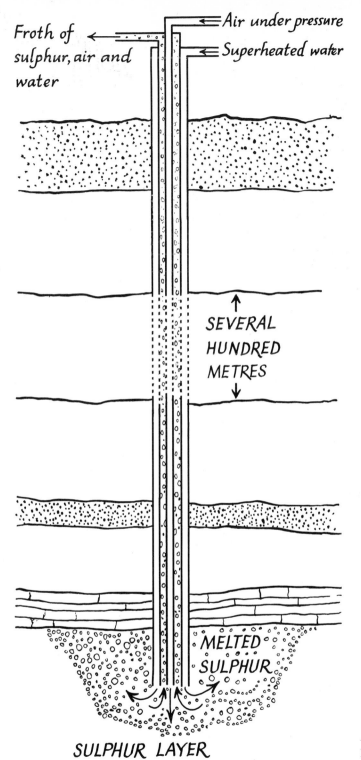

Froth of sulphur, air and water

Air under pressure

Superheated water

SEVERAL HUNDRED METRES

MELTED SULPHUR

SULPHUR LAYER

Figure 27.3 The Frasch process

with sulphur and inhibitors, it becomes much harder and resistant to wear and to the action of oxygen. Compare the properties of a piece of car tyre with those of a piece of soft india rubber.

Sulphur is used in agriculture as a **fungicide**. Powdered sulphur is used for dusting crops to protect them from diseases produced by fungi. It is also used in medicine, ointments containing sulphur being of value in the treatment of skin infections. The 'sulphur candle' is used to disinfect greenhouses: on burning, sulphur dioxide gas is given off, and this compound is an excellent disinfectant.

27.4. The allotropes of sulphur

Like carbon, and many other elements, sulphur exists in different forms in the same physical state – that is it has allotropes.

Investigation 27a. The preparation of rhombic sulphur

★ Warning – this investigation should only be carried out by the teacher: carbon disulphide is a very inflammable substance. Its vapour, mixed with air, may explode on slight heating.

Take a small quantity of 'flowers of sulphur', put it into a test-tube and add about four times the volume of carbon disulphide. Carbon disulphide is an evil-smelling liquid. It is also very inflammable. No naked flames should be allowed near. There is also a danger of the heavy vapour collecting in sinks and other hollows and forming an explosive mixture with air.

Shake the test-tube for some time, so that the flowers of sulphur come into close contact with the carbon disulphide, and then filter into a small evaporating dish. Put the dish in a warm place so that the very volatile liquid may evaporate. Do **NOT** heat the dish. When all, or nearly all, of the liquid has evaporated, examine the residue. What shape are the crystals? If possible, put a drop or two of the liquid on a microscope slide and allow it to evaporate under the microscope, watching as the crystals form.

Explosive   Flammable

Toxic

Investigation 27b. The preparation of monoclinic sulphur

Break up some roll-sulphur and about one-third fill a 150 mm × 19 mm test-tube with it. Prepare a *double* filter-paper cone, by folding *two* filter papers together in the usual way. When the cone is completed, fit a paper-clip so as to hold it together.

Now, gently heat the sulphur in the test-tube until it melts to a pale-coloured liquid. If the sulphur is heated too strongly, it will turn a dark-red colour. If this happens, the test-tube should be allowed to cool a little.

★ Warning – the liquid in the cone will be very hot and it should be handled carefully.

Hold the filter-paper cone in a pair of crucible tongs and pour the amber-coloured liquid sulphur into it.

When the liquid has cooled sufficiently to form a thin crust, take off the paper-clip and open up the paper cone. Drain off the extra liquid. Note the shape of the crystals which will be found growing inwards from the solid crust.

Investigation 27c.   The preparation of plastic sulphur

As in the last investigation, melt some roll-sulphur in a test-tube. This time, however, heat the sulphur until it boils. Heat slowly, and note all the changes that take place. What, in particular, do you notice about the 'stickiness' (viscosity) of the liquid sulphur as the temperature rises?

When the sulphur is boiling, pour it into some 300 cm³ of cold water in a beaker – that is cool it *very rapidly*. Pick the sulphur out of the water and examine it carefully. Rub it between your fingers. What happens? Note its elastic properties.

The three different forms of sulphur, prepared in Investigations 27a, 27b and 27c, differ from each other in crystal form. Nevertheless, they are all just sulphur, and nothing else. The remarkable series of changes seen when the element is heated are indicated in the following diagram:

★ Warning – the sulphur vapour may well ignite as the liquid is poured out.

Flammable

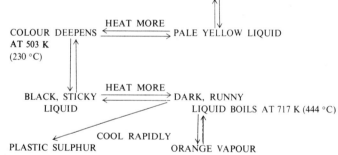

Sulphur atoms go round in groups of eight, so the sulphur molecule is S_8. The temperature at which the change from the rhombic form to the monoclinic form takes place, and vice versa, is known as the **transition temperature**.

27.5.   The chemical properties of sulphur

As has already been stated, sulphur is a very reactive element.

Investigation 27d.   Some simple properties of sulphur

For the following investigations, sulphur in either the roll form, or as flowers of sulphur, may be used.

a. Place a very small quantity of sulphur on one end of a wood splint, or in a deflagrating spoon. Heat the sulphur in the bunsen flame and note what happens. *Carefully* smell the vapours that are formed, taking care not to inhale them.

b. Make a mixture of some iron filings with about three times their bulk of sulphur, and put some of the mixture into a small test-tube.

Heat the mixture, gently at first, and then more strongly, observing all that happens. Finally, heat very strongly. Allow the tube to cool thoroughly, wrap it in a sheet of paper and break it open.

Remove some of the residue and examine it carefully. Put a small part of the residue into a test-tube and add a little dilute hydrochloric acid. Note the smell of the gas which is formed.

Put a little of the original unheated mixture of iron filings and sulphur into another test-tube and add a little dilute hydrochloric acid. Compare the smell of the product with that obtained from the heated mixture.

c. Put a small quantity of sulphur in the bottom of a test-tube. Take a piece of thin copper foil, of such a size that it may be wedged in the upper part of the test-tube (see Figure 27.5).

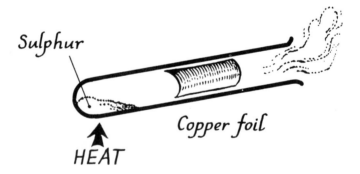

Figure 27.5 Heating copper foil in sulphur vapour

Heat the sulphur until it vaporizes and allow the vapour to pass over the foil. Now, heat the foil as well. Note all that happens. When the reaction is over, cool the tube and remove the solid product that has been formed from the copper. Compare the properties of this product with those of the original copper foil.

When sulphur burns in air or oxygen it combines with oxygen to form **sulphur dioxide** gas:

$$S(s) + O_2(g) \rightarrow SO_2(g)$$

We shall learn more about this gas and its properties in Section 27.7 and Investigations 27e and 27f.

When sulphur is heated with many elements, it combines with them, forming **sulphides**. For example, with iron and copper:

$$Fe(s) + S(g) \rightarrow FeS(s)$$
$$2Cu(s) + S(g) \rightarrow Cu_2S(s)$$

When hydrogen gas is bubbled through boiling sulphur, hydrogen sulphide gas (see Section 27.8) is formed:

$$H_2(g) + S(l) \rightarrow H_2S(g)$$

27.6. The manufacture of sulphur dioxide gas

Sulphur dioxide gas is required on a very large scale for the manufacture of sulphuric(VI) acid. For this purpose, it is made by two main methods:

a. By burning sulphur (obtained from the underground sources in the U.S.A., or from Sicily):

$$S(l) + O_2(g) \rightarrow SO_2(g)$$

b. By roasting a sulphide ore, such as iron pyrites, in a current of air:

$$4FeS_2(s) + 11O_2(g) \rightarrow 2Fe_2O_3(s) + 8SO_2(g)$$

27.7. The large-scale uses of sulphur dioxide gas

As well as for making sulphuric acid(VI), sulphur dioxide has many other uses.

a. It is used in the food industry as a **preservative** to prevent the growth of moulds and bacteria.
b. It is used to prepare solutions of sodium and calcium hydrogensulphate(IV). The solutions are used for the **bleaching of wood-pulp** in paper manufacture.
c. It is used for the **bleaching** of fine fabrics, such as silk, wool, straw and other materials, too delicate for chlorine bleaching.

Investigation 27e. The laboratory preparation of sulphur dioxide gas

In the laboratory, sulphur dioxide gas is usually made by the action of a dilute acid, normally hydrochloric, on a sulphate(IV). Set up the apparatus shown in Figure 27.6.

Put some crystals of sodium sulphate(IV) into the flask. The U-tube should be filled with large pieces of calcium chloride. Passing the gas over this substance will dry it. Make sure that the open end

★ *Warning – this investigation should be carried out in a fume-cupboard or in a well-ventilated room; care should be taken not to inhale the gas.*

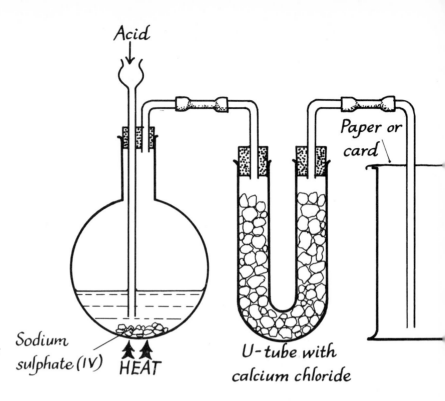

Figure 27.6 Apparatus for the laboratory preparation of sulphur dioxide gas

Acid

Paper or card

Sodium sulphate (IV)

HEAT

U-tube with calcium chloride

Toxic

Harmful

of the gas-jar is covered with a piece of card, wood, or paper, so that draughts of air will not blow the gas out of the jar as it is being collected.

By delivering the heavy gas to the bottom of the gas-jar, the gas will build up in the jar from the bottom upwards, and will displace the air in the jar as it does so.

When all is ready, start heating the flask. Continue heating until a steady supply of the gas is seen to come off, and then remove the source of heat. If this is not done, the gas will come off at an ever-increasing rate and there is a danger that the contents of the flask will boil over.

The equation for the reaction is:

$$Na_2SO_3(s) + 2HCl(aq) \rightarrow 2NaCl(aq) + SO_2(g) + H_2O(l)$$

It is difficult to tell when a jar is full, but as soon as a piece of moist blue litmus paper, held near the mouth of the jar, is reddened, replace the jar with a fresh one, taking care to cover the first jar with the usual greased glass disc.

When several gas-jars have been collected, the properties of the gas may be tested.

For the purposes of testing, it is more convenient to have a small cylinder of compressed sulphur dioxide available. Such cylinders, with a simple release valve, are available for a relatively small sum from the usual chemical suppliers.

Investigation 27f. The properties of sulphur dioxide gas

First, carry out the usual COWSLIPS tests (see Investigation 9e). Take particular care when smelling the gas. Notice that it leaves a bitter taste in the mouth. Open a jar under water to test the solubility, or, alternatively, if you are using a cylinder of the gas, fit the cylinder with a rubber tube and a glass delivery tube so that a stream of gas may be bubbled through some water in a beaker. Try the action of the gas on moist blue litmus paper. Put a lighted splint into a jar of the gas. Does the gas catch on fire? Does the splint go on burning?

Toxic

Harmful

Try bubbling the gas through a solution of potassium dichromate(VI) to which a little dilute sulphuric(VI) acid has been added. What happens? Try, also, bubbling the gas through a solution of potassium manganate(VII) to which, again, a little dilute sulphuric(VI) acid has been added. What happens?

The reactions with potassium dichromate(VI) and with potassium manganate(VII) show that sulphur dioxide is a **reductant (reducing agent)**. It can lower the proportion of oxygen in other substances. For this reason, it is used as a bleaching agent for materials such as wood-pulp and straw. As the bleaching is by removal of oxygen from the coloured substances, there is a tendency for the bleached material to pick up oxygen again from the air and to regain its colour. For this reason, paper that has been bleached with sulphur dioxide will gradually turn yellow. The change is accelerated if the paper is exposed to sunlight. You will have seen old newsprint that has yellowed in this way.

27.8. Hydrogen sulphide gas

Hydrogen sulphide is a very poisonous gas and considerable care must be taken when it is being prepared or used in the laboratory. The gas destroys the sense of smell and it is possible for the concentration to build up without its presence being noticed, despite its initially very noticeable – and unpleasant – smell.

The gas is normally prepared by the action of a dilute acid, usually hydrochloric acid, on a sulphide. Iron(II) sulphide is often used:

$$FeS(s) + 2HCl(aq) \rightarrow FeCl_2(aq) + H_2S(g)$$

★ *Warning – hydrogen sulphide is a very poisonous gas and great care must be taken not to let too much get into the air.*

301

★ *Warning – the large-scale preparation should only be done by the teacher and a fume-cupboard must be used.*

On the very small scale, a little of the gas may be prepared by putting a small piece of iron(II) sulphide in a test-tube and then adding a little dilute hydrochloric acid.

On the larger scale, the usual apparatus (see Figure 27.7) for the preparation of a gas by the action of a liquid on a solid is used. The apparatus should be set up in the fume-cupboard.

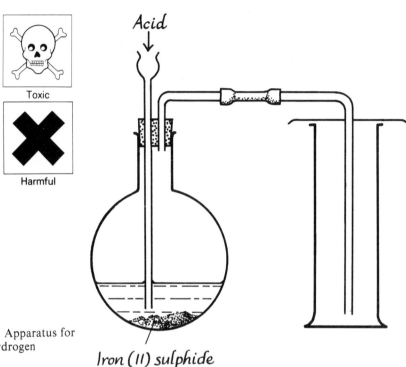

Toxic

Harmful

Figure 27.7 Apparatus for preparing hydrogen sulphide gas

Hydrogen sulphide is a useful gas in qualitative analysis, that is in finding out what elements are present in an unknown substance, so it is often convenient to have a supply of the gas 'on tap'. For this purpose, a **Kipp's apparatus** – commonly known as a 'Kipp' – is used. A diagram of a Kipp's apparatus is given in Figure 27.8.

When the tap is opened, the weight of the acid in the top bulb forces gas out through the tap. When the acid level rises in the middle bulb, until it reaches the solid material, the acid reacts with the solid, and more gas is formed. When the tap is now closed,

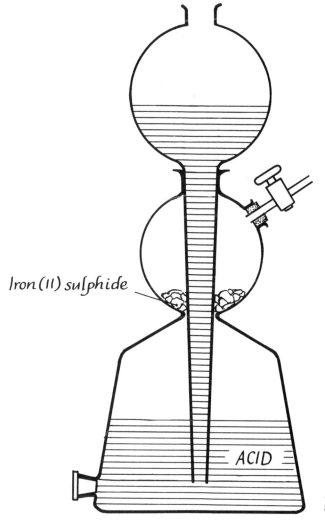

Iron (II) sulphide

ACID

Figure 27.8 A Kipp's apparatus

the gas pressure in the middle bulb forces the acid away from the solid and up into the top bulb.

Kipp's apparatus may be used to prepare any gas which is made by the action of a cold liquid on a solid. Other gases frequently made in this type of apparatus are carbon dioxide (from hydrochloric acid and marble chips) and hydrogen (from metallic zinc and hydrochloric acid).

Investigation 27h. The properties of hydrogen sulphide gas

The hydrogen sulphide for these tests is best obtained from a Kipp's apparatus.

Again, first carry out the usual COWSLIPS tests. In particular, try the effect of the gas on a piece of moist blue litmus paper.

Also, fill a test-tube or a small gas-jar with the gas and apply a light to the open end. Note the deposit of yellow solid that forms near the region where the gas ignites. What is this solid?

Soak a piece of filter paper in a solution of lead(II) ethanoate or lead(II) nitrate(V) and hold it in the gas. What happens? The characteristic reaction is often used to identify hydrogen sulphide gas. The reaction is a good example of the use of the gas in qualitative analysis:

$$Pb^{2+}(aq) + H_2S(g) \rightarrow PbS(s) + 2H^+(aq)$$

To show further the use of the gas in qualitative analysis, prepare small quantities of approximately 1 M solutions of the following substances: copper(II) sulphate(VI), lead(II) nitrate(V), zinc sulphate(VI), nickel(II) sulphate(VI) and manganese(II) sulphate (VI). Divide each solution into three parts (fifteen test-tubes in all). Leave the first set of five as it is. To the second set of five tubes, add 1 cm³ of dilute hydrochloric acid. To the third set of five, add a large pinch of solid ammonium chloride, followed by about 1 cm³ of ammonia solution (2 M).

Now, add a solution of hydrogen sulphide gas in water to each of the tubes. What happens in each case?

The use of hydrogen sulphide gas under different conditions of acidity and alkalinity may, or may not, bring about the precipitation of the sulphides of certain metals. The colours of these precipitates, together with the conditions under which they are produced, can help us to identify those metals.

TABLE 27.2. THE PRECIPITATION OF METAL SULPHIDES

Metal Sulphide	Acid Conditions	Alkaline Conditions
Mercury(II)	Black	Black
Copper(II)	Brown	Brown
Manganese(II)	—	Pink
Cadmium(II)	Yellow (if acid is *very* dilute)	Yellow
Nickel(II)	—	Black
Zinc	—	White
Tin(II)	Brown-black	{ Alkaline solutions of these
Lead(II)	Black (sometimes red)	last two cannot be made

Table 27.2 lists some of the sulphides and the conditions under which they are precipitated.

Test your understanding

1. Where is sulphur found free in nature?
2. Name, and give the chemical formulae for, FOUR naturally occurring sulphur compounds.
3. Describe, using a labelled diagram, how sulphur is extracted by the Frasch process.
4. State, and explain, THREE large-scale uses of sulphur.
5. Name the three main allotropes of sulphur and describe how you could obtain samples of each in the laboratory, starting from flowers of sulphur.
6. Outline the changes that take place when flowers of sulphur are slowly heated in a test-tube from room temperature until the sulphur finally boils.
7. What happens when: (a) sulphur is burned in air, (b) copper metal foil is heated in sulphur vapour, (c) iron filings are mixed with sulphur and the mixture is heated. Give the chemical equations for the changes that occur.
8. Describe, giving a fully labelled diagram of the apparatus you would use, how you would prepare and collect several gas-jars of sulphur dioxide gas in the laboratory. Name the chemicals you would use and give the equation for the reaction.
9. What are the main large-scale uses of sulphur dioxide gas?
10. What changes do you see take place when sulphur dioxide gas is bubbled through (a) acidified potassium dichromate(VI) solution, (b) acidified potassium manganate(VII) solution?
11. Paper which has been bleached by the use of sulphur dioxide will turn yellow or brown in the course of time. Why is this?
12. Describe, giving a labelled diagram of the apparatus you would use, how you would prepare and collect some hydrogen sulphide gas in the laboratory. Name the chemicals you would use and give the equation for the chemical change taking place.
13. Why is it essential that all experiments using hydrogen sulphide gas should be carried out in an efficient fume-cupboard?
14. What do you *SEE* happening when hydrogen sulphide gas is: (a) burned at the mouth of a test-tube in which it is being made, (b) bubbled into a solution containing an acid solution of copper(II) sulphate(VI), (c) bubbled into a solution containing an acid solution of manganese(II) sulphate(VI)? What would happen if the last solution was then made alkaline?
15. Why does silver 'tarnish' in air?

Chapter 28

Sulphuric(VI) Acid and the Sulphates(VI)

28.1. The importance of sulphuric(VI) acid

Sulphuric(VI) acid is the most widely used substance in the chemical industry. In the United Kingdom, the annual production is of the order of 3 500 000 tonnes.

The largest single use is in the manufacture of phosphate fertilizers, some 1 000 000 tonnes being used annually for this purpose. Synthetic fibres and plastics require 400 000 tonnes. Sulphate of ammonia takes 250 000 tonnes. Steel 'pickling' needs 120 000 tonnes and oil refining some 60 000 tonnes.

28.2. The manufacture of sulphuric(VI) acid

Today, nearly all the sulphuric(VI) acid produced is made by the **Contact process**. This process was first introduced in 1825 in Bristol by Peregrine Phillips, a vinegar manufacturer, but its large-scale use did not come until much later.

In outline, the process consists of making sulphur dioxide gas and of combining it with more oxygen to make sulphur trioxide. The sulphur trioxide is then reacted with water to produce sulphuric(VI) acid. It is the conversion of the sulphur dioxide to trioxide, which requires a catalyst, that constitutes the 'contact' step, and it is this that gives the name to the whole process.

28.3. The production of the sulphur dioxide

The easiest method of producing sulphur dioxide for use in the Contact process is to burn sulphur in air:

$$S(l) + O_2(g) \rightarrow SO_2(g)$$

The burning takes place in long, slowly rotating cylinders. The sulphur is fed in at one end, is melted by the heat, spreads out over the surface of the rotating cylinder, and burns in a current of air that is blown through at the same time (see Figure 28.1).

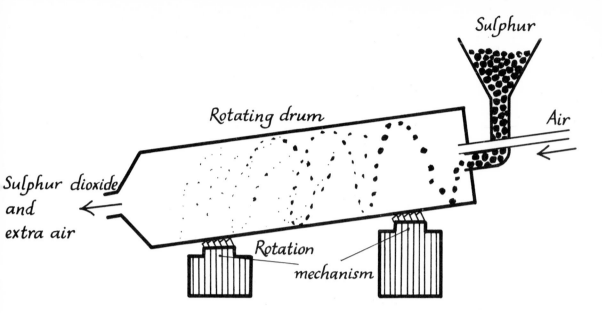

Figure 28.1. A sulphur burner

This method is by far the most economical, providing that the supplies of natural sulphur are freely available. However, the present known sources are rapidly being depleted and, as a result, the price of the free element has risen. This means that other sources for the sulphur dioxide have to be used.

One of these sources is iron pyrites. This may be roasted in a current of air to liberate sulphur dioxide:

$$4FeS_2(s) + 11O_2(g) \rightarrow 2Fe_2O_3(s) + 8SO_2(g)$$

Some sulphur dioxide is also obtained as a by-product in the extraction of certain metals, such as zinc, from sulphide ores:

$$2ZnS(s) + 3O_2(g) \rightarrow 2ZnO(s) + 2SO_2(g)$$

However, in the United Kingdom, after the use of free sulphur, the most important source of sulphur dioxide is the mineral **anhydrite** ($CaSO_4$), which occurs abundantly. A mixture of anhydrite with coke, sand and ashes, the last providing aluminium oxide, is heated in a long, rotating kiln, exactly the same as that used in the manufacture of cement (see Figure 32.2). The heating is done in the same way, by blowing in powdered coal and air. The issuing gases contain some 9 per cent of sulphur dioxide. The residue left is cement clinker (see Section 32.4).

28.4. Purifying the gases for the Contact process

No matter what method has been used for making the sulphur dioxide gas, it has to be cleaned and purified, together with the accompanying air, before it is allowed to meet the catalyst.

307

Purification is done by washing the gases with concentrated sulphuric(VI) acid, or by **electrostatic precipitation**. In both methods, the gases are first cooled in heat exchangers.

In the case of washing with acid, the gases are passed up a tower, down which the acid is being sprayed. The tower is packed with some inert material, such as pieces of earthenware, so that a large surface of the liquid is exposed to the gases (see Figure 28.2).

In electrostatic precipitation (see Figure 28.3), the gases are passed through a box-like chamber. From the top of this chamber, wires are suspended, which are charged electrically to a very high voltage. The dust particles in the gases become charged and are attracted to the wires of opposite charge; they are collected on those wires. When the wires are shaken, the voltage first being removed, the dust, now collected into lumps, is deposited on the floor of the chamber, from which it may be withdrawn.

The principle of electrostatic precipitation is frequently used to remove dust particles from smoke, for example, at power stations,

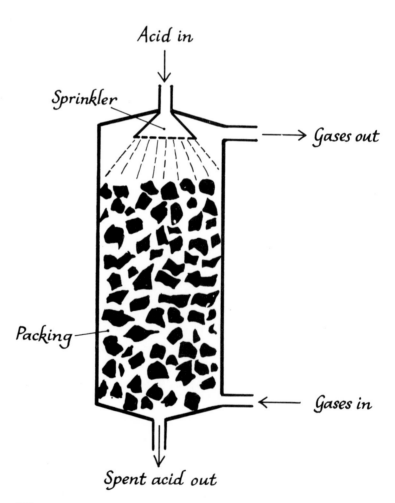

Figure 28.2. Washing the gases for the Contact process

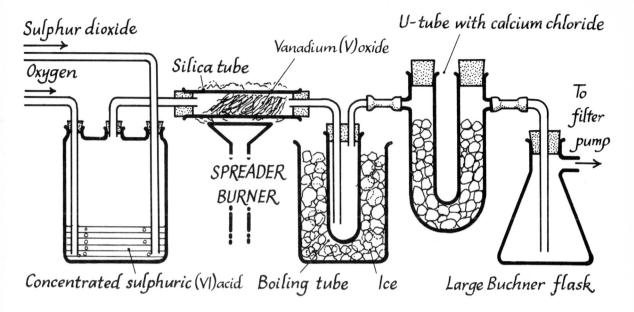

Sulphur dioxide

Oxygen

Silica tube

Vanadium (V) oxide

U-tube with calcium chloride

To filter pump

SPREADER BURNER

Concentrated sulphuric (VI) acid   Boiling tube   Ice   Large Buchner flask

oxide to make a thick cream. Transfer the cream to an evaporating dish and heat, first to dryness, and then as strongly as possible. The resulting solid will be found to break up into lumps on removing from the dish and the pieces may be used to pack the silica tube.

Before use, the whole apparatus must be thoroughly dried. This is best done by allowing a gentle stream of oxygen to pass through the silica tube, while it is being strongly heated. Moisture will condense in the tubes on the cool side away from the oxygen source and may be removed by using a filter paper on a glass rod.

If, now, a slow stream of the mixed gases is passed through the silica tube while it is being heated moderately, but not too strongly, white crystals of sulphur trioxide should be formed in the cooled boiling tube.

When sufficient of the product has been made, turn off the sulphur dioxide supply and allow the apparatus to cool while still keeping a slow stream of oxygen running through it.

When the apparatus is cool, detach the boiling tube. As you do this, you will probably see thick, white fumes forming in the air near the outlet tube of the apparatus. These fumes are due to the reaction of the sulphur trioxide with moisture in the air.

To show the very vigorous nature of the reaction between sulphur trioxide and water a small quantity of the sulphur trioxide may be added, with great care, to some water. The resulting solution contains sulphuric(VI) acid.

To confirm the presence of sulphuric(VI) acid in the solution, carry out the test for a sulphate(VI) ion. This is done by adding a little dilute hydrochloric acid, followed by some barium chloride

Figure 28.4 The laboratory preparation of sulphur trioxide

★ *Warning – take very great care when opening up the apparatus and when adding the product to water.* **NEVER** *add water to the product.*

solution. The formation of a dense white precipitate of barium sulphate(VI) shows the presence of sulphate(VI) ions:

$$SO_4{}^{2-}(aq) + Ba^{2+}(aq) \rightarrow Ba^{2+}SO_4{}^{2-}(s)$$

28.7. The properties of sulphuric(VI) acid

Pure, concentrated sulphuric(VI) acid is a dense oily liquid. It is commonly called **oil of vitriol**. When the concentrated acid is diluted with water, a very great quantity of heat is evolved. So much heat is given out that it is dangerous to add water *to* sulphuric(VI) acid: *the acid must always be added to the water*. If water is added to the acid, the heat produced may be sufficient to bring the water to its boiling-point and change it to steam, thus causing a sudden expansion which may throw the acid out of its container. If the acid is added to the water, there will be a sufficent bulk of water to take up the heat safely.

The heat produced when water reacts with sulphuric(VI) acid is due to the reaction:

$$H_2SO_4(l) + water \rightarrow H^+(aq) + HSO_4{}^-(aq)$$

Concentrated sulphuric(VI) acid has properties which differ from the properties of the diluted acid. When diluted, the acid properties are apparent; when concentrated, the compound is an oxidizing agent.

Corrosive

Harmful

★ *Warning – Investigation 28b.2 must only be carried out by the teacher.*

★ *Warning – Investigations 28b.3 and 28b.4 must only be carried out under direct supervision of the teacher.*

Investigation 28b. The properties of concentrated sulphuric(VI) acid

1. Put about 30 mm depth of water into a 150 mm × 19 mm test-tube. Add a *small* quantity of concentrated sulphuric(VI) acid, drop by drop. Note the rise in temperature of the tube and its contents.
2. Put about 5 g of sugar into a 100 cm³ beaker. Stand the beaker on a tile Add *two drops* of water and then add about 5 cm³ of concentrated sulphuric(VI) acid. Immediately cover the beaker with a larger beaker (400 cm³ would be a good size). See Figure 35.1 for the arrangement. A vigorous reaction should start. Note what happens.

    See Investigations 35c and 35d for details of the properties of carbohydrates; this will explain what happens.
3. Into another 150 mm × 19 mm test-tube, put a small quantity of powdered carbon (charcoal). Add about 10 mm depth of concentrated sulphuric(VI) acid and heat the tube.

    Carefully note the smell. Test the gas which is evolved by drawing some into a teat pipette and by then bubbling it through some potassium dichromate(VI) solution in another test-tube.

4. Repeat Investigation 28b.3, using a little powdered sulphur in place of the carbon.

Investigation 28b.2 illustrates the action of concentrated sulphuric(VI) acid as a **dehydrating agent**. Sugar (sucrose) is a **carbohydrate** with the formula $C_{12}H_{22}O_{11}$. When it is treated with the concentrated acid, the elements of water are removed, leaving the carbon:

$$C_{12}H_{22}O_{11}(s) - 11H_2O(g) \rightarrow 12C(s)$$

The test with potassium dichromate(VI) used in Investigations 28b.3 and 28b.4 shows that sulphur dioxide gas has been liberated in both cases. This shows that the acid is acting as an **oxidising agent**, and is itself being reduced in the process:

$$C(s) + 2H_2SO_4(l) \rightarrow CO_2(g) + 2SO_2(g) + 2H_2O(g)$$
$$S(s) + 2H_2SO_4(l) \rightarrow 3SO_2(g) + 2H_2O(g)$$

It is not possible to detect the carbon dioxide given off in the first of these reactions by using the usual lime-water test, as the sulphur dioxide present at the same time prevents the formation of the white precipitate of chalk.

Thus, we see that the acid can act in three ways: as an acid, when dilute; and as a dehydrating agent and an oxidizing agent, when concentrated.

28.8. The salts of sulphuric(VI) acid (the sulphates(VI))

Both hydrogen atoms in the molecule of sulphuric(VI) acid are replaceable. Two series of salts can be made, for example:

$$Na^+OH^-(aq) + H_2SO_4(aq) \rightarrow Na^+HSO_4^-(aq) + H_2O(l)$$
$$2Na^+OH^-(aq) + H_2SO_4(aq) \rightarrow Na_2^+SO_4^{2-}(aq) + 2H_2O(l)$$

The first series, containing the HSO_4^- radical, are known as the **hydrogensulphates(VI)**; the second series, with the SO_4^{2-} radical, are the **sulphates(VI)**.

Investigation 28c. Making the two sodium salts of sulphuric(VI) acid

Fill a burette with either 2 M or 1.5 M sulphuric(VI) acid. The exact concentration of the acid does not matter.

Use a test-tube (150 mm × 19 mm) to measure a volume of 2 M or 3 M sodium hydroxide solution into a beaker and add one or two drops of a suitable indicator solution, methyl orange for example.

Harmful

Corrosive

With continual stirring, run acid into the alkali until a permanent colour change shows that the alkali has been neutralized. Note the volume used.

Now, take two 250 cm³ beakers and into each put a test-tube full of the sodium hydroxide solution. This time, do *not* add any indicator, but, to the first beaker, add exactly the same volume of acid as was used in the first part of the investigation. To the second beaker, *add exactly twice this volume* of the acid.

Evaporate the water from both solutions until crystallization occurs. In the case of the hydrogensulphate(VI) (second beaker), it will be found necessary to remove almost all the water before crystals can be formed.

Drain the water away from the crystals, transfer them to some clean blotting paper, and dry them. Compare the crystals obtained.

28.9. Sodium sulphate(VI) and sodium hydrogen-sulphate(VI)

The crystals of sodium sulphate(VI) contain ten molecules of water of crystallization and have the empirical formula $Na_2SO_4 . 10H_2O$. They are sometimes known by their common name of **Glauber's salt**. The crystals are efflorescent and fall to a white powder when left exposed to the air, losing all their water of crystallization. Sodium sulphate(VI) has found a use in medicine as a purgative.

The hydrogensulphate(VI) gives a strongly acid solution when dissolved in water (see Investigation 15k). It is used in many lavatory cleaners, where its acid properties are of value in dissolving away any deposits of lime or chalk that may have been formed.

28.10. Other important sulphates(VI)

Ammonium sulphate(VI)

This substance has been met in the chapters on nitrogen and ammonia and the ammonium salts. It is the most important artificial nitrogenous fertilizer. For details of its preparation, properties and uses, see Sections 25.3, 25.8, 24.11 and Investigations 25e and 25h.

Calcium sulphate(VI)

This substance occurs naturally in many forms. Among the more important of these are **gypsum** ($CaSO_4 . 2H_2O$) and **anhydrite** ($CaSO_4$). The use of the latter as a source of sulphur dioxide gas for the manufacture of sulphuric(VI) acid has been described in Section 28.3.

When gypsum is carefully heated, it loses part of its water of crystallization, and leaves a powder:

$$2CaSO_4.2H_2O(s) \rightarrow (CaSO_4)_2.H_2O(s) + 3H_2O(g)$$

The powder, when treated with water, re-forms the gypsum and, in doing so, sets to a hard mass. The powder is known as **Plaster of Paris**. It is very useful for making **casts** of objects, and for 'setting' broken bones, so that they may be held rigidly in position while the bone-tissues are re-uniting.

Investigation 28d. Making a plaster cast

Prepare a mould of the object that is desired to be copied. This may be done by using a soft modelling-wax, and by pressing the object into it. A key might be used as a model for this purpose.

Make a paste, by stirring some Plaster of Paris with water until it is of a thick, but still pourable, consistency. Pour the paste into the mould and leave it to set for some hours. The casting may then be removed from the mould.

When making large casts, such as casts of footprints, it is usual to include some strengthening material in the cast. Thin strips of wood are added while the paste is being poured.

Magnesium sulphate(VI)

The preparation of this salt is described in Investigations 15d, 15e and 15f. The crystals have the formula $MgSO_4.7H_2O$ and are known as **Epsom Salts**. Like sodium sulphate(VI) they have found a use in medicine as a purgative.

Copper(II) sulphate(VI)

This substance has already been met many times. Investigations 15e and 15f give its preparation. The crystals have the formula $CuSO_4.5H_2O$ and the common name is **blue vitriol**.

Iron(II) sulphate(VI)

This salt is best prepared by warming metallic iron with dilute sulphuric(VI) acid (Investigation 15d), taking care to exclude air as far as possible. The crystals have the formula $FeSO_4.7H_2O$ and the common name is **green vitriol.**

Investigation 28e. The action of heat on iron(II) sulphate(VI)

Put some iron(II) sulphate(VI) crystals into a small (75 mm × 10 mm) hard-glass tube and heat them, gently at first, and then more strongly. Note all the changes that take place. Finally, heat *very strongly*. The residue that is left is iron(III) oxide. It is a fine hard powder which is used by jewellers to polish precious stones and

Harmful

315

metals. For this reason, it is sometimes called **jeweller's rouge** (jeweller's red).

$$FeSO_4.7H_2O(s) \rightarrow FeSO_4(s) + 7H_2O(g)$$
(green)   (white)

$$2FeSO_4(s) \rightarrow Fe_2O_3(s) + SO_2(g) + SO_3(g)$$
(red)

The white fumes of sulphur trioxide, liberated during the later stages of the heating, can be clearly seen.

Zinc sulphate(VI)

This has the common name of **white vitriol**, and has the formula $ZnSO_4.7H_2O$. It is used in medicine in ointments for the treatment of skin infections.

Barium sulphate(VI)

This is the white precipitate produced in the test for a sulphate(VI) (see Investigation 28a). It is an extremely insoluble substance. It finds a practical use in medicine, in X-ray photography.

When it is desired to take an X-ray photograph of the stomach or the intestines, it is necessary to make these show up on the plate by using special methods. Being soft tissues, they would not show up at all without special techniques. The technique used is that of the **barium meal.**

For the barium meal, the patient is given a gruel which contains barium sulphate(VI). Barium metal has a high atomic number, and barium sulphate(VI) has a high density. As the gruel passes through the stomach and the intestines, it adheres to the walls. An X-ray photograph will now show up the outline of the alimentary system, as the dense barium sulphate stops the X-rays. Barium compounds are *very poisonous*, but the sulphate(VI) is so insoluble that it passes through the digestive system without being absorbed at all.

Test your understanding

1. 'Sulphuric(VI) acid is one of the most widely used substances in industry.' Comment on this statement and gives examples of the use of sulphuric(VI) acid.
2. Most sulphuric(VI) acid is made by the 'Contact' process. This process starts with sulphur dioxide gas. Outline THREE ways that might be used for preparing this substance industrially.
3. In many industries where smoke is produced, the smoke is prevented from polluting the atmosphere by using the process of electrostatic precipitation. How does this process work?
4. What catalysts are available for use in the 'Contact' process? What advantages and/or disadvantages does each have?
5. What is 'oleum'?

6. How would you test a solution for the presence of the sulphate(VI) ion?
7. What happens when concentrated sulphuric(VI) acid is added to water? What ions are present in the solution so formed? Why is it very dangerous to add water to the concentrated acid?
8. Give ONE example of how concentrated sulphuric(VI) acid can act as a dehydrating agent.
9. Give ONE example of how concentrated sulphuric(VI) acid can act as an oxidant.
10. Starting from dilute sulphuric(VI) acid and sodium hydroxide solution, describe how you would prepare some crystals of (a) sodium sulphate(VI), (b) sodium hydrogensulphate(VI), in the laboratory.
11. Give the correct chemical name and the formula for each of the following substances: (a) anhydrite, (b) Glauber's salt, (c) Plaster of Paris, (d) gypsum, (e) Epsom Salts.
12. What is the chief use of ammonium sulphate(VI)? How is it made?
13. Sulphuric(VI) acid is known commonly as 'oil of vitriol'. What are (a) white vitriol, (b) blue vitriol, (c) green vitriol? Give their formulae.
14. What happens when iron(II) sulphate(VI) crystals are heated (a) gently, (b) very strongly. Give the equations for the changes that take place.
15. What is a 'barium meal'?
16. Starting from copper(II) oxide, how would you prepare some dry crystals of copper(II) sulphate(VI)-5-water in the laboratory?
17. How would you use some Plaster of Paris to make a cast of a footprint?

Chapter 29

The Chemistry of Salt

One of the most plentiful and widespread minerals in the world is common salt. This substance, known chemically as sodium chloride, is essential to all forms of animal life. About seventy-five million tonnes of salt are used in the world each year; much of this (about one-third) is obtained by the evaporation of sea-water. The remainder of the salt is extracted from underground deposits and is known as **rock salt**. Figure 29.1 shows the location of salt deposits in the world.

29.1. The salt deposits

We are quite used to the presence of salt in the sea, but might wonder how it came to be deposited inland. Geologists tell us that many parts of the world that are now land were, at one time, under the seas and oceans. At some time in the past, the land

Figure 29.1 The world's salt deposits

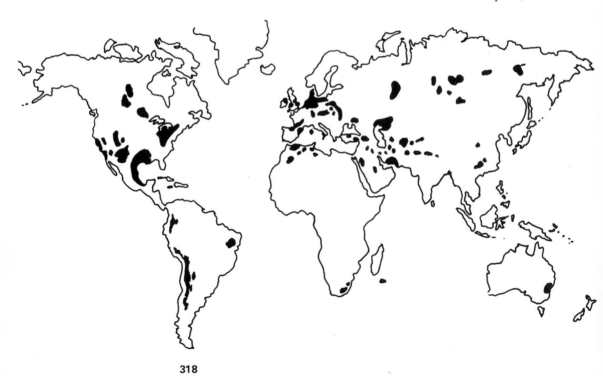

masses rose from the sea, trapping large inland lakes of salt water. As the water evaporated from these inland lakes, large deposits of salt were left behind.

This process is happening in many places today, the best known example being the *Dead Sea*, where the water is so dense that it is possible to sit in the sea without sinking. Evaporation in this way has led to large areas of salt, for example the *Utah salt flats* in the United States of America. In other parts of the world, this salt has become covered with other deposits. A large underground deposit of salt is found in Cheshire; this deposit supplies most of the needs of the United Kingdom.

In Cheshire, the salt is obtained by washing it to the surface as a solution in water. This strong solution is then evaporated to dryness. In other parts of the world, for example at Stassfurt in Germany, the salt is mined in a similar way to coal.

Where inland deposits are not available, salt is obtained by the evaporation of sea-water. The amount of salt in sea-water varies from place to place, but the average content of the oceans is about 3.5 per cent. Some land-locked seas, for example the Mediterranean and Baltic Seas, contain less salt. In temperate and tropical regions the evaporation is done by heat from the sun.

29.2. The properties of sodium chloride

We have already learned quite a lot about sodium chloride, but this is a good opportunity to collect our ideas together. We have seen that sodium chloride is a crystalline solid, the crystals being cubic in shape (see Figure 2.5). It is soluble in water (see Table 2.1), and this solution conducts electricity. The solid melts at the high temperature of 1 076 K (803 °C).

In Chapter 17, we saw that molten sodium chloride allowed the passage of an electric current, and that this method was used in the manufacture of sodium metal. At the same time, chlorine gas is produced. We will now investigate the effect of passing an electric current through a solution of sodium chloride.

Investigation 29a. Electrolysis of sodium chloride solution

For this investigation, the apparatus used in Investigation 12b should be used; this is shown in Figure 12.3. The wide glass tube should be filled with a strong sodium chloride solution, commonly called brine. Connect the pencil leads to a 6-volt direct current supply, as in the previous investigation, and observe what happens.

What gas do you think is given off at the cathode (negative lead)? How could you test this gas?

Toxic

Examine the tube over the anode (positive lead) carefully. Do you notice any coloration in the tube? Holding a piece of white paper behind the tube may make it easier for you to answer this question. The tube may take some time to fill with gas, but, when it does, remove the tube and hold a piece of moist blue litmus paper in the tube. What happens to the litmus paper? Can you recognize the smell?

Toxic

In this electrolysis, the products are different from those obtained when electricity is passed through melted sodium chloride (see Investigation 17a). Since water is present, in addition to the sodium ions and the chloride ions, $H^+(aq)$ ions and hydroxide ions $OH^-(aq)$ are present. Since the $H^+(aq)$ ions are less active than the sodium ions, hydrogen is discharged at the cathode in preference to the sodium ions. This may be confirmed by applying a lighted splint to the mouth of the tube that was placed over the cathode. At the anode, both oxygen and chlorine are liable to be discharged, but, since the solution is concentrated, most of the gas collected will be chlorine. It has the familiar smell associated with swimming baths, and turns blue litmus paper red and subsequently bleaches it. It may take some time for enough gas to be collected in the anode tube since chlorine dissolves in the brine.

Chlorine is manufactured on the large scale by the electrolysis of brine.

29.3. The uses of sodium chloride

If chlorine is manufactured using the same apparatus as that used in Investigation 29a, the resulting solution would be a mixture of sodium chloride and sodium hydroxide. Commercially, the electrolysis of brine is carried out in such a way as to remove the sodium hydroxide from the sodium chloride solution.

Investigation 29b.   Preparation of sodium hydroxide solution

★ *Warning – since this investigation requires the use of mercury, it must be done as a demonstration.*

Take a U-tube and pour in sufficient mercury to fill the bend at the bottom of the tube (see Figure 29.2). Carefully pour a solution of sodium chloride into one limb and some distilled water into the other limb. Since mercury is a dense liquid, the water and sodium chloride solution will float on the mercury. Lower a carbon rod into the sodium chloride solution and connect this to the positive terminal of a 6-volt direct current supply. Lower another carbon rod into the distilled water until it *just dips into the mercury* at the bottom. Connect a lead from this carbon rod, via a bulb or ammeter (3 A), to the negative terminal of the supply. A few drops of universal indicator should be put into the distilled

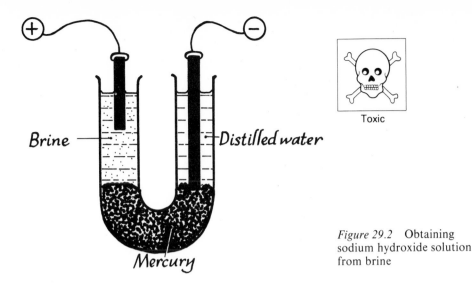

Brine — Distilled water

Toxic

Mercury

Figure 29.2 Obtaining sodium hydroxide solution from brine

water. After a few minutes, it should be possible to observe a number of changes. What are they?

The cell most commonly used in modern industry for the electrolysis of sodium chloride solution uses principles similar to those used in Investigation 29b. The cell used is the **Castner – Kellner** (or moving mercury) cell, which is illustrated in Figure 29.3.

Mercury is made to flow across the bottom of the sloping floor of the cell. The mercury serves as the cathode and the sodium ions are attracted to it. The sodium is able to dissolve in the mercury

Figure 29.3 The Castner – Kellner cell

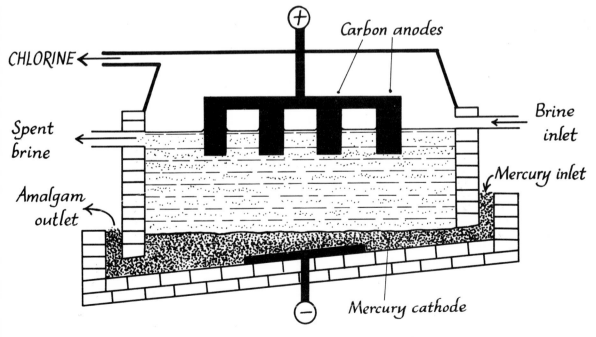

CHLORINE

Carbon anodes

Spent brine

Brine inlet

Amalgam outlet

Mercury inlet

Mercury cathode

to form an amalgam (see Chapter 18). Chlorine is given off at the anode as before (see Investigation 17a). The sodium/mercury amalgam is now mixed with water. The sodium reacts, liberating hydrogen gas from the water:

$$2Na(s) + 2H_2O(l) \rightarrow 2NaOH(aq) + H_2(g)$$

The water becomes a solution of sodium hydroxide, an important chemical. As we shall learn in Chapter 30, much of the hydrogen produced is combined with the liberated chlorine to form hydrogen chloride gas. Hence, this industrial process enables brine to be used to obtain chlorine, hydrogen, sodium hydroxide and hydrogen chloride.

Compare the results that you observed in Investigation 29b with the industrial process. You should find that the products are the same and that the apparatus used in the investigation is similar to the Castner–Kellner cell.

Figure 29.4 The uses of salt

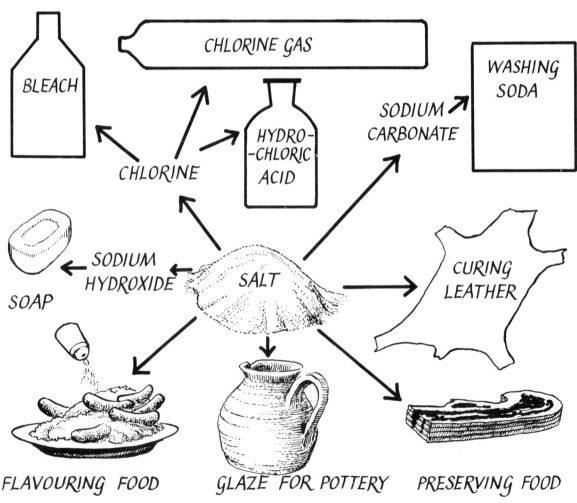

We have already seen that sodium chloride is the source of metallic sodium (see Section 17.4) and is used in the manufacture of sodium carbonate and sodium hydrogencarbonate (see Section 22.8). We are also familiar with the use of salt for preserving and for flavouring food. Salt taken in this way is an essential part of man's diet: a shortage of salt in the body often causes 'cramp'. In addition to these uses it is used for glazing pottery and for preserving animal skins. It is also used to make snow and ice melt (see Investigation 1k).

Test your understanding

1. Why is common salt so widely distributed over the earth?
2. Why do you think that certain land-locked seas, like the Baltic, contain less salt than the average for the oceans?
3. What tests would you apply to confirm that a gas was chlorine?
4. Why must strong brine solution be used to obtain chlorine by the electrolysis of brine?
5. What is the special feature of the Castner–Kellner cell for the electrolysis of brine?
6. If you heated brine, would it boil (a) below 373 K (100°C), (b) at 373 K (100°C) or (c) above 373 K (100°C)?

Chapter 30

Hydrochloric Acid and the Chlorides

In Chapter 29, we discovered that both hydrogen and chlorine gases were liberated when electricity is passed through a solution of sodium chloride. We learned that this process was used for the manufacture of hydrogen chloride. Such a method is not suitable for the preparation of hydrogen chloride in the school laboratory. We can, however, obtain hydrogen chloride from the same compound, namely sodium chloride.

30.1. The liberation of hydrogen chloride

We have learned that hydrogen chloride is a gas and that it forms salts. We shall now examine the effect of heating some salts of hydrogen chloride in the presence of a strong acid.

Investigation 30a. The action of sulphuric(VI) acid on chlorides

★ *Warning – these experiments must be carried out with extreme care.*

Corrosive

Harmful

Take a hard-glass test-tube (Pyrex type) and place a little sodium chloride in the tube. With a teat pipette add just enough concentrated sulphuric(VI) acid to cover the sodium chloride. Place the teat pipette in a beaker or sink, do *not* leave it on the bench. If there appears to be no action, place the test-tube in a suitable holder and gently warm the bottom of the tube in a non-luminous flame.

What do you observe happening at the mouth of the test-tube? Hold a piece of moist blue litmus paper at the mouth of the test-tube. What happens? Hold a piece of dry blue litmus paper at the mouth of the test-tube. What happens? Does the result of the second test with litmus paper differ from the result of the first test?

Rinse a glass rod with distilled water so that it is quite clean, and dip it into some silver nitrate(V) solution so that a drop of the solution forms on the rod. Hold the rod at the mouth of the tube. After a moment, examine the rod carefully. What has happened to the silver nitrate(V) drop?

Take the stopper off a bottle of ammonia solution and gently

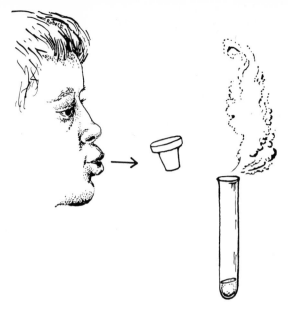

Figure 30.1 Mixing ammonia gas with hydrogen chloride

blow ammonia gas from the stopper across the top of the tube, as shown in Figure 30.1. What do you see?

This investigation may be repeated with the chlorides of some other metals.

In this investigation you have prepared some hydrogen chloride gas. Whenever you heat a metal chloride with concentrated sulphuric(VI) acid, the sulphuric(VI) acid displaces the more volatile hydrogen chloride from its salt, leaving a metal hydrogen-sulphate(VI).

$$Cl^-(s) + H_2SO_4(l) \rightarrow HSO_4^-(s) + HCl(g)$$

When the hydrogen chloride gas leaves the tube and mixes with the air, it fumes. This is because the gas reacts with the water vapour in the air to form an acid, hydrochloric acid. This action was explained in Section 7.2. You will have noticed that the moist blue litmus paper turned red almost at once, whereas the dry litmus paper took some time to change colour. The reason for this difference is that the hydrogen chloride does not behave as an acid until it dissolves in water.

The silver nitrate(V) solution on the glass rod turns milky, because the hydrogen chloride dissolves in the solution, producing hydrochloric acid, and reacts to form silver chloride. Silver chloride is one of the few chlorides that are insoluble in water. This is the basis of the test that is used for chlorides:

$$AgNO_3(aq) + HCl(aq) \rightarrow AgCl(s) + HNO_3(aq)$$

When ammonia gas is blown into the hydrogen chloride gas, white fumes of ammonium chloride are formed:

$$NH_3(g) + HCl(g) \rightarrow NH_4Cl(s)$$

325

The tests that we have used are suitable for testing a gas to see if it is hydrogen chloride. The method that we have used can be adapted for the production of larger quantities of hydrogen chloride.

Investigation 30b. The laboratory preparation of hydrogen chloride and hydrochloric acid

We have already examined many of the properties of hydrogen chloride and, since it is a dangerous corrosive gas, we shall now make a solution of the gas in water for further examination. Your teacher will probably arrange to collect some more hydrogen chloride for two special experiments.

Set up the apparatus shown in Figure 30.2(a).

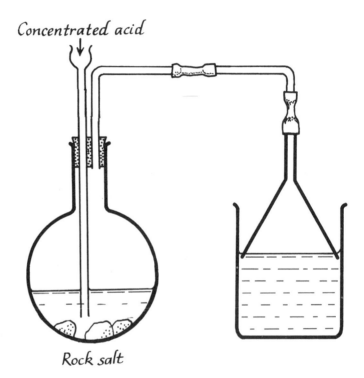

Concentrated acid

Rock salt

Figure 30.2(a) The preparation of a solution of hydrogen chloride gas

Corrosive

Harmful

Place a few lumps of rock salt in the flask. Arrange the funnel so that its edge is just below the surface of the water in the beaker. This is done since the gas is very soluble in water.

When all is ready, pour concentrated sulphuric(VI) acid down the thistle funnel until the acid level is just above the bottom end of the thistle funnel; this ensures that the gas leaves the flask through the delivery tube and not through the thistle funnel.

Your teacher will arrange an apparatus so that some dry hydrogen chloride gas can be collected. The flask is set up in the same way as in Figure 30.2, but the delivery tube passes to the container in which the gas is to be collected. Since hydrogen chloride is heavier than air (it has a relative molecular mass of 36.5), it is collected by downward delivery. The arrangement of the apparatus is shown in Figure 30.2(b).

The solution of hydrogen chloride in water that is obtained from this investigation should be retained for examination later. Meanwhile, a number of experiments should be carried out with the dry gas.

Collect a flask full of the gas and repeat the 'fountain experiment' (see Figure 25.3). This time, however, colour the water in the bottom container with universal indicator or blue litmus solution.

Make a solution of hydrogen chloride in *dry* methylbenzene. The methylbenzene should be dried by standing overnight with some calcium chloride in the bottle. The hydrogen chloride should then be bubbled through the methylbenzene. About 20 cm³ of methylbenzene should be sufficient. Divide the solution of hydrogen chloride in methylbenzene into two equal parts and place these two

★ *Warning – these experiments are best carried out as demonstrations by the teacher.*

Flammable

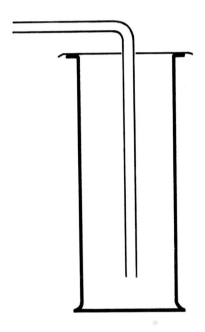

Figure 30.2(b) Collection of hydrogen chloride gas

parts in two *dry* test-tubes. To one portion add a small crystal of silver nitrate(V). Observe this tube. Add a little distilled water and then set aside for a few minutes. Meanwhile, dip a piece of *dry* blue litmus paper into the other portion. Observe the colour of the litmus paper and then add a little blue litmus solution to the solution in methylbenzene, shake it and stand the tube aside for a few minutes.

The fountain experiment shows that hydrogen chloride is very soluble in water. One volume of water will dissolve about 450 volumes of hydrogen chloride gas at room temperature. The solution that is obtained is acidic. We see that when hydrogen chloride is dissolved in methylbenzene the solution is not acidic in reaction, nor does the silver nitrate(V) solution turn milky. If, however, the methylbenzene is allowed to stand, the litmus solution will turn red and the silver nitrate(V) will turn milky as hydrogen chloride gas dissolves in water in preference to the methylbenzene.

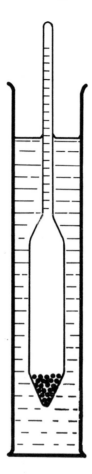

Figure 30.3 Using a hydrometer to find the density of the acid

30.2. Hydrochloric acid

In Investigation 30b, you prepared a solution of hydrogen chloride in water and this solution was found to give an acidic reaction with litmus. This solution is known as hydrochloric acid. We shall now use the solution prepared in Investigation 30b to examine some of the properties of hydrochloric acid.

Carry out the following tests on the hydrochloric acid obtained in Investigation 30b.

Corrosive   Harmful

1. Take a small portion of the solution in a test-tube and add an indicator, such as litmus.
2. Repeat this with another small portion of solution and with universal indicator. Use the colour chart on the bottle, or one that you have made (see Figure 13.5), to assess the strength of the acid.
3. Place a little solid calcium carbonate in the bottom of a test-tube. Use a teat pipette to add some of the acid. Test the reaction of any gas produced, with lime-water.
4. Place a little of the acid in a test-tube and drop in a small piece of zinc or magnesium ribbon.
5. Warm a little of the liquid with a little solid copper(II) oxide in a test-tube.
6. Place about 10 cm³ of your liquid in a small glass beaker, drop in a piece of litmus paper, and add sodium hydroxide solution, a little at a time, until the litmus paper just turns blue. Evaporate the solution to dryness.

Tests 1 to 6 all show the acid nature of the solution that you have prepared in Investigation 30b. Try to write chemical equations for as many of the reactions as you can. Then continue with the other tests.

7. Place a little of the hydrochloric acid in a test-tube and add a little dilute nitric(V) acid, followed by a few drops of silver nitrate(V) solution. Observe what happens and then stand the tube aside for a while. After it has stood for about ten minutes, again examine the tube. Now pour away most of the contents and add dilute ammonium hydroxide solution to the remainder.
8. Pour your acid into a narrow cylinder, such as a measuring cylinder, and lower a hydrometer into the acid. It may be necessary to collect all the acid from the class, in order to get enough acid for the hydrometer to float. Record the reading on the hydrometer. Your teacher may also use the hydrometer to find the density of the acid from a bottle of concentrated acid.

Hydrogen chloride, as we have discovered, forms an acidic solution in water. Universal indicator shows that it has a low pH value, showing that it is a strong acid. Like all strong acids, it liberates carbon dioxide from carbonates:

$$CaCO_3(s) + 2HCl(aq) \rightarrow CaCl_2(aq) + CO_2(g) + H_2O(l)$$

When it is added to zinc or magnesium, hydrogen gas is liberated:

$$Mg(s) + 2HCl(aq) \rightarrow MgCl_2(aq) + H_2(g)$$

When it is added to a base or an alkali, a salt is formed:

$$CuO(s) + 2HCl(aq) \rightarrow H_2O(l) + CuCl_2(aq) \text{ (blue copper(II) chloride)}$$
$$NaOH(aq) + HCl(aq) \rightarrow H_2O(l) + NaCl(aq) \text{ (sodium chloride)}$$

In test 7, we used the standard laboratory test for the chloride ion, Cl^-. The substance to be tested is dissolved in distilled water and a little nitric(V) acid is added; the solution is said to be acidified with nitric(V) acid. (If the substance should be insoluble in water, it should be dissolved in dilute nitric(V) acid.) The addition of silver nitrate(V) solution causes a precipitate of white silver chloride:

$$Ag^+(aq) + Cl^-(aq) \rightarrow AgCl(s)$$

If this precipitate is allowed to stand in sunlight, it begins to turn mauve. The effect of light on silver salts is used in photographic work. The silver chloride may be dissolved in dilute ammonium hydroxide solution, forming a complex ion. A similar result was observed in Investigation 11e.

Since hydrochloric acid is a solution of a gas, hydrogen chloride, in water, it is not possible to obtain a pure sample of the acid. Com-

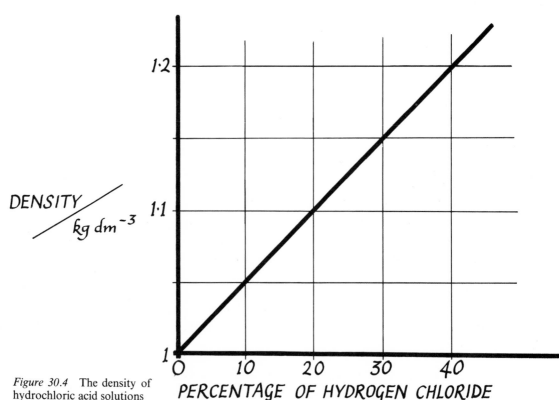

DENSITY / kg dm^{-3}

PERCENTAGE OF HYDROGEN CHLORIDE

Figure 30.4 The density of hydrochloric acid solutions

mercial concentrated hydrochloric acid contains 36 per cent hydrogen chloride gas, by mass.

The most convenient way to find the concentration of a solution of hydrochloric acid is to find its density. Figure 30.4 shows how the density of a solution of hydrogen chloride varies with concentration. Use this graph to find the concentration of your solution, using the result that you obtained in Investigation 30c, test 8. Use the graph to determine the concentration of the concentrated hydrochloric acid, from the density found by your teacher.

30.3. The chlorides of metals

In an earlier investigation in this chapter, we prepared two chlorides, namely sodium and copper(II) chloride; in Investigation 15g, we prepared iron(III) chloride by the direct combination of iron with chlorine. Let us, now, examine some of the properties of these chlorides.

Investigation 30d. The solubility of chlorides

Take a number of test-tubes and approximately half fill them with water, distilled if possible. To each tube add one of the chlorides listed in Table 30.1. Use about as much as will cover 10 mm at the end of a splint or spatula. Cork the tubes and shake them. Make a copy of Table 30.1 and place a tick against those chlorides that will dissolve in cold water. If the chloride does not

TABLE 30.1.   THE SOLUBILITY OF CHLORIDES

Chloride	Soluble in		Action on Litmus
	Cold Water	Hot Water	
Sodium			
Potassium			
Calcium			
Lead(II)			
Copper(II)			
Iron(III)			
Mercury(I)			
Magnesium			
Silver			
Zinc			

Toxic

Harmful

331

dissolve, gently warm the tube to see if the chloride will dissolve in hot water; if it does, place a tick in the column headed 'hot water'.

Do not pour the solutions away since we shall use them in the next investigation.

You may, of course, use other chlorides of metals. Also, it is not necessary for every pupil to use all the chlorides; the results of a class investigation could be used.

Investigation 30e. The action of chloride solutions on litmus

Dip a small piece of blue litmus paper into each solution. If it does not change colour, dip in a small piece of red litmus paper. Record the results of this test in the column headed 'action on litmus' on your copy of Table 30.1. Write acidic, basic or neutral, according to the results that you observe. When you have done this, look at the list of metals in Table 16.1 and see if the reaction with litmus has any connection with the order of these metals.

Toxic

Harmful

Most of the chlorides are quite soluble in cold water. This is because the **chloride ion** is a very soluble ion. Only lead(II), silver and mercury(I) chlorides do not dissolve in cold water. Lead(II) chloride, however, is appreciably soluble in hot water.

Some of the solutions of chlorides in water give an acidic re-action to litmus. This occurs because the metals are not strong metals and the metal ions form hydroxides with the hydroxyl ions from water, leaving an excess of $H^+(aq)$ ions. These $H^+(aq)$ ions, as we know, are the cause of acidic properties. For example:

$$H_2O(l) \rightarrow H^+(aq) + OH^-(aq)$$
$$Fe^{3+}(aq) + 3OH^-(aq) \rightarrow Fe(OH)_3(s)$$

Notice that this occurs with metals that are low in the electro-chemical series.

We can learn a lot about metals by a study of their chlorides. One way in which we find out something about the elements is to compare the melting-points of their chlorides. Table 30.2 gives a list of the melting-points of some chlorides. Look at these values and the position of the elements in the periodic table. If you have not got a large periodic table on the wall of your laboratory or classroom, you will find one in this book (Table 3.2).

You will notice that elements towards the bottom left-hand corner of the periodic table have chlorides with high melting-points. As we learned, in Chapter 7, this is a sign of ionic bonding. You may also have noticed that the elements in the same position, the bottom left-hand corner, are also at the top of the electro-chemical series (see Table 16.1). They are also, of course, the elements that form basic oxides.

332

TABLE 30.2   THE MELTING-POINTS OF SOME CHLORIDES

Chloride	Melting-point
Sodium	1 076 K (803 °C)
Potassium	1 043 K (770 °C)
Barium	1 228 K (955 °C)
Calcium	1 048 K (775 °C)
Carbon	250 K ($-$ 23 °C)
Silicon	203 K ($-$ 70 °C)
Magnesium	988 K (715 °C)

Test your understanding

1. Why is hydrochloric acid manufactured by the electrolysis of sodium chloride rather than by the action of sulphuric acid on a chloride?
2. Why is pure hydrogen chloride not acidic, whilst a solution of it in water is a strong acid?
3. Explain what has happened to the ammonium chloride shown in Figure 30.5.

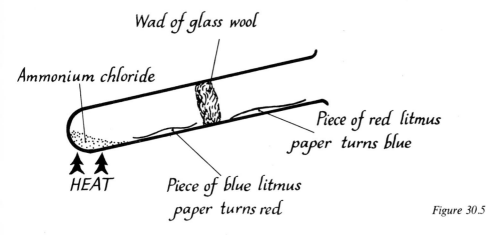

Wad of glass wool

Ammonium chloride

Piece of red litmus paper turns blue

HEAT

Piece of blue litmus paper turns red

Figure 30.5

4. Why should you avoid inhaling hydrogen chloride gas?
5. Why do you think hydrogen chloride is so soluble in water?
6. Write the chemical equation for the action of dilute hydrochloric acid on solid sodium carbonate.
7. Name two metals that will not liberate hydrogen from hydrochloric acid.
8. How would you show that a sample of distilled water that you had made did not contain dissolved chlorides?
9. Why are silver salts used in making photographic films?
10. The label on a commercial sample of hydrochloric acid says that its density is 1.15 kg dm^{-3}. What is the percentage of hydrochloric acid in the solution?
11. How concentrated is the hydrochloric acid in a fresh bottle of concentrated acid?

12. You are given a mixture of silver, lead(II) and potassium chlorides. Suggest a method by which you could obtain reasonably pure sample of silver and lead(II) chlorides from this mixture.
13. Would you expect strontium chloride ($SrCl_2$) to have a high or a low melting-point?
14. A new element X has been discovered; the melting-point of its chloride is 913 K (640 °C). Will its oxide be basic or acidic?

Chapter 31

Chlorine and the Halogens

We found that chlorine was liberated at the anode during the electrolysis of brine (see Investigation 29a). We subsequently learned that this method was used on the large scale for the manufacture of chlorine. We also learned that a large proportion of the chlorine is converted to hydrogen chloride gas, by burning hydrogen in the chlorine gas (see Section 29.3).

31.1. Chlorine in the laboratory

It is more convenient in the laboratory to obtain chlorine gas by the oxidation of concentrated hydrochloric acid rather than by the electrolysis of brine. Chlorine is, however, a *very poisonous gas*. (It has been estimated that one part of chlorine in 50 000 parts of air may cause injury.) For this reason, chlorine should never be prepared in the laboratory in larger quantities than is absolutely necessary. Your teacher will prepare some chlorine gas and demonstrate some of its properties to you. You will then use a solution of chlorine in water to investigate some other properties of chlorine for yourself.

Investigation 31a. The preparation of chlorine

Assemble the apparatus shown in Figure 31.1. Place sufficient potassium manganate(VII) crystals in the flask to just cover the bottom of the flask. Place some concentrated hydrochloric acid in the tap-funnel; the use of a tap-funnel allows the rate of production of chlorine to be carefully controlled. The screw clip is provided to prevent the escape of chlorine gas to the atmosphere. When the clip is closed, the gas cannot enter the empty gas jar, but is made to bubble into a solution of sodium hydroxide, which will dissolve most of the gas. When the clip is open, the water pressure in the jar with the sodium hydroxide solution prevents escape through this jar and chlorine gas is collected in the empty gas jar.

★ *Warning – this investigation should only be carried out by an experienced teacher.*
A fume-cupboard must be used.

Corrosive

Harmful

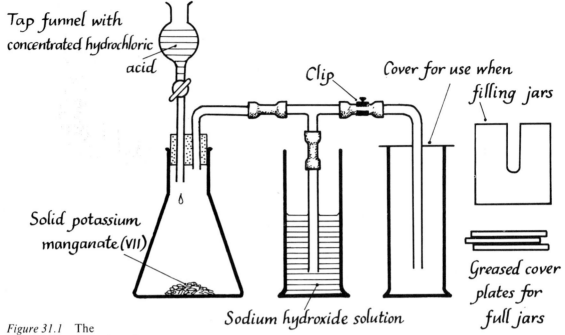

Tap funnel with
concentrated hydrochloric
acid

Clip

Cover for use when
filling jars

Solid potassium
manganate (VII)

Sodium hydroxide solution

Greased cover
plates for
full jars

Figure 31.1   The
laboratory preparation of
chlorine

Corrosive

Harmful

When all is ready, allow a few drops of concentrated hydrochloric acid to fall on to the potassium manganate(VII) and observe what happens.

Chlorine is liberated from hydrochloric acid by all oxidizing agents; the advantage of potassium manganate (VII) is that it does not require heating. The essential reaction may be represented by the equation:

$$2HCl(aq) + [O](s) \rightarrow H_2O(l) + Cl_2(g)$$

The symbol for oxygen is placed in square brackets since this oxygen has to be liberated from a suitable chemical and is not 'free' oxygen. The full chemical equation for the reaction that you have observed is:

$$16HCl(aq) + 2KMnO_4(s) \rightarrow 2KCl(aq) + 2MnCl_2(aq) + 5Cl_2(g) + 8H_2O(l)$$

When a gas jar is full of gas, cover the open end with a well-greased cover plate. The first jar or two will contain air and should be allowed to empty in the *fume-cupboard*. After a while the jars will fill with a light-green gas. As each jar is collected it should be covered with a greased cover plate. Four jars of chlorine will be required for the demonstrations, but it would be well to have an extra jar.

Investigation 31b. Some properties of chlorine gas

Toxic

Explosive

Your teacher will show you some experiments with chlorine gas; you must observe what happens and try to explain the changes that occur.

1. Take a small piece of Dutch metal in some tongs. (Dutch metal is an alloy of copper and zinc which has been beaten into a thin sheet; it is used as a substitute for gold plate in lettering, etc.) Remove the cover plate from a gas jar of chlorine and lower the Dutch metal into the gas jar. What happens? What reaction has taken place? (It may help you to look back to Investigation 15g.)

2. Connect a hydrogen cylinder to a jet as shown in Figure 31.2. Collect a small test-tube of hydrogen and ignite it. If it burns quietly, you may proceed with the experiment. If it explodes, allow a little more hydrogen to escape and repeat the burning test.

★ *Warning – A safety screen should be placed between the class and the apparatus.*

As soon as the hydrogen gas will burn quietly, and therefore safely, light the hydrogen escaping from the jet.

Remove the cover plate from the top of another jar of

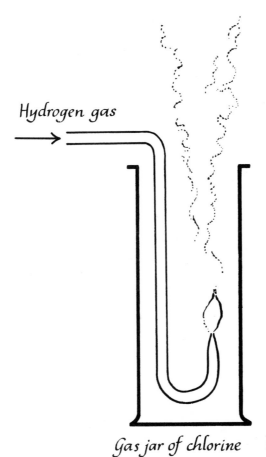

Hydrogen gas

Gas jar of chlorine

Figure 31.2 Burning hydrogen gas in chlorine

337

Flammable

Toxic

chlorine and carefully lower into the jar the burning jet of hydrogen gas. What happens to the hydrogen flame? What do you think is produced? (It may help if you blow a little ammonia gas across the top of the gas jar, as shown in Figure 30.1.)

3. For this experiment you will need a little turpentine; this is a hydrocarbon having the formula C_8H_{18}. Dip a piece of filter paper into the turpentine. Hold the piece of paper in tongs and lower it into a gas jar of chlorine. What happens? Name the products of the reaction.

4. Tip into the fourth jar of chlorine some powdered quicklime, about as much as would fill a 150 mm × 19 mm test-tube. What happens to the chlorine gas in the jar?

In these experiments, we have seen some of the more striking properties of chlorine gas and those that can be performed with the gas alone. Later, you will examine some other properties using a solution of chlorine in water; this solution can be prepared with the apparatus used in Investigation 31a. The screw clip should be closed and the jar containing sodium hydroxide solution should be replaced with one containing water.

Investigation 31c. Experiments with a solution of chlorine in water

Toxic

We have just seen some of the reactions shown by chlorine gas. Many of the other properties of chlorine may be demonstrated by the use of a solution of chlorine in water, known as **chlorine water**. The use of chlorine water for this work has the advantage that large amounts of chlorine do not escape into the laboratory. Your teacher will give you some chlorine water and you should place a little into a test-tube for each of the following experiments.

1. Place a piece of blue litmus paper on a watch glass. Use a teat pipette to drop a few drops of the chlorine water on to the litmus paper. Carefully observe what happens. There are two changes that you should observe.

2. Place about 5 cm³ of potassium bromide solution in a small test-tube. Add about five drops of the chlorine water to the solution in the test-tube. What do you observe? Can you suggest what the name of the new substance in the tube is?

3. Repeat the last experiment using a solution of potassium iodide in place of the solution of potassium bromide.

31.2. The properties of chlorine

From the experiments that you and your teacher have performed, you will have formed some good ideas about the properties of chlorine. Let us now examine these a little more closely.

We have seen that chlorine is a yellow-green gas with a *smell* that is quite characteristic (swimming baths). We have seen that it is *heavier* than air; it is collected by 'downward delivery'. The gas is quite *soluble* in water (2.5 volumes of the gas dissolve in 1 volume of water) and the solution may be used to 'store' chlorine. When treated with litmus solution (or moist litmus paper) the gas gives an *acid reaction* at first, then the colour is *bleached* from the paper, that is it is turned white. This shows that the solution of chlorine in water is acidic and also bleaching in nature. A mixture of two acids is formed: the familiar hydrochloric acid and a weaker acid, **chloric(I) acid**:

$$H_2O(l) + Cl_2(g) \rightarrow HCl(aq) + HOCl(aq)$$

The chloric(I) acid decomposes to form hydrochloric acid and the oxygen that is given up combines with the dye to form a colourless compound:

$$HOCl(aq) + dye \rightarrow HCl(aq) + (dye + oxygen)$$
$$(coloured) \qquad (colourless)$$

If chlorine water is left in sunlight, it will slowly lose oxygen to form hydrochloric acid. Notice that if chlorine is used as a bleaching agent it must be moist and that hydrochloric acid is produced by the chlorine in the water. Hence the object that has been bleached by the chlorine must be thoroughly rinsed to remove the acid, since this would 'rot' the object. Further, any excess of chlorine must be removed by a suitable chemical. Some substances, such as straw and paper, are so delicate that chlorine must not be used for bleaching; sulphur dioxide is often used instead (see Section 27.7).

Since chlorine dissolves in water to form a mixture of acids, it will react with most alkaline substances to form salts. We have seen two examples of this. In the preparation of chlorine, a safety-trap of sodium hydroxide solution was used to dissolve any surplus chlorine. This solution will slowly be converted to a solution of sodium chloride and **sodium chlorate(I)**, the sodium salt of chloric(I) acid:

$$2NaOH(aq) + Cl_2(g) \rightarrow NaCl(aq) + NaOCl(aq) + H_2O(l)$$

Sodium chlorate(I) decomposes on heating to form sodium chloride. It is dangerous to attempt to concentrate a solution of sodium chlorate(I). The quicklime that was dropped into a gas-jar of chlorine also combined with the chlorine (you should have noticed that the colour of the chlorine disappeared from the gas-jar). The compound that is formed does not appear to have a fixed formula, rather it appears to behave as a mixture of substances; its chemical behaviour can be represented, however, as a compound of calcium that is part chlorate(I) and part chloride:

$$Ca\!\!<^{OCl}_{Cl}$$

This substance is known as **bleaching powder**.

We have already learned that chlorine is a very poisonous gas. We have seen that although chlorine will not burn, it will support the combustion of many substances. Iron(III) chloride was made in Investigation 15g by heating iron wire in chlorine, and we have seen that Dutch metal bursts into flames when it is dropped into a jar of chlorine. The products of combustion are, of course, metal chlorides. We have also seen that hydrogen and hydrogen-containing compounds, like turpentine, will burn readily in chlorine. The hydrogen forms hydrogen chloride:

$$H_2(g) + Cl_2(g) \rightarrow 2HCl(g)$$

which will form fumes of ammonium chloride if mixed with

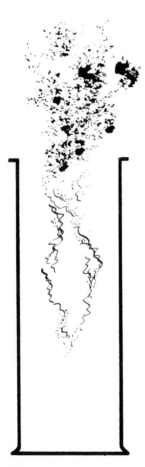

Figure 31.3 Turpentine burning in chlorine

340

ammonia gas. Similarly, the hydrogen in the turpentine combines with chlorine, liberating carbon:

$$C_8H_{18}(l) + 9Cl_2(g) \rightarrow 8C(s) + 18HCl(g)$$

The burning of hydrogen in chlorine is the method used in the industrial manufacture of hydrochloric acid.

In the last two investigations that you carried out with chlorine water, you saw the *family* behaviour of the halogen elements. Chlorine will displace **bromine** and **iodine** from solutions of their salts:

$$Cl_2(g) + 2NaBr(aq) \rightarrow 2NaCl(aq) + Br_2(aq)$$
$$Cl_2(g) + 2KI(aq) \rightarrow 2KCl(aq) + I_2(aq)$$

The *red liquid* bromine that is liberated colours the water orange and you may have seen a small drop of bromine at the bottom of the tube. The released iodine colours the water yellow and sometimes *blackish crystals* of solid iodine may be seen.

31.3. Other products from electrolysis

We have learned that chlorine is obtained by the electrolysis of sodium chloride solution. You may have noticed that, in your experiment, it took some time for the chlorine to appear at the anode. You will realize, now, that the chlorine was dissolving in the solution. You have seen that chlorine will dissolve in alkali to form chlorate(I). This method is used to prepare sodium chlorate(I) on a large scale. If a cold weak sodium chloride solution is electrolysed and the chlorine liberated at the anode is stirred into the solution, a dilute solution of sodium chlorate(I) is formed. If a strong hot solution of sodium chloride is used, a different product, **sodium chlorate(V)**, is obtained.

Chlorate(V) is an extremely powerful oxidizing agent and *must be handled with great care*. In particular, it must neither be ground nor mixed with other substances, but see page 97.

Investigation 31d. Oxygen from potassium chlorate(V)

Place a small amount of pure potassium chlorate(V) in a small test-tube and heat it strongly in a bunsen flame. Test the gas evolved for oxygen.

Explosive

31.4. A family of elements

We have already seen that if chlorine is passed through a solution of a bromide or through a solution of an iodide, the chlorine forms a salt taking the place of either the bromine or the iodine in the compound. This example must have suggested to you that chlorine, bromine and iodine are similar elements. These three elements are, in fact, very similar to each other and this family-like resemblance was observed very early in the study of chemistry. These three elements, together with two others, are known as the **halogens** (which means salt formers). Look at your periodic table and find the names of the other two elements. You have seen the elements chlorine, bromine and iodine, but you are unlikely to see the other two in a school laboratory. The one which is heavier than iodine is *radioactive* (see Chapter 37), and hence decays very quickly to other elements. Can you suggest why you will not find a sample of the element lighter than chlorine in the school? (If you cannot think of an answer to this question, look at the results of Investigation 31c, tests 2 and 3.)

It is easy to find out the names of the similar elements, because we have learned that elements that have similar properties have similar atomic structures. The periodic table lists elements with similar structures in vertical columns called *groups* (see Chapter 40). The halogens are found in group VII; they have seven electrons in their outer shells. The two other halogens are fluorine and astatine.

We know, from our experiments, that halogens are reactive elements; in particular, we have seen that chlorine combines very readily with other elements. We have also seen that the lighter the halogen, the more reactive it seems to be towards other elements; chlorine displaces bromine and iodine from their salts. We should not be very surprised to learn that fluorine is extremely reactive. This is why we shall not find it, as an element, in the school laboratory: it combines so strongly with other elements. Table 31.1 gives a comparison of the properties of the halogen elements.

TABLE 31.1. THE PROPERTIES OF THE HALOGENS

Element	Relative Atomic Mass	Colour	Physical State at Laboratory Temperature	Melting-point
Fluorine	19	Pale yellow	Gas	53 K (-220 °C)
Chlorine	35.5	Yellow-green	Gas	172 K (-101 °C)
Bromine	80	Red-brown	Liquid	266 K (-7 °C)
Iodine	127	Violet-black	Solid	387 K (114 °C)
Astatine	210	—	Solid	About 570 K (about 300 °C)

Test your understanding

1. What is the raw material from which chlorine is manufactured?
2. Why is chlorine only prepared in small quantities at school?
3. What do you think is mixed with the chlorine obtained by heating hydrochloric acid with an oxidizing agent that is not present when the cold acid reacts with potassium manganate(VII)?
4. What would you expect to be formed if a few hot iron filings were sprinkled into a gas-jar of chlorine?
5. Candle wax is a compound of carbon and hydrogen. What would happen if a lighted candle was placed in a jar full of chlorine?
6. What test, other than colour and smell, would you apply to a gas that you suspected was chlorine, in order to confirm your suspicion?
7. Suggest a reason for the use of chlorine in swimming baths.
8. If you used some chlorine water for bleaching a piece of material, what must you do afterwards? Why do you have to do this?
9. Make a list of substances that contain chlorine that you might expect to find at home.
10. What is the name given to a family of elements?
11. What properties of the halogens cause them to be regarded as a chemical family?
12. Why do the halogens possess similar properties?
13. What would you *EXPECT* to happen if you passed some fluorine into a solution of potassium chloride?

Chapter 32

Silicon

As we learned in Section 3.11, next to oxygen silicon is the most abundant element in the earth's crust. Think for a moment about the surface of the earth. What is it made of? If the centre of the earth is full of a white-hot liquid, as many geologists believe it is, can you think of any special properties that the surface of the earth must possess?

32.1. Silicon in the world

As you will have realized, most of the earth (including the sea bed) is covered with clay, sand and rocks. **Clays** are complex compounds of silicon, oxygen and metals; **sand** is silicon dioxide and most **rocks** contain silicon compounds. Table 32.1 gives a list of some naturally occurring compounds of silicon.

TABLE 32.1. NATURALLY OCCURRING SILICON COMPOUNDS

Sand
Quartz
Mica
Asbestos
Granite
All forms of clay

You will have realized that the surface of the earth must be resistant to heat, if the centre of the earth is a very hot liquid.

Investigation 32a. The effect of heat on compounds of silicon

Obtain some of the silicon compounds listed in Table 32.1 (*do not use asbestos*) and heat them in crucibles, or on tin lids. (You will be able to obtain some mica from the supports for electric iron or toaster elements.) Can you observe any definite change, other than the loss of a little water on drying?

You will not be very surprised with your result, if you think of some of the uses that are made of the substances listed in Table 32.1.

Sand is used to form moulds for casting metals.

Mica, as we have seen, is used to support electric elements that become red-hot in use.

Asbestos is used for heat-resisting mats and fire-proof suits.

Clay is moulded and then heated to a very high temperature, to set it, in making pottery. You may have had some of your own pottery 'fired' in the school craft-room.

32.2. Silica

Silica is the name given to the oxide of silicon, silicon dioxide. It occurs naturally as sand and quartz. **Quartz** is a very pure form of silica and, when perfectly formed, exists as colourless six-sided crystals. It is sometimes coloured by traces of other minerals and, because of their attractive appearance and hardness, these forms are known as 'semi-precious' stones. The violet-blue *amethyst* and dark red *jasper* are examples of such stones. Another form of quartz that is valuable because of its hardness is *agate*, the substance used in the construction of knife-edges on scientific balances.

The most common form of silica, in nature, is sand. Although more common, it is not as pure as quartz; you would guess this from the wide range of colours that are found. The purest form of sand is known as silver sand.

Another form of silica is *kieselguhr* (or diatomaceous earth). This very fine powder is the sediment from the skeletons of dead diatoms, the small marine plants forming part of the plankton. Silicon also occurs in other plants; in more remote parts of the world, silica from plants is used in the same way that we use silica from minerals.

Investigation 32b. Making silica

Place a few cubic centimetres of water-glass (a solution of sodium silicate(IV) in a test-tube. Using a dropping pipette, add a few drops of concentrated hydrochloric acid. Note what happens.

Pour off the liquid that remains and scrape the residue on to an iron tray and heat it, carefully at first but more strongly later. Since it requires heating for rather a long time, your teacher may collect all the solid from the class and heat it together.

Corrosive

Harmful

Investigation 32c. Another form of silica

Place some more water-glass in a test-tube, about the same amount as before. This time slowly sprinkle into the tube some sodium hydrogensulphate(VI). What happens this time?

In Investigation 32b, the strong acid, hydrochloric acid, displaced a weaker acid of silicon from the solution of its salt. The acid which is formed is of doubtful composition, but we call it **silicic acid**. Since the composition is not definite, we cannot write a chemical equation so we will use a word equation:

sodium silicate(IV)(aq) + hydrochloric acid(aq) → sodium chloride(aq) + silicic acid(s

when heated, the silicic acid loses water to form silicon dioxide

silicic acid(s) → silicon dioxide(s) + water(l)

Can you think of another example of an element with the same valency as silicon (4) that behaves in the same way; that is, one which forms salts from which the acid can be displaced by a stronger acid, and of which the acid loses water on heating to form an oxide?

In Investigation 32c, the form of silica that you made is called **silica gel**. You will have noticed the jelly-like nature of the substance formed. Silica gel is used as a drying agent in the laboratory and on the industrial scale; it is also used to remove sulphur from crude oil.

Silicon occurs in group IV of the periodic table (see Table 3.2), below carbon. Silicon dioxide is a solid, however, and not a gas like carbon dioxide. This is because silica is an example of a giant molecule, like diamond (see Section 6.7). Each silicon atom is covalently bonded to four oxygen atoms, and each oxygen atom is similarly bonded to two silicon atoms. Figure 32.1 shows one structure of silica; compare this with the structure of diamond (see Figure 6.14).

32.3. Clay, bricks and pottery

Clay is formed when rocks containing silicon are 'weathered'; that is when they are broken down into small particles by the action of wind and water. To be known as clay, the product of weathering of rocks must possess three properties:
  a. it must be 'sticky' enough to be moulded;
  b. it must retain the shape into which it is moulded, on drying; and
  c. when heated to a high temperature, it must form a hard solid mass of the same shape.

In Investigation 32a, we saw how a silicate, water-glass, reacted with an acid. Rocks are also weathered more quickly by acidic water, for example water which contains dissolved carbon dioxide or water from a peat bog. In Investigation 32b, we saw how silica, under certain conditions, formed a jelly-like substance, silica gel,

346

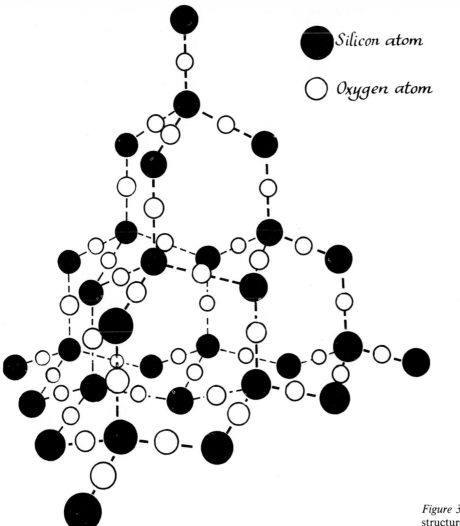

Silicon atom

Oxygen atom

Figure 32.1 **One structure of silica**

that was capable of absorbing water. These two investigations should help us to understand how clays come to be formed.

Bricks may be made from almost any form of clay. The clay is 'puddled' with water to ensure that it has the correct moisture content for moulding. It is then shaped to the required form, either by hand or by a press, and set aside to dry. When the brick has dried off, it is put into a large brick kiln and heated to a high temperature for a long time. When cold, the bricks are hard and brittle. In warm parts of the world some buildings are made from bricks that have been dried in the sun and have not been fired. The colour of the bricks will depend upon several factors, two of which are the temperature of the kiln and the metals that are in the clay.

Pottery is usually made from special blends of clays, but it can

347

be made from local clay. The process used is very similar to that used in making bricks, except that the firing is carried out at a higher temperature, about 1 300 K (about 1 000 °C). **China** is made from a very pure form of clay known as **china clay** or **kaolin**. This white clay is a hydrated mixture of aluminium oxide and silica.

Bricks and pottery are porous. In order to prevent the seepage of water through the pores of the material it must be sealed. This is done by painting another silicate mixture over the surface and refiring. After this second firing the surface is covered with a glass-like layer which is not porous. This is known as **glazing** and is similar to the manufacture of glass which is described later. In some parts of the world the glazing material is obtained from the ashes of straw and bamboo.

32.4. Cement and concrete

In this chapter, we have learned that silicon dioxide is an acidic oxide; we have also used a compound of this acidic oxide with a basic oxide, namely sodium silicate(IV). Let us now examine what happens when silicon dioxide reacts with another basic oxide.

In Chapter 22, you made some lime mortar and learned how it 'set' and later deteriorated. Walls that have been built using lime mortar require 're-pointing' after a time. A more permanent mortar may be made by using **cement**. There are many different types of cement, but the one in most common use is called 'Portland' cement. It has been given this name since it sets to a hardness similar to Portland stone (a type of limestone).

Portland cement is made by heating a liquid slurry made by adding water to finely ground chalk or limestone, mixed with clay. The mixture is fed into the top end of a long sloping, cylindrical cement kiln. This is some 100–120 metres in length and is rotated slowly (see Figure 32.2). Powdered coal and air are blown in at

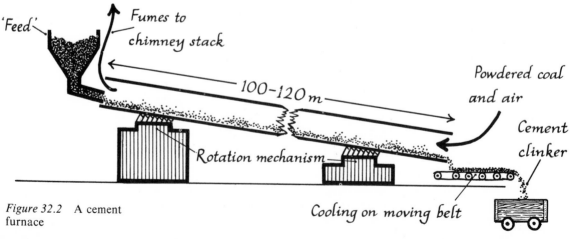

Figure 32.2 A cement furnace

348

the lower end and the burning coal provides the heat needed for the reaction.

Complex chemical changes take place and cement clinker is formed. A large part of this clinker is made up of compounds of lime (CaO) and silicon dioxide. The clinker is mixed with some calcium sulphate(VI) (so that the product will not set too quickly), and the mixture is ground to a fine powder.

When water is added to the cement, a further reaction occurs and a hard mass is formed. When it is mixed with sand or gravel, and water is added, the resulting hard mass is known as **concrete**. Concrete can be very strong, especially if it is formed around a lattice of steel rods or bars. When formed in this way it is known as reinforced concrete.

Investigation 32d.   Comparison of lime and cement mortars

Mix thoroughly one part of powdered slaked lime and three parts of sand. Add water, a little at a time, until a thick paste is formed. Mould this into a shape and set it aside for a few days. This paste is lime mortar.

Repeat the experiment using Portland cement instead of slaked lime in the mixture. After moulding this cement mortar to a shape, set it aside with the lime mortar. Examine the two forms of mortar after a few days and compare their hardness.

Mortars are used to joint bricks and stones in buildings. Bricks, above damp-course level, are commonly 'laid' with a mortar of one part Portland cement, one part lime and five parts sand.

32.5. Glass

Commercially made glass requires a very high temperature in the manufacturing process; we shall attempt to make a glass at a lower temperature.

Investigation 32e. Making glass

Using a pestle and mortar, crush together two parts by mass of lead(II) oxide with two parts by mass of sodium hydrogen-carbonate and one part by mass of silica (the residue from Investigation 32b or some silver sand). Place this mixture in a metal crucible (a steel thimble or thin water pipe fitting is suitable). Heat the mixture strongly inside a suitable crucible furnace.

Stir the molten mixture with a piece of *steel* rod (for example a knitting needle). When molten, draw out a string of the melt from the crucible, or use a hot steel spoon to pour some on to a warm surface.

Toxic

349

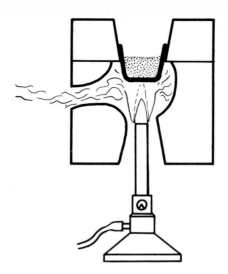

Figure 32.3 A furnace for making glass

You should have managed to obtain some low melting-point glass. You might like to try colouring this glass by adding small amounts of cobalt, copper or manganese oxides at the melting stage. The effect of these is best observed by obtaining a bead of glass from the melt. This may be done by dipping in a piece of nichrome wire shaped as shown in Figure 32.4.

The glass in most common use is usually known either as **soft glass** or as **soda glass**. This is made from sand, limestone and sodium carbonate. It is used for making bottles, window glass and glass laboratory ware of the *thinner* type. It can be melted easily in the hot part of a bunsen flame. A slight green colour seen in this glass is due to the presence of iron oxide impurity.

The *thicker* laboratory glass is usually made of a mixture which contains sodium, boron and aluminium with silicon. It is often known as **boro-silicate glass**; glass of this type is usually marked with a trade name ('Pyrex' is one of the most common). These boro-silicate glasses expand less on heating than ordinary soda glass; hence they withstand changes in temperature better than the softer glasses.

32.6. Water-glass

You have already used water-glass in your study of silicon. Since you have learned that water-glass is sodium silicate(IV) and

Figure 32.4 A former for a glass bead

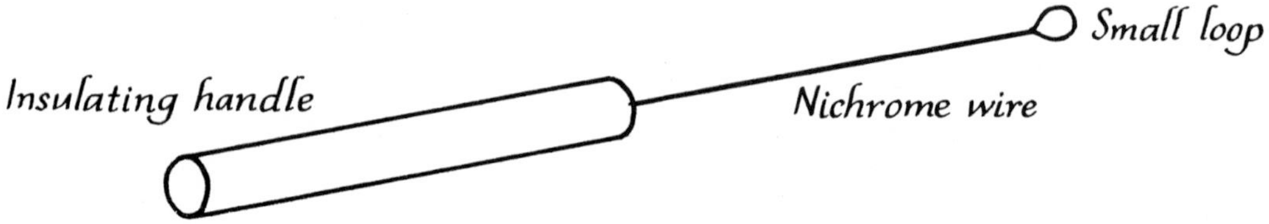

Insulating handle   Nichrome wire   Small loop

that it is soluble in water, you may have wondered why sodium silicate(IV) is used in the making of glass. The other silicates present in glass almost prevent the sodium silicate(IV) from dissolving. Even so, a little of the sodium silicate(IV) will dissolve, especially at high temperatures. If a glass is going to be used continuously at high temperatures, potassium carbonate is used in its manufacture, rather than sodium carbonate.

We can use the soluble property of water-glass in an interesting way.

Investigation 32f. A chemical garden

Take a 400 cm³ beaker and half fill it with warm water, about 330 K (about 60 °C). Stir into this about 100 cm³ of water-glass. Drop into the beaker one or two small crystals of the sulphate(VI) of the following metals: copper(II), nickel(II), chromium(III), cobalt(II), manganese(II) and iron(II). Cover the top of the beaker with a piece of cardboard or a clock-glass, and stand it where it can remain undisturbed for two or three days. Since this experiment does not involve heating, a jam jar could be used in place of a beaker (see Figure 32.5).

Harmful

Observe any changes that occur in the beaker. All the sulphates (VI) that you placed in the beaker are soluble in water and the colours of the sulphates(VI) are due to their metal ions. Consider the statements made in the previous sentences when you try to answer the following questions:

a. What new substances have been formed in the beaker?
b. Are these new substances soluble?
c. Why have they taken on this shape?

Try to write the word equation for the change of the cobalt(II) sulphate(VI).

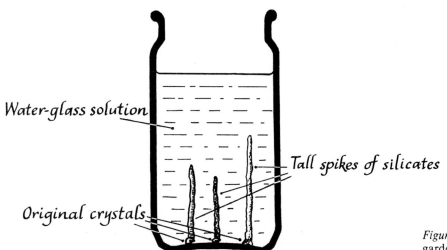

Water-glass solution

Tall spikes of silicates

Original crystals

Figure 32.5   A chemical garden

351

Whilst sodium silicate(IV) is soluble in water, the other silicates are not. Therefore, when copper ions in the copper(II) sulphate(VI) crystal are surrounded by silicate(IV) ions from the water-glass, copper(II) silicate(IV) (an insoluble compound) forms at the surface of the copper(II) sulphate(VI) crystal:

copper(II) ions(aq) + silicate(IV) ions(aq) → copper silicate(IV)(s)

32.7. Silicones

The **-Si-O-Si-O-** linkage can form a long chain. However, in some cases a silicon atom may be joined to a methyl group of atoms, CH_3 (see Chapter 33), in which case the silicon atoms will not form a linkage through an oxygen atom to another silicon atom. Some of the compounds that can be obtained from silicon in this way are shown in Figure 32.6.

You will see that, by careful control, a wide range of different compounds can be formed – either very large cross-linked structures or chain polymers. All these polymers are known as **silicones**.

32.8. The properties of silicones

Since the silicones are modified forms of the silica structure, we could expect them to possess the stable nature of silica: many silicones are able to withstand a temperature of about 470 K (about 200 °C) for several years.

Investigation 32g. The use of silicone grease

Apparatus fitted with ground glass joints or with glass taps requires lubricating with a grease. Tap-funnels and burettes are examples of such pieces of apparatus.

The greasing may be done with a *carbon based* grease, such as 'Vaseline', or with a *silicone* grease. Take a piece of metal about 100 mm by 50 mm and support one end in a clamp so that the other end is about 100 mm above a bunsen burner. Close to the clamped end, place a small blob of 'Vaseline' and a similar sized blob of silicone stop-cock grease. Light the bunsen burner and adjust the flame so that the tip of a semi-luminous flame just touches the metal sheet. Watch carefully to see which melts first. Can you suggest any reason for the difference?

When you have completed this first experiment, remove the bunsen burner and allow the metal to cool. When it is cold, turn it round, so that the end with the grease is now furthest from the clamp. Replace the bunsen burner underneath the grease samples this time. Observe what happens and gradually increase the heat until both greases have changed. Which grease changed first?

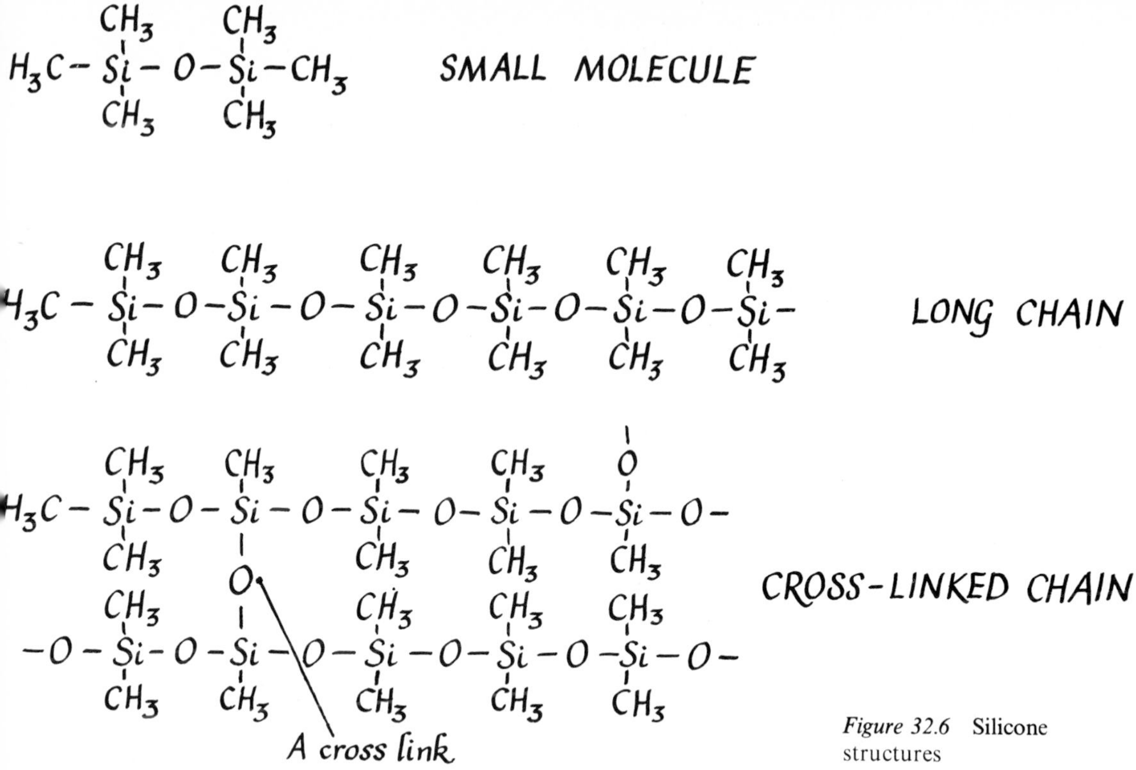

$$H_3C - \underset{\underset{CH_3}{|}}{\overset{\overset{CH_3}{|}}{Si}} - O - \underset{\underset{CH_3}{|}}{\overset{\overset{CH_3}{|}}{Si}} - CH_3 \qquad \textit{SMALL MOLECULE}$$

$$H_3C - \underset{\underset{CH_3}{|}}{\overset{\overset{CH_3}{|}}{Si}} - O - \underset{\underset{CH_3}{|}}{\overset{\overset{CH_3}{|}}{Si}} - O - \underset{\underset{CH_3}{|}}{\overset{\overset{CH_3}{|}}{Si}} - O - \underset{\underset{CH_3}{|}}{\overset{\overset{CH_3}{|}}{Si}} - O - \underset{\underset{CH_3}{|}}{\overset{\overset{CH_3}{|}}{Si}} - O - \underset{\underset{CH_3}{|}}{\overset{\overset{CH_3}{|}}{Si}} - \qquad \textit{LONG CHAIN}$$

LONG CHAIN

CROSS-LINKED CHAIN

A cross link

Figure 32.6 Silicone structures

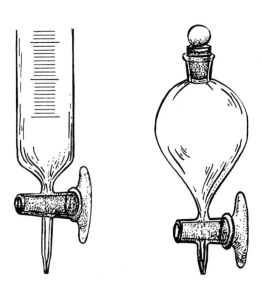

Figure 32.7 Glass apparatus requiring grease

What do you think is left on the metal where the silicone grease was?

In this investigation, we see that silicone greases do not melt as easily as carbon greases. They remain of almost the same consistency over a wide range of temperatures. This is another property that is due to the silica-type structure that they possess.

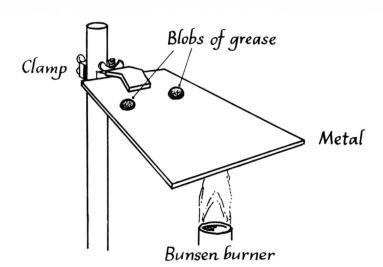

Figure 32.8 Comparison of greases

When strongly heated, the carbon greases burn, but the silicone grease leaves a light-coloured residue of silica.

Investigation 32h. Silicone-treated surfaces

We shall now carry out a number of simple tests on surfaces to see how treatment with silicones can be used to help in our life.

1. Take a small piece of impact-adhesive plastic ('Fablon' is of this type). Remove the paper backing and stick the plastic to a piece of brown paper. Now try to remove the brown paper. Compare this with removing the original backing paper.
2. Take the backing paper from the last experiment and fix a piece of self-adhesive acetate tape (for example 'Sellotape') to the paper. How well does it stick? Try a similar piece of tape on a piece of brown paper.
3. Make up an ordinary fruit jelly according to the maker's instructions. Pour some of the jelly into a 'non-stick' baking case and some into a similar untreated mould. Allow to set and see which 'turns-out' more easily.
4. Repeat experiment 3 above with a cake mixture, if possible. Bake as usual and 'turn-out' the cakes.

Surfaces can be treated with silicones to prevent 'sticking'. We have seen some uses of this treatment; others include the surface coating of frying-pans and saucepans. Metal surfaces treated in this way are usually a light grey in colour.

Investigation 32i.   Another surface treatment with silicones

Take two similar pieces of old washed cloth (a discarded sheet would do quite well). Treat one piece with a silicone water-

354

proofing solution. (This can usually be obtained from shops that deal in camping supplies. Some is for painting on the fabric with a brush, but it can be obtained in an aerosol pack. 'Fabsil' is a solution of this type.) Allow the proofing to dry and then spray water on the two pieces of fabric. Compare the results.

A similar investigation should also be carried out with a masonry water-repellent. Small pieces of clay roofing tiles are suitable surfaces for treatment in this way.

In this investigation we have seen that silicones are used to make surfaces 'water-repellent' (see Figure 32.9).

Flammable

Drop of water spreads on untreated surface

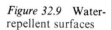

Drop of water on treated surface

In addition to the uses that we have examined, silicone water-repellent treatment is used to protect many electrical systems from damage that would be caused if they became wet. For example, large glass and porcelain insulators used to carry high voltage cables are treated with a silicone so that water will 'run-off' the surface. Many polishes have a silicone added for its water-repellent properties. These silicone polishes are also easier to use than the older wax polishes.

In the same way that cake moulds are treated with silicones to make them 'non-stick', large industrial moulds, such as those used for making car tyres, are also treated. Silicones are also used to completely enclose small and delicate electrical components.

Figure 32.9 Water-repellent surfaces

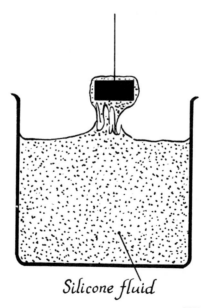

Silicone fluid

Figure 32.10 Encapsulating electrical components

1. What special property do you associate with compounds of silicon?
2. List some forms of silica.
3. Why is silicon dioxide a solid whilst carbon dioxide is a gas?
4. What is the difference between the clay used for bricks and that used for china?
5. State TWO ways in which pottery can be treated to prevent water seeping through.
6. What is the difference between lime mortar and cement mortar? What is the 'long-term' advantage of using cement mortar?
7. Why can Pyrex glass apparatus be made thicker than soft glass apparatus?
8. How is glass coloured?
9. Why do silicone compounds possess special properties?
10. Make a list of some uses of silicones in daily life.

Chapter 33

Introduction to Organic Chemistry

In the course of our study of chemistry, we have already learned quite a lot about the element carbon. We learned, in Chapter 20, that a carbon atom was able to join to other carbon atoms to form chains. This property is unique to carbon and leads to the formation of homologous series of compounds (see Section 20.5). Table 20.1 gives a list of the first six members of the alkane hydrocarbons.

33.1. The preparation of alkanes in the laboratory

We will now make an alkane in the laboratory and investigate some of its properties.

Investigation 33a. The preparation of methane

Take a little *anhydrous* sodium ethanoate (about 10 mm high in a 19 mm diameter test-tube). Mix this with about the same bulk of soda-lime, add a 'pinch' of iron filings, and grind the mixture together. Place about half of this mixture in a hard-glass (Pyrex) test-tube and fit the tube with a delivery tube, leading to a beaker, as shown in Figure 33.1.

Flammable

Sodium ethanoate is the sodium salt of ethanoic acid. If you have not any anhydrous sodium ethanoate, the hydrated form can be dehydrated for use. (To do this, heat the ethanoate gently in a crucible until it appears to melt. Stir the substance thoroughly and continue heating until no more steam is given off.)

Heat the test-tube sufficiently strongly to cause a steady stream of bubbles to escape from the delivery tube. Allow the first few bubbles to escape and then collect some of the escaping gas in a test-tube of water, as shown in Figure 33.1. When the tube is full of gas, remove it and cautiously apply a lighted splint to the open end. If the gas burns quietly, it is free of air. If it does not burn quietly, repeat the test. When the escaping gas that is collected burns quietly, collect a further four test-tubes of the gas. Leave

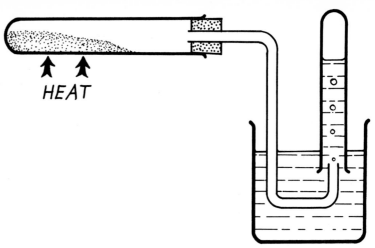

Figure 33.1 The preparation of methane

Flammable

these tubes, open end downwards, in the beaker, for use later. When you have collected four tubes of gas, stop heating. As soon as the gas stops escaping from the delivery tube, disconnect the delivery tube from the test-tube by pulling out the bung. (This stops the cold water in the beaker being drawn up into the hot test-tube as the gas in the test-tube contracts on cooling.)

Now carry out the following tests on the gas that you have collected:

1. Turn the first tube of gas so that its open end points upwards and, at once, apply a lighted splint. As we already know, the gas burns, but what sort of flame does it burn with? Look carefully at the walls of the test-tube after the gas has burnt. Can you see anything that was not there before? What does the type of flame suggest about the amount of carbon in the gas? Can you test the gas left after burning to show that there was some carbon in the methane? Why did you apply the lighted splint *at once*?

2. Have a cork ready which fits the test-tube with the methane in it. Take the test-tube and turn it open end upwards. With a teat pipette, add a few drops of bromine water and put the cork in the tube as soon as possible. Shake the test-tube. Does anything happen in one or two minutes?

3. Into the third tube of gas, put a few drops of potassium manganate(VII) solution, in the same way as the bromine water was added in test 2. The potassium manganate(VII) solution should be very dilute and alkaline to litmus. (This solution can be prepared by taking two drops of a stock solution in a test-tube and adding water until it is possible to see through the solution. Then add two drops of sodium carbonate solution. Use a few drops of this solution.) Can you observe any change when the potassium manganate(VII) solution is shaken with methane gas?

The fourth tube is a reserve, in case you need to repeat any of these tests. Keep it until you have checked your observations.

The methane gas burns with a pale blue flame. There are no signs of soot and there is little or no yellow colour in the flame, indicating that there is not a very high proportion of carbon atoms in the compound. If there had been a lot of carbon, the flame would have been yellow and sooty. The walls of the tube are covered with a fine mist of condensation, showing that hydrogen is contained in methane. If we shake a few drops of lime-water in the tube it turns milky, indicating the production of carbon dioxide from the carbon in the methane. Hence methane contains both carbon and hydrogen. There should have been little or no change with bromine water and potassium manganate(VII) solution. Remember this; we shall use the result later.

Now that we know something about the gas produced we can write the equation:

$$CH_3COONa(s) + NaOH(s) \rightarrow CH_4(g) + Na_2CO_3(s)$$

Soda-lime is quicklime that has been slaked with sodium hydroxide solution; it acts chemically as sodium hydroxide.

Ethanoic acid (sodium ethanoate, you will remember, is a salt of ethanoic acid) is one of the acids in the homologous series known as **alkanoic acids.** We learned about some other members of this series and their uses in soap manufacture (see Chapter 23). Figure 33.2 shows the arrangement of atoms in some of these acids.

Only one of the hydrogen atoms has *acidic* properties. Those joined directly to a carbon atom are *not acidic*.

Sodium salts of the other acids can be used to make other

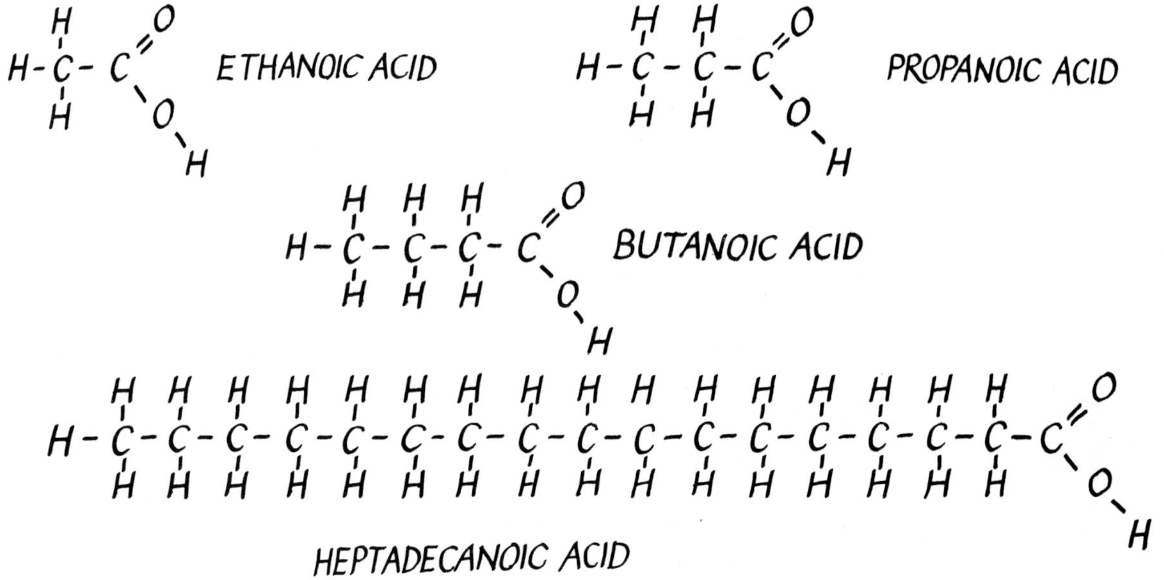

Figure 33.2 Some alkanoic acids

HEPTADECANOIC ACID

members of the alkane series. For example, sodium propanoate will produce ethane:

$$CH_3CH_2COONa(s) + NaOH(s) \rightarrow CH_3.CH_3(g) + Na_2CO_3(s)$$

Investigation 33b The properties of another alkane

Your teacher will assemble the apparatus shown in Figure 33.3, used to collect some test-tubes of another alkane, **butane** (C_4H_{10}).

You will then be given the test-tubes of the gas and you should apply the same tests to the butane as you used to test the methane in Investigation 33a.

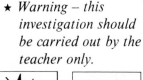
Explosive   Flammable

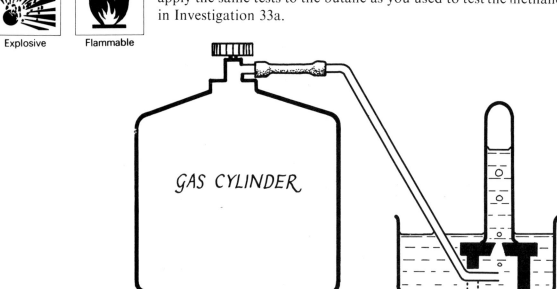

GAS CYLINDER

Figure 33.3   The collection of butane

Compare the results that you obtained for the butane with the results for methane. We shall refer to the results of this investigation when we have completed Investigation 33e.

Investigation 33c. The properties of a liquid alkane

★ *Warning – this investigation should be carried out by the teacher only.*

Flammable   Harmful

Place a few drops of either liquid pentane or hexane in an evaporating basin. Apply a lighted splint to the liquid. Shake a few drops of the liquid with some bromine water and with an alkaline solution of potassium manganate(VII). Again compare the results with those obtained previously.

33.2. The nature and uses of alkanes

As we have seen in our investigations, alkanes are not very reactive compounds. In a plentiful supply of air, alkanes burn to form water and carbon dioxide:

$$CH_4(g) + 2O_2(g) \rightarrow CO_2(g) + 2H_2O(l)$$
$$2C_4H_{10}(g) + 13O_2(g) \rightarrow 8CO_2(g) + 10H_2O(g)$$

Hence they are used mainly as fuels. Methane is the main component of natural gas, which is the main fuel gas used in the United Kingdom. Butane, as we have seen, is used for supplies of gas for portable appliances, for example camping stoves and lighters. (Propane is used in a similar way for larger users who do not have a town gas supply available.) Although butane and propane are gases at normal temperatures, at increased pressure they can be stored as liquids. In liquid form they occupy a much smaller volume. When these gases are supplied for use as fuels it is usual to add another type of gas, in small amounts, to render the flame more easily visible; because of this added gas, your results with butane may differ from those observed with methane and hexane or pentane.

Higher members of the alkane series that are normally liquids are found in various petroleum fuels, whilst solid alkanes are used in candles.

Alkanes will also react with halogens in the presence of sunlight. An atom of a halogen – for example, chlorine – replaces a hydrogen atom in the alkane:

$$CH_4(g) + Cl_2(g) \rightarrow HCl(g) + CH_3Cl(g) \text{ chloromethane}$$

The reaction is known as **substitution** and may continue until all the hydrogen atoms have been replaced by halogen atoms:

$$CH_3Cl(g) + Cl_2(g) \rightarrow HCl(g) + CH_2Cl_2(g) \text{ dichloromethane}$$
$$CH_2Cl_2(g) + Cl_2(g) \rightarrow HCl(g) + CHCl_3(l) \text{ trichloromethane}$$
$$CHCl_3(l) + Cl_2(g) \rightarrow HCl(g) + CCl_4(l) \text{ tetrachloromethane}$$

33.3. The homologous series

In the last set of investigations, namely 33a, 33b and 33c, we have examined the properties of various alkanes. We find that they all display, if pure, similar properties: we might say that they form a 'family' of compounds.

Look at Table 20.1, which lists the alkanes, and try to find out what is the common difference between one alkane and the next. Now make a list of the alkanoic acids given in Figure 33.2, in order of ascending relative molecular mass. Look at the first three on this list in the same way that you looked at the alkane series. Do you see the same common difference?

As we learned in Chapter 20, a series of compounds with similar properties is known as a homologous series. The second member of such a series has one more CH_2 than the previous one, hence its relative molecular mass is 14 greater than the previous member of the series.

TABLE 33.1. SOME EXAMPLES OF HOMOLOGOUS SERIES

Alkanols	Chloroalkanes
Methanol (CH_3OH)	Chloromethane (CH_3Cl)
Ethanol (CH_3CH_2OH)	Chloroethane (CH_3CH_2Cl)
Propanol ($CH_3CH_2CH_2OH$)	Chloropropane ($CH_3CH_2CH_2Cl$)
Butanol ($CH_3CH_2CH_2CH_2OH$)	Chlorobutane ($CH_3CH_2CH_2CH_2Cl$)

Do you notice any similarity in names of these compounds and compounds that you have used before?

Since all the alkanols contain an **OH** group, they will have similar properties. Their boiling-points will rise steadily with the increase in relative molecular mass.

33.4. Chains and rings

When carbon joins with itself, it forms either chains of atoms or rings of atoms. The stable ring form is six atoms in a hexagonal arrangement.

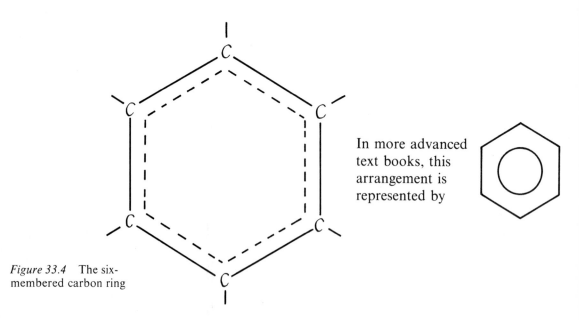

In more advanced text books, this arrangement is represented by

Figure 33.4 The six-membered carbon ring

Many of the compounds that are based on this six-membered ring belong to a group of compounds that are known as **aromatic** compounds, since many of them have a pleasing smell. Where have you seen this six-membered ring of carbon atoms before?

Figure 33.5 shows the arrangement of atoms in some well-known carbon compounds.

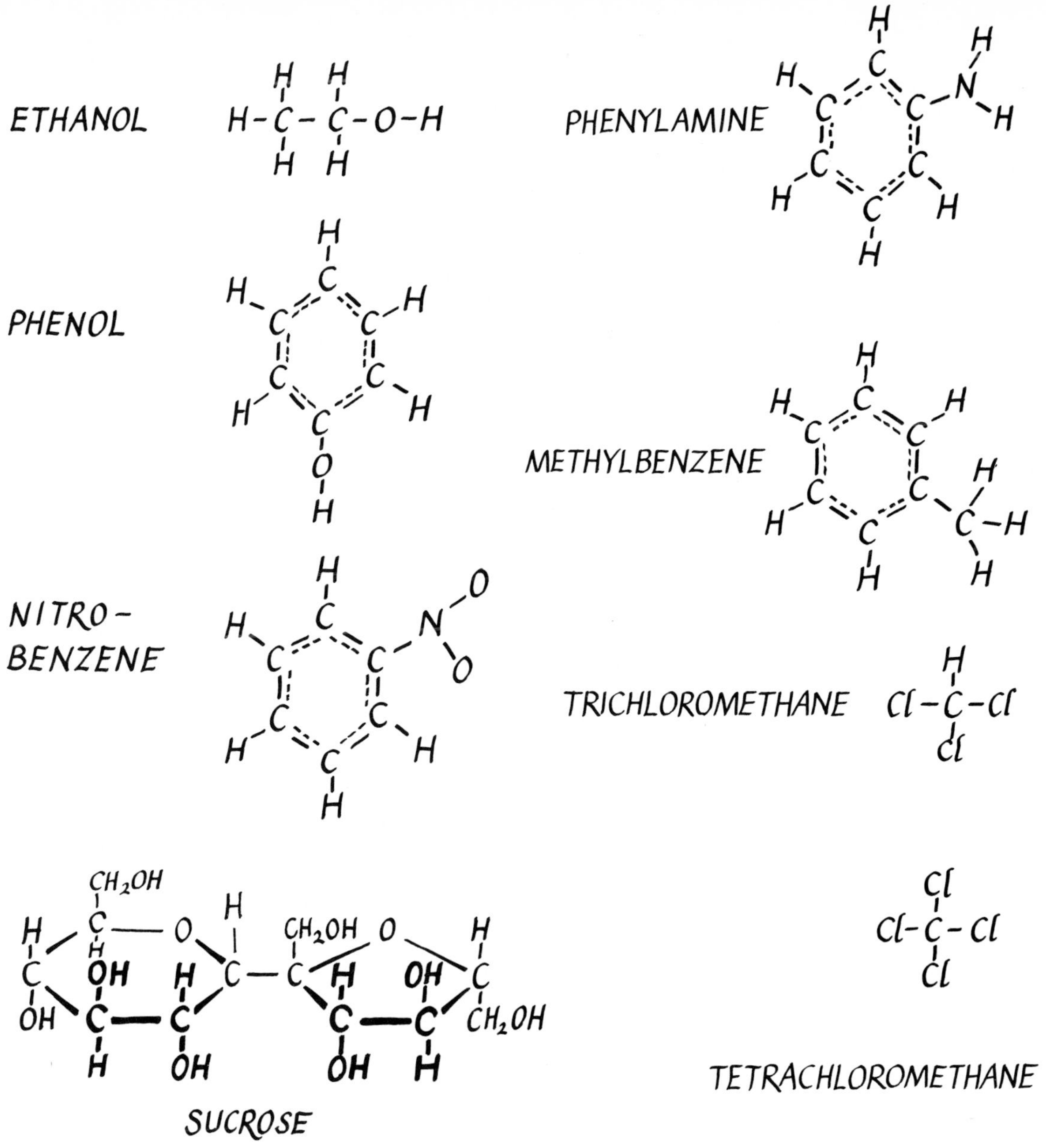

ETHANOL

$$H-\underset{\underset{H}{|}}{\overset{\overset{H}{|}}{C}}-\underset{\underset{H}{|}}{\overset{\overset{H}{|}}{C}}-O-H$$

PHENYLAMINE

PHENOL

NITRO-
BENZENE

METHYLBENZENE

TRICHLOROMETHANE $Cl-\underset{\underset{Cl}{|}}{\overset{\overset{H}{|}}{C}}-Cl$

SUCROSE

$Cl-\underset{\underset{Cl}{|}}{\overset{\overset{Cl}{|}}{C}}-Cl$

TETRACHLOROMETHANE

Figure 33.5 The structure
of some carbon compounds

33.5. Isomerism

In Chapter 20, you learned that the same number of carbon and
hydrogen atoms could have different arrangements and that the
various compounds were said to be isomers. The isomers discussed
in Chapter 20 were those where the atom arrangements differed,
but the compounds had the same properties, the only difference
being small changes in melting- and boiling-points. Isomers of
butane are shown in Figure 33.6.

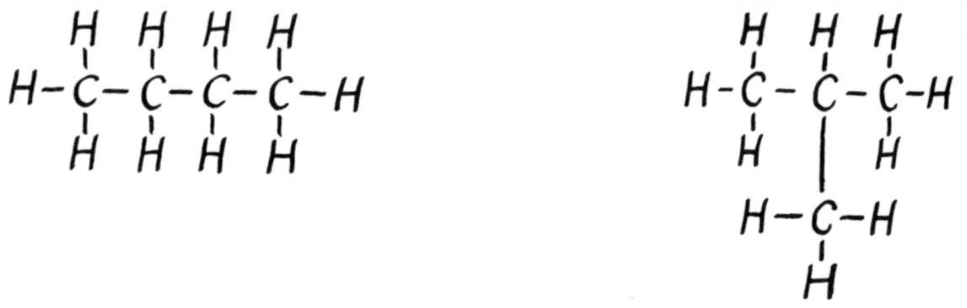

Figure 33.6 Isomers of butane

$$H-\overset{\overset{\displaystyle H}{|}}{\underset{\underset{\displaystyle H}{|}}{C}}-\overset{\overset{\displaystyle H}{|}}{\underset{\underset{\displaystyle H}{|}}{C}}-\overset{\overset{\displaystyle H}{|}}{\underset{\underset{\displaystyle H}{|}}{C}}-\overset{\overset{\displaystyle H}{|}}{\underset{\underset{\displaystyle H}{|}}{C}}-H$$

$$H-\overset{\overset{\displaystyle H}{|}}{\underset{\underset{\displaystyle H}{|}}{C}}-\overset{\overset{\displaystyle H}{|}}{\underset{\underset{\displaystyle H-\overset{\overset{\displaystyle |}{C}}{\underset{\underset{\displaystyle H}{|}}{}}-H}{|}}{C}}-\overset{\overset{\displaystyle H}{|}}{\underset{\underset{\displaystyle H}{|}}{C}}-H$$

This property of isomerism can, however, also lead to different chemical properties. **Propanone** and **propanal** both have the same molecular formula, C_3H_6O.

Investigation 33d. A comparison of propanal and propanone

Your teacher will give you a few cubic centimetres of each of the two substances. Make sure that you label the tubes so that you will know which liquid is in each tube.

1. Carefully smell each of the liquids. You may well recognize the smell of one or both liquids.
2. Test the two substances with Schiff's reagent.
3. Take two drops of potassium manganate(VII) solution, add distilled water until it is a pale mauve colour, and then add two drops of dilute sulphuric(VI) acid. Use this to test the two isomers.

As a result of the tests you have used, you can tell that the two substances are different in their properties. The arrangement of the atoms in these two compounds is shown in Figure 33.7.

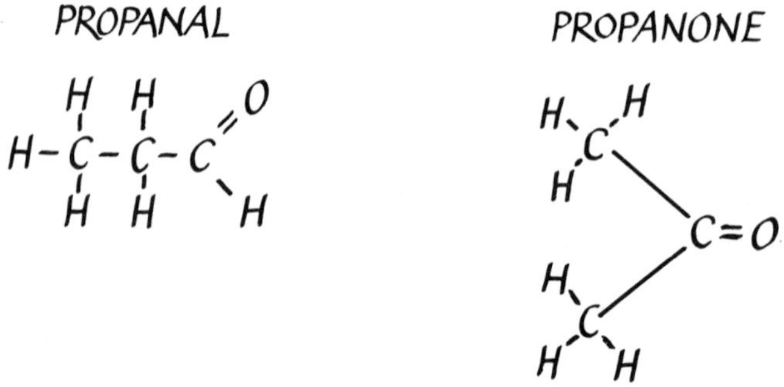

PROPANAL

PROPANONE

Figure 33.7 The structure of propanal and propanone

33.6. Some other hydrocarbons

In Section 33.1, we made a study of one type of hydrocarbon (carbon – hydrogen compounds) known as alkanes. We shall now examine some other hydrocarbons.

Investigation 33e. The preparation of ethene

Take a hard-glass test-tube (150 mm × 19 mm) and pour into it

364

enough ethanol to come about 20 mm up the tube. Now drop in a little 'Rocksil' fibre and, with a stirring rod, mix this with the ethanol. Continue adding Rocksil and stirring until all the ethanol has been soaked up in the fibre. Now, support the tube in a clamp so that the open end points slightly downwards. Place a small pile of pumice lumps in the tube, about half-way between the Rocksil and the mouth of the tube. Fit the test-tube with a delivery tube leading to a beaker and a collecting tube, as shown in Figure 33.8.

Flammable

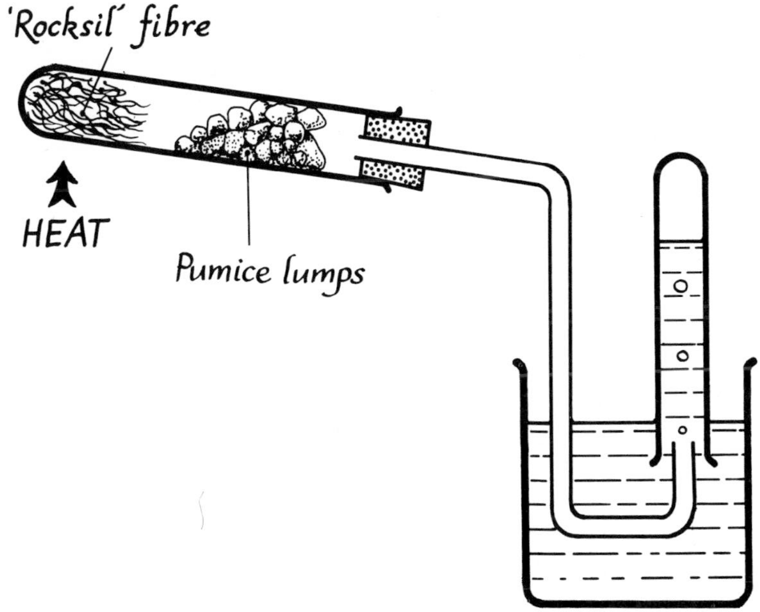

Figure 33.8   The preparation of ethene

Heat the pumice with a bunsen burner. The heat will cause some of the ethanol to evaporate and pass down the tube over the heated pumice. Discard the first two tubes of gas collected; then collect four further tubes of gas. Test the gas in these tubes in the same way that you tested methane in Investigation 33a. If necessary, the Rocksil fibre may be gently heated to drive off some ethanol vapour. However, this heating should be kept to a minimum.

Compare the results of your tests on ethene with the results that you observed with methane, butane, pentane and hexane. You should notice some differences (there may be some similarity with the results of the impure sample of butane). This gas, ethene, is a member of a different series of hydrocarbons, known as alkenes.

Investigation 33f. The properties of a liquid alkene

Your teacher will provide you with a little hexene; this is a liquid of the same homologous series as ethene. Test it in the same way that the pentane (or hexane) was tested in Investigation 33c.

Flammable

Again compare the results with those in Investigations 33a, 33b, 33c and 33e.

You should have noticed a number of differences between alkanes and alkenes which are best summarized in a table.

TABLE 33.2. THE DIFFERENCES BETWEEN ALKANES AND ALKENES

Property	Alkane	Alkene
Burning	Burns with an almost colourless flame	Burns with a rather yellow flame
Action on bromine water	None	Decolourizes
Action on alkaline potassium manganate(VII) solution	None	Turns green and then brown

These differences in properties suggest to us that: alkenes contain relatively more carbon than alkanes (see Table 33.3); and alkenes are more reactive than alkanes.

TABLE 33.3. THE CARBON/HYDROGEN RATIO IN ALKANES AND ALKENES

Alkane	Carbon/Hydrogen Ratio	Alkene	Carbon/Hydrogen Ratio
Ethane (C_2H_6)	1:3	Ethene (C_2H_4)	1:2
Butane (C_4H_{10})	1:2.5	Butene (C_4H_8)	1:2
Hexane (C_6H_{14})	1:2.3	Hexene (C_6H_{12})	1:2

You will see that the ratio in the alkenes is constant at 1:2 and hence the general formula (see Section 20.5) for alkenes is C_nH_{2n}.

You will probably wonder about the formula C_2H_4, if you remember that the normal valency of carbon is four and that of hydrogen is one. Chemists believe that, in alkenes, two carbon atoms are joined together by two pairs of electrons, rather like oxygen atoms in the oxygen molecule (see Figure 6.2(b)).

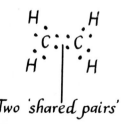

Figure 33.9 The structure of ethene

366

This double bond is responsible for the reactivity of alkenes. For example, with bromine water the bromine adds on to the ethene to form 1,2-dibromoethane:

$$
\begin{array}{c}
\text{H} \qquad \text{H} \\
\diagdown \quad \diagup \\
\text{C}=\text{C} \quad (g) + Br_2(aq) \rightarrow \\
\diagup \quad \diagdown \\
\text{H} \qquad \text{H}
\end{array}
\qquad
\begin{array}{c}
\text{H} \; \text{H} \\
| \quad | \\
\text{Br}-\text{C}-\text{C}-\text{Br}(l) \\
| \quad | \\
\text{H} \; \text{H}
\end{array}
$$

Because of this special property, alkenes are very important starting materials in modern industry (see Chapter 36).

33.7. Ethyne

Another type of hydrocarbon, known as **ethyne,** can be prepared in the laboratory from calcium dicarbide.

Investigation 33g. The properties of ethyne

Prepare some ethyne as described in Investigation 21a. Collect some tubes of the gas and use them as described below. Discard the first two tubes of gas, and burn the gas in the third tube. If the gas in the third tube burns successfully, add a little bromine water to the fourth tube of gas and shake. (If the gas in the third tube does not burn, continue collecting gas until it does burn and then carry on with the tests.)

What can you say about the carbon/hydrogen ratio in ethyne?

Like ethene and the other alkenes, ethyne combines readily with bromine. You will also see that the flame from burning ethyne is very smoky. This suggests that the carbon/hydrogen ratio is even higher than in ethene. Ethyne has the formula C_2H_2; the ratio is $1:1$.

Explosive

Flammable

Harmful

The reaction of water with calcium dicarbide may be represented as:

$$CaC_2(s) + 2H_2O(l) \rightarrow H\text{-}C \equiv C\text{-}H(g) + Ca(OH)_2(s)$$

You will see that ethyne is represented as having a 'triple bond' between the two carbon atoms. Compounds with multiple bonds between carbon atoms, like ethene and ethyne, are said to be **unsaturated**. They add halogen atoms when mixed with a halogen supply to satisfy the four valencies of carbon:

$$
\text{H}-\text{C} \equiv \text{C}-\text{H}(g) + 2Cl_2(g) \rightarrow
\begin{array}{c}
\text{H} \quad \text{Cl} \\
| \quad | \\
\text{Cl}-\text{C}-\text{C}-\text{Cl}(l) \\
| \quad | \\
\text{Cl} \quad \text{H}
\end{array}
$$

367

Ethyne used to be used for lights on bicycles and motor cars; now it is used, like ethene, in the manufacture of many modern materials.

Test your understanding

1. Is methane heavier or lighter than air?
2. How does the flame of burning methane differ from a candle flame?
3. 1,2-ethanedioic acid has the formula:

$$
\begin{array}{c}
\quad\quad O \\
\quad\quad \parallel \\
C-O-H \\
\mid \\
C-O-H \\
\quad\quad \parallel \\
\quad\quad O
\end{array}
$$

Are both of the hydrogen atoms acidic?
4. Complete the equation for the reaction of sodium pentanoate and soda lime on heating:

$$CH_3CH_2CH_2CH_2COONa(s) + NaOH(s) \rightarrow$$

What is the name of the alkane obtained in this reaction?
5. Which of the following formulae is not in the same homologous series:

$$CH_3NH_2; \quad C_2H_5NH_2; \quad C_3H_7NH_2; \quad C_4H_8NH_2; \quad C_5H_{11}NH_2?$$

6. What are alkanes used for?
7. The arrangement of atoms in ethanol is:

$$
\begin{array}{c}
\quad H \quad H \\
\quad \mid \quad \mid \\
H-C-C-O-H \\
\quad \mid \quad \mid \\
\quad H \quad H
\end{array}
$$

Try to write the arrangement of atoms in an isomer of ethanol.
8. What is the cause of 'unsaturation' in the alkenes?
9. What test could you use to show that a colourless organic liquid was not an alkene?
10. Why is ethyne such a good starting material in the chemical industry?
11. Without attempting to give practical details of any process, suggest how ethanol might be obtained cheaply from ethene.

Chapter 34

Ethanol

In the previous chapter, we made a study of some very simple compounds of carbon. We saw how carbon atoms have the ability to form quite large molecules by bonding covalently to each other. Living matter is made up of compounds of carbon; this will be realized when you remember the way in which charcoal is made. The compounds that we studied in the last chapter are rather unreactive, and we shall obtain a better idea of the importance of carbon compounds from the study of a more reactive substance.

34.1. Ethanol from natural materials

The compound **ethanol** (or to give it its common name, alcohol) is an important substance in the link between very big molecules, like **starch**, and the simpler compounds, water and carbon dioxide. Ethanol has been prepared for hundreds of years from natural **sugars** and starchy materials. Brewers and distillers continue to use these processes today.

Investigation 34a. Fermentation of a sugar

Take about 25 cm³ of a 20 per cent solution of glucose and place it in a flask (250 cm³ for preference). Add to this about one gram of **yeast** and cover the opening of the flask with a piece of paper, as shown in Figure 34.1. Do not use a cork or other tight closure.

Stand the flask by a radiator or in some other warm place; a temperature of about 300 K (about 25 °C) is best. At the end of the first or second day, remove a sample of gas from the neck of the flask and try to identify the gas. After four days, filter the product and collect the filtrate. Note the smell of the filtrate.

Place about 10 cm³ of the filtrate in a filter tube arranged for distillation as shown in Figure 34.2 and distil. Note the temperature at which the liquid is collected in the receiver tube. When you have collected about 2 cm³ of liquid in the receiver tube, stop heating. Examine the distillate noting its colour and smell. Try to burn a little of the distillate in an evaporating basin.

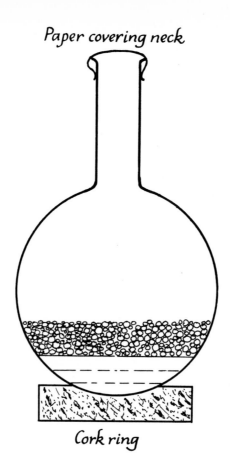

Paper covering neck

Cork ring

Figure 34.1 The
fermentation of glucose

The product is ethanol – you will probably have recognized it by its smell. Ethanol boils as 351 K (78 °C).

The reaction that you have used is the last stage of the process used in the **brewing industry**. The yeast that you added acted in a similar way to a catalyst although it is a *living organism*. Living organisms act in this way because they produce organic catalysts, known as **enzymes**.

In the brewing industry, starch is converted into sugar by the enzyme, **diastase**. This enzyme is found in germinating barley. The sugar, **maltose**, which is formed in this process is then converted into a simpler sugar, **glucose**, by one of the other enzymes in yeast, **maltase**. Then a second enzyme in yeast, **zymase**, converts the glucose into ethanol and carbon dioxide. It is this last process that we carried out in Investigation 34a:

$$C_6H_{12}O_6(aq) \rightarrow 2C_2H_5OH(aq) + 2CO_2(g)$$

You should have been able to detect the carbon dioxide in your investigation. The froth or 'head' produced on the liquid is due to the effervescence of the liquid.

370

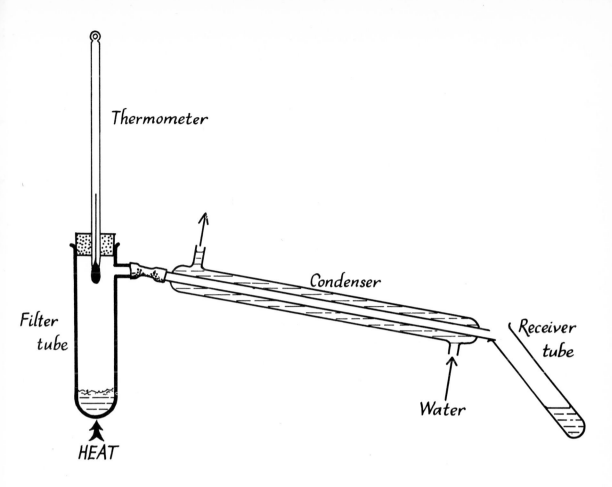

Figure 34.2 Distillation of the filtrate

When the juice of grapes is fermented, the first stage is not used and the enzymes necessary for the fermentation are found on the outside of the grape skin. Figure 34.3 illustrates, in diagram form, the formation of ethanol from starch.

In the manufacture of alcoholic drinks the type of starchy material used largely determines the flavour of the drink produced.

The alcoholic content of drinks is indicated by **proof number**. This proof number is approximately twice the percentage of ethanol in the drink. Hence a drink that contains 35 per cent ethanol would be said to be 70° proof.

Figure 34.3 The conversion of starch to ethanol

STARCH $\xrightarrow{\text{DIASTASE}}$ MALTOSE $\xrightarrow{\text{MALTASE}}$ GLUCOSE $\xrightarrow{\text{ZYMASE}}$ ETHANOL
(a sugar) (Simpler sugar) + CARBON
DIOXIDE

GRAPE JUICE

34.2. Industrial preparation of ethanol and its uses

Ethanol has a wide range of uses in everyday life, in addition to its use in alcoholic drinks. Hence a cheap source of ethanol is needed to supply these requirements. Although, at one time, it was made from potatoes, it is now manufactured synthetically from simple compounds. One of these is the combination of water and ethene in the presence of a catalyst:

$$H_2C = CH_2(g) + H_2O(g) \rightarrow H_3C\text{-}CH_2\text{-}OH(g)$$

Most of the ethanol which is made by this process is '**de-natured**', that is about 5 per cent **methanol** is added to make it unsuitable for drinking. Methanol, CH_3OH, is a poisonous alkanol. Other substances, including a purple dye, are sometimes added – this is so in the case of the well-known '**methylated spirits**'.

Ethanol is used as a solvent and as a fuel. As *surgical spirit*, it is sold for medical use and it is used in perfumes and other toilet goods. It is also used in the manufacture of ethanoic acid.

34.3. Ethanoic acid

Ethanoic acid may be made by the oxidation of ethanol.

Investigation 34b. Laboratory oxidation of ethanol

Flammable

Take a test-tube (150 mm × 19 mm) and place enough ethanol in the tube to come about 10 mm up the tube. Push into the tube enough 'Rocksil' fibre to absorb the ethanol. Clamp the tube in an almost horizontal position and place a pile of black copper(II) oxide about half-way down the tube, as shown in Figure 34.4.

Heat the copper(II) oxide with a bunsen flame. Some of the heat will cause the ethanol to vaporize and pass up the tube over the heated oxide. Test the gas escaping from the tube with a piece of moist blue litmus paper.

The method that you have used is the basis of an industrial

Figure 34.4 Laboratory oxidation of ethanol

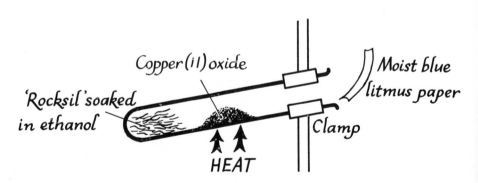

method. Examine the oxide left at the end of the experiment. Can you see any change?

Investigation 34c. Oxidation of ethanol by air

The cheapest source of oxygen is the air. On the manufacturing scale, air is used to oxidize ethanol. This process may be attempted in the laboratory in the following way. Take two filter tubes and a 200 mm length of glass tube, about 20 mm internal diameter. Arrange the apparatus as shown in Figure 34.5, and connect the

Explosive   Flammable

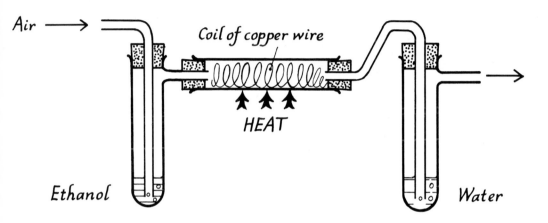

Figure 34.5 The industrial preparation of ethanoic acid

outlet from the second tube to a water vacuum pump. Alternatively, the inlet to the first tube could be connected to an air pump, for example a bicycle pump.

Heat the copper wire in the central tube and allow air to pass through the ethanol in the first filter tube. After about fifteen minutes, stop heating and stop the air flow. Disconnect the apparatus as soon as possible to prevent liquid being drawn up into the hot tube.

Test the product in the second filter tube with litmus and smell it. The product is ethanoic acid:

$$C_2H_5OH(g) + O_2(g) \rightarrow CH_3COOH(g) + H_2O(g)$$

In organic compounds, the only hydrogen with acid properties is the one in the − COOH group.

Solutions of ethanol containing less than 10 per cent ethanol will oxidize to ethanoic acid if left open to the air. The catalyst in this natural oxidation is an enzyme introduced by bacteria from the air. **Vinegar** is manufactured by this process from weak wines. Solutions of ethanol containing more than 15 per cent ethanol do not oxidize in this way, since the stronger ethanol solution kills the bacteria.

373

34.4. Acids and alkanols

You will have noticed that ethanol (and all the alkanols listed in Table 33.1) contains an – OH group of atoms. You may have wondered if it reacts in a similar way to alkalis. Let us now investigate this possibility.

Investigation 34d. The action of an acid on an alkanol

Take a boiling tube and place in it about 5 cm³ of ethanol. Test this with a piece of red litmus paper. Is it alkaline? Now carefully add to the ethanol, with continual stirring, 5 cm³ of glacial ethanoic acid (*this is concentrated acid*).

Stand the tube in a large beaker of cold water and arrange it so that the mouth of the tube does not point at anyone, if possible arrange for it to point at the wall. Using a teat pipette add, *one drop at a time*, about 1 cm³ of concentrated sulphuric(VI) acid. Stir the contents of the tube after every two or three drops of acid have been added. Allow the tube to stand for a few minutes until it is cool.

When the tube is cool, carefully dry the outside of the tube and hold it in a test-tube holder. Warm the mixture gently for a few minutes. *Do not allow it to boil*. After heating, allow it to cool and then carefully smell the vapour escaping from the tube. Compare the smell with that of the original ethanol and ethanoic acid. Has the smell changed?

Investigation 34e. The reaction between methanol and 2-hydroxybenzoic acid

2-hydroxybenzoic acid is a solid organic acid. Place a little of the solid in a test-tube and add 2 cm³ of methanol. Arrange the tube as in Investigation 34d. Again add some concentrated sulphuric(VI) acid, this time about ten drops. As before, dry the tube and gently warm it. Do you recognize the smell of the new substance in the tube?

We have seen from these last two investigations that although alkanols are not alkaline, they do react with acids. The concentrated sulphuric(VI) acid acts as a catalyst to the reaction. The new compounds that are formed are known as **esters,** and the process is called **esterification**. In Investigation 34d you made some ethyl ethanoate:

ethanoic acid + ethanol → ethyl ethanoate + water

$$CH_3COOH(l) + C_2H_5OH(l) \rightarrow CH_3COOC_2H_5(l) + H_2O(l)$$

In Investigation 34e, you made some methyl 2-hydroxybenzoate, this is commonly known as **oil of wintergreen**.

2-hydroxybenzoic acid + methanol → methyl 2-hydroxybenzoate + water

$$C_6H_4(OH)COOH(s) + CH_3OH(l) \rightarrow C_6H_4(OH)COOCH_3(l) + H_2O(l)$$

Hence, we see that in general terms:

organic acid + alkanol → ester + water

34.5. The properties of esters

Investigation 34f. Hydrolysis of an ester

Place about 5 cm³ of distilled water in a test-tube and add to it about 3 cm³ of ethyl ethanoate. Drop in a small piece of red and a small piece of blue litmus paper, shake the tube and then allow it to stand. Examine carefully the contents of the tube for about two minutes. What do you see?

Now, fit the tube with a cork carrying a length of glass tubing (about 150 mm long and about 10 mm in diameter). Support the tube in a beaker of water and heat the beaker until the mixture in the test-tube begins to boil. The purpose of the vertical glass tube is to allow the vapour escaping from the tube to cool and condense; this condensed liquid then runs back into the test-tube. This process is known as **refluxing**.

After about ten minutes of boiling, remove the test-tube from the hot water and allow it to stand. Can you see any difference in the contents of the tube? Remove the bung and length of glass

Flammable

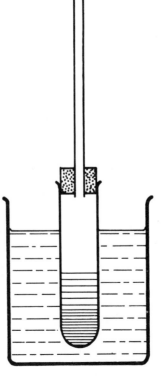

Figure 34.6 Boiling under reflux

375

tubing and drop in two new pieces of litmus paper (one red and one blue). Is there any change?

From this investigation, we learn that esters are liquids which do not easily dissolve in water and that most of them are less dense than water, that is most esters will float on the surface of water. On boiling, the water reacts with the ester, and the two layers that were present initially disappear. The result of the action of the water, on boiling, appears to produce an acid.

On warming with water, the esterification is reversed:

$$\text{ester} + \text{water} \rightarrow \text{acid} + \text{alkanol}$$

This interaction of water with a compound is known as **hydrolysis**. A special form of hydrolysis, known as saponification, is used in the manufacture of soap (see Investigation 23b).

We have also discovered that esters have a sweet fruity smell. Because of this many of them are used for artificial flavourings; for example ethyl butanoate ($C_3H_7COOC_2H_5$) has a smell like pineapples.

Test your understanding

1. What is the special property of carbon that makes it important in living matter?
2. To what group of organic compounds does glucose belong?
3. What are 'enzymes'?
4. What is produced by the fermentation of glucose?
5. What is 'proof number'?
6. What two letters at the end of the name of an organic compound indicate that it is an alkanol?
7. How is commercial ethanol treated to make it unsuitable for drinking?
8. Do you think that methanol has a higher or lower boiling-point than ethanol?
9. How may ethanoic acid be manufactured from ethanol?
10. What are the characteristics by which you might recognize a substance as an ester?
11. Write the equation for the formation of methyl ethanoate. What would you have to add to this reaction in order to make the change occur fairly quickly?
12. What happens when an ester is boiled with water?
13. Are you likely to find an ester in your home? If so, what might the label on the bottle say?

Chapter 35

Larger Organic Molecules

In the last chapter, we saw how a fairly simple organic molecule, ethanol, could be obtained from a larger molecule, starch, and how even the ethanol could be broken down into simpler molecules, carbon dioxide and water. In this chapter, we shall examine some of these larger molecules in greater detail. Although they are complex and difficult to experiment with, they are very important in our life. We shall consider the naturally occurring molecules; in the next chapter, you will learn something about man-made large molecules.

35.1. Starches and sugars

If you remember how the brewer makes his ethanol, you will realize that starch and sugar are closely related compounds.

Investigation 35a.   Starch into sugar

Place a small piece of *dry* bread in your mouth and chew it, *but do not swallow* it. Keep it in your mouth for about three minutes. Do you notice any change in the taste? Your saliva contains an enzyme, **ptyalin**, which starts the digestive processes of the body. What do you think ptyalin does to the starch in the bread?

If this investigation is to be fair one, you should not have eaten any sweets for a little time. Why?

Investigation 35b. What happens in the mouth?

The previous investigation showed us something about a very important process that occurs in the mouth; but it is not a very good investigation since we cannot control the experiment very well. Now we will try to examine the action of ptyalin with a little more care.

For this investigation, you will need a 1 per cent suspension of

starch in water; about 2 cm³ will be sufficient. Put about half of this starch suspension into a test-tube and add enough warm water to about half fill the tube. Now add about five drops of a solution of iodine in potassium iodide solution. What happens?

If you have been eating recently, wash out your mouth with a little warm water. Now, from a *clean* test-tube, pour into your mouth a test-tube full of warm water. Swill it round your mouth and then let it run into a clean beaker. About half fill a test-tube with the water that you have collected in the beaker. Add the remaining starch solution to the saliva-water mixture in the test-tube and stir well. Allow the tube to stand for about five minutes and then drop in the solution of iodine in potassium iodide solution, as before. Is there any difference in the result?

When you add iodine to starch, a dark blue colour is produced. This is a chemical test for starch (or for iodine); in this investigation, we are using it as a test for starch. In Investigation 35a, you noticed that a piece of bread became sweet after it had been in your mouth for some time. The enzyme, ptyalin, was converting the starch in the bread into sugar. In the second investigation, you were able to see that, after a few minutes, the ptyalin had converted all the starch to a new substance, since the iodine did not give a blue colour. From the result of Investigation 35a we should assume that this is sugar (a test for sugar is given in Section 35.7).

Investigation 35c.   What do starch and sugar contain?

For this investigation, you will need a number of small, cheap tubes (75 mm × 10 mm is a suitable size). Place a little granulated sugar at the bottom of one tube and heat it in a bunsen flame, gently at first and then more strongly. Can you recognize any gas or vapour escaping from the tube? If you think you know what is produced, try a chemical test.

Continue heating until there does not appear to be any further change in the tube. Then set the tube aside to cool. When it is cold, examine the residue in the tube to see what it is.

Repeat this investigation with samples of starch, flour, paper, straw, wood, custard powder and cotton wool. You may not be able to do all of these, but your teacher will probably arrange that at least one member of the class does a sample of each.

Investigation 35d. A further examination of the nature of sugar

Place a little granulated sugar in the bottom of a 100 cm³ beaker. Stand this beaker in a larger beaker (250 cm³). Now take a tall-form beaker (1 dm³) and check that it can be placed, open end downwards, over the other two beakers. When you have made sure that the apparatus can be screened, pour about 25 cm³ of concentrated sulphuric(VI) acid on the sugar in 100 cm³ beaker and, *at once*, replace the large beaker, as shown in Figure 35.1.

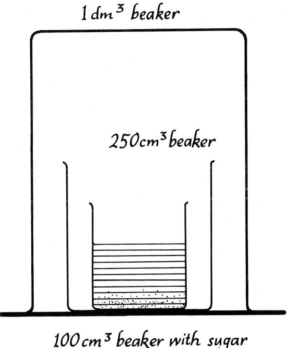

1 dm³ beaker

250 cm³ beaker

100 cm³ beaker with sugar and concentrated acid

Corrosive

Harmful

Figure 35.1 A safe arrangement of the apparatus

Try to answer these questions:
a. What is produced in the 100 cm³ beaker?
b. Is any heat generated?
c. What is the function of the sulphuric(VI) acid? (You may need to look up the properties of concentrated sulphuric(VI) acid in Chapter 28.)

When the reaction has died down, the apparatus can be dismantled. This should be done with care since it is possible that there are drops of concentrated sulphuric(VI) acid on the beakers and the bench. The beakers and bench should be washed, at once, with a plentiful quantity of water. The solid residue, in the 100 cm³ beaker, should be tipped into a sink full of water and washed well. It may then be examined by the pupils.

In the last two investigations we have seen that sugar and the other substances can be converted to carbon, the black residue, and water. The water should have been confirmed by its action on anhydrous copper(II) sulphate(VI). The concentrated sulphuric(VI) acid is used as a dehydrating agent and a considerable amount of heat is evolved.

Hence, we see that all these substances may be broken down into water and carbon. They must, therefore, contain the elements hydrogen, oxygen and carbon. Further, the elements hydrogen and oxygen are present in the same ratio that they are present in water (that is two atoms of hydrogen to one atom of oxygen). We could then express the composition of these substances by the formula $C_x(H_2O)_y$ where x and y are whole numbers. These substances, like sugar and starch, are known as **carbohydrates**.

35.2. Constant change

We have learned so far that sugar and starch are carbohydrates. These substances are of vegetable origin. You learned, in Chapter 21, how green plants convert carbon dioxide and water, by the process known as photosynthesis, into simple sugars, like glucose $C_6H_{12}O_6$, and then into more complex sugars and starch. The stages in this plant-building process are very complex and are not yet fully understood, but we believe that starch acts as an energy store for the plant. Certain plants, cane sugar, sugar-beet and carrots, store their energy as the sugar, sucrose ($C_{12}H_{22}O_{11}$), which is the sugar which we commonly use. Starch is a much more complex molecule than sucrose; its formula is usually written $(C_6H_{10}O_5)_n$, where n is a large number (about 100). Some of the starch is converted into an even larger molecule, **cellulose**, which forms the basic component of plant cell walls. Cellulose is thought to have a formula $(C_6H_{10}O_5)_n$, where n, this time, is about 1 000.

Certain animals are able to digest cellulose, but man is not able to do so. It is thought that one reason for man being unable to digest cellulose is that the molecule is ten times as big as starch.

You have also learned, in Chapter 34, how starch and sugar can be converted into ethanol, which in turn can be burnt to carbon dioxide and water. All these changes are summarized in Figure 35.3.

35.3. Proteins

In the last section, we learned something about the importance of chains of carbon, hydrogen and oxygen atoms in building up large molecules. In this section we shall examine the way in which another chain-like molecule is built up from simpler parts. The

✦ Represents $C_6H_{10}O_5$ group

• Simple sugar like glucose

❘ Sugar like sucrose

Starch

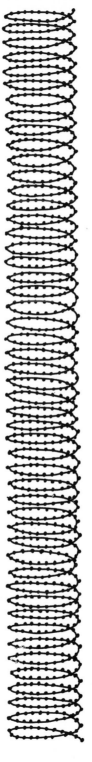

Cellulose molecule

Figure 35.2 A comparison of the size of sugar, starch and cellulose molecules

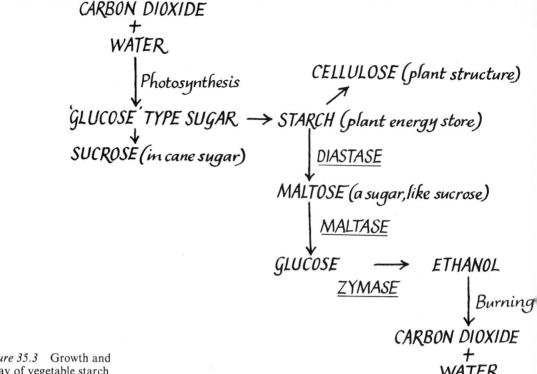

Figure 35.3 Growth and decay of vegetable starch

large chains that we shall study are known as **proteins**. You will probably have heard of the importance of proteins in your diet, and you will know that meat and eggs are good sources of protein.

Investigation 35e. What is present in a protein?

Corrosive

Harmful

Take a small test-tube and put into the tube about 1 cm³ of egg white. Using a spatula, put into the tube a little soda-lime. (If soda-lime is not available, a pellet of sodium hydroxide could be used.)

Now, heat the test-tube in a bunsen flame. Do you recognize the gas evolved? If you are in doubt, hold a piece of moist red litmus paper in the escaping gas. Keep the tubes and contents for the next investigation.

You can repeat this test with some other protein foods – for example, a small piece of meat or cheese. It is interesting to notice which protein foods break up most readily. Are they the ones that are easily digested? Two other sources of proteins that you might test in this way are clippings from your finger nails and some hairs.

Investigation 35f. Another element in proteins

Take the tubes from the last investigation and add about 5 cm³ of dilute hydrochloric acid to the residue in the tube. Warm the tube gently and hold a piece of moist lead(II) ethanoate paper at the mouth of the tube. (If you have no lead(II) ethanoate paper, a substitute may be made by dipping some strips of filter paper into lead(II) nitrate(V) solution.) What happens to the test paper? What does this tell you about the gas escaping from the tube? (See Investigation 27h.)

Toxic

This test should be applied to all the protein residues. Do you get the same result in all cases?

From these investigations we have seen that:

a. all proteins contain nitrogen; the gas given off in Investigation 35e was ammonia gas (see Investigation 25b).

b. many proteins contain sulphur; the gas given off in Investigation 35f was hydrogen sulphide.

35.4. Where does a protein come from?

We saw, in the last section, that all proteins contain nitrogen and, in Chapter 24, you learned about the nitrogen cycle and the importance of nitrogen to all life. We have learned that organic compounds are composed of a chain of carbon and hydrogen atoms to which may be attached special groups of atoms which give the compound its special properties. For example, the series of alcohols all contain the −OH group. There is a series of compounds similar to ammonia which is known as the **amine** series.

TABLE 35.1. SOME AMINES

Amine	Formula
Aminomethane	CH_3NH_2
Aminoethane	$C_2H_5NH_2$
Aminopropane	$C_3H_7NH_2$

You will see from Table 35.1 that these compounds are ammonia in which one atom has been replaced by a carbon − hydrogen group. Like ammonia, these compounds are bases. The carbon − hydrogen chain may have other special groups of atoms attached at some other place. We are interested, in this section, in the compounds that have an amino group and also an acid group, − COOH (see Section 34.3). Such compounds are known as amino acids. Figure 35.4 represents an amino acid structure.

Plants are able to convert the nitrogen-containing compounds taken in from the soil into amino acids. The amino acids have a very special property. The amino group, which is basic, of one

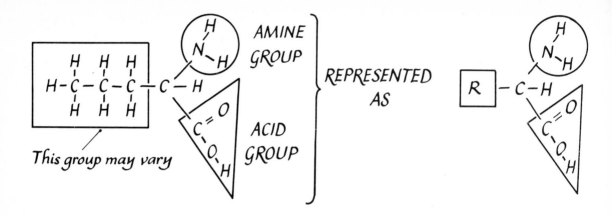

Figure 35.4 An amino acid

amino acid can react with the acid group of another amino acid, with the elimination of water. This process is known as **condensation** and a bigger molecule is formed. This process can repeat itself; in this way a protein is formed. (You will perform a similar process to that in which a long protein is formed when you make nylon in Chapter 36.)

35.5. Any protein?

By breaking down proteins into their component amino acids, scientists have found that there are twenty-four amino acids that may be used to make a protein. How many proteins can be made? Let us try a simple investigation.

Investigation 35g

On a piece of paper, see how many different combinations of the letters *A*, *B* and *C* you can make. You may use either two or three letters and you may also use the same letter more than once. However, the same order of letters in reverse may not be counted. For example: *ABA* and *BB* would count, but *ABC* and *CBA* count as only one combination.

You will find that even from three letters you can make quite a number of combinations.

You might like to repeat this investigation using the letters in the word *REACT*. Since there are five letters there will be many more combinations.

You may now ask: 'Can all the possible combinations exist as proteins?' The answer is that a very large number of different combinations occur, but the combinations are controlled. Your combinations of the letters in *REACT* would be controlled if the only combinations permitted were words that could be found in the dictionary.

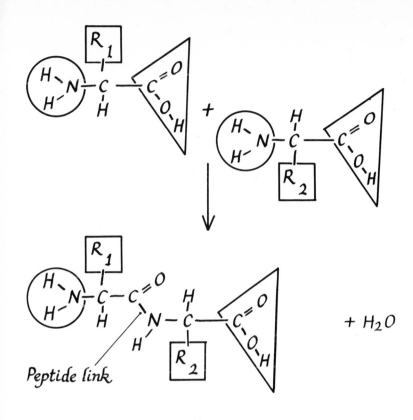

Peptide link

$+ H_2O$

THEN

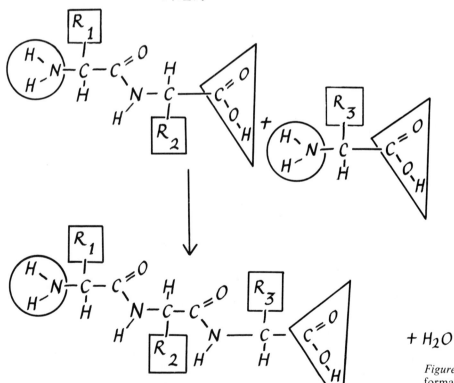

$+ H_2O$

Figure 35.5   The
formation of a protein

385

This control is exercised by a *template*. Let us see how templates manage to control the formation of proteins from amino acids.

Investigation 35h. A model of a protein

For this investigation you will need some cardboard and Sellotape. The work to be done would best be accomplished if you work in pairs.

One partner will need a strip of cardboard 200 mm long and 20 mm wide. With a ruler and pencil, divide this into squares with 20 mm sides, as shown in Figure 35.6(*a*). Then, using a pair of scissors, cut out shapes from one edge of this strip to the patterns shown. You have ten squares to cut out and may use any combination of the four patterns shown.

The second partner should cut out several pieces of cardboard to the patterns shown in Figure 35.6(*b*). When these have been cut out, fix a small piece of Sellotape to the pieces of cardboard as shown.

The long strip of cardboard represents the template and the small pieces of cardboard represent amino acid molecules.

Arrange the pieces of cardboard that represent the amino acids to fit the template as shown in Figure 35.6(*c*). When the pieces are in place, join them together with the pieces of Sellotape. When you have finished joining this up, you will be able to lift the chain of cardboard pieces away from the cardboard template, as shown in Figure 35.6(*d*).

The chain of pieces of cardboard now represents a protein. In your investigation, the protein model only contains ten amino acid models, but, in practice, a protein would contain a hundred or more amino acids. Further, we have allowed only four possible amino acids in our models, instead of the twenty-four found in real proteins.

Compare the model of a protein that you have made with others in the class, to see if anyone else has made the same model that you have made.

You will see that the template (which is known as a nucleic acid) is quite specific, and will only allow one type of protein to be made.

The system is rather like a lock and key – only a key of the right pattern will fit the lock.

35.6. Fats

So far, in this chapter, we have studied carbohydrates (foods that liberate energy) and proteins (body-building foods). We will now study a third form of food, **fats**.

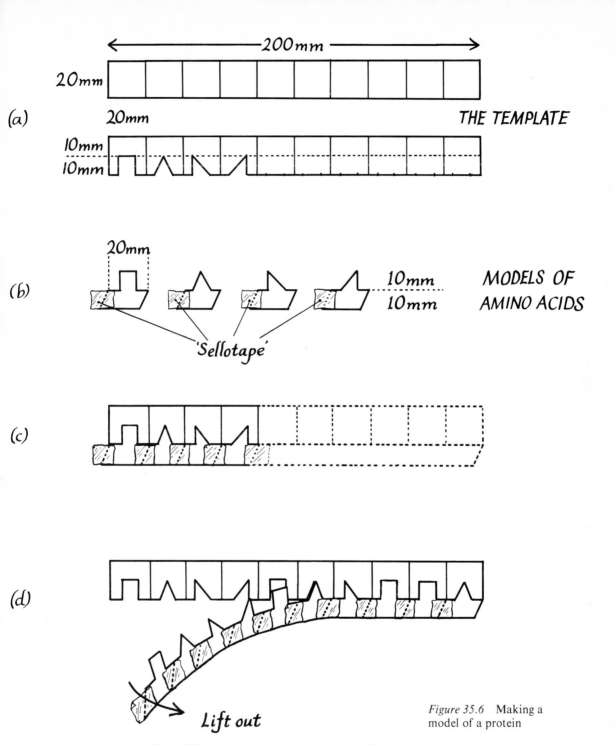

20mm
200mm
THE TEMPLATE

20mm
10mm
10mm

20mm

(b)

10mm
10mm
MODELS OF
AMINO ACIDS

'Sellotape'

(c)

(d)

Lift out

Figure 35.6 Making a
model of a protein

Fats provide our bodies with a reserve energy store. Unlike the
carbohydrates, that are broken down rapidly to provide energy,
the fats are only partially broken down at first. This first breakdown of fats yields an alkanol and some acids. The alkanol is
propane-1,2,3-triol, commonly known as **glycerine**.

A fat is a solid at room temperature; oils are liquid at room temperature. You will realize that there is little difference between oils and fats except for this. Let us see what is present in a typical fat and a typical oil, and see if we can find out why one is a liquid and one is a solid.

Propane-1,2,3-triyl trioctadec-9-enoate is a typical oil. It is a colourless liquid which freezes at 267 K (-6 °C). **Propane-1,2,3-triyl trioctadecanoate** is a typical fat; it is colourless and melts at 344 K (71 °C). Figure 35.7 shows the arrangement of atoms in these

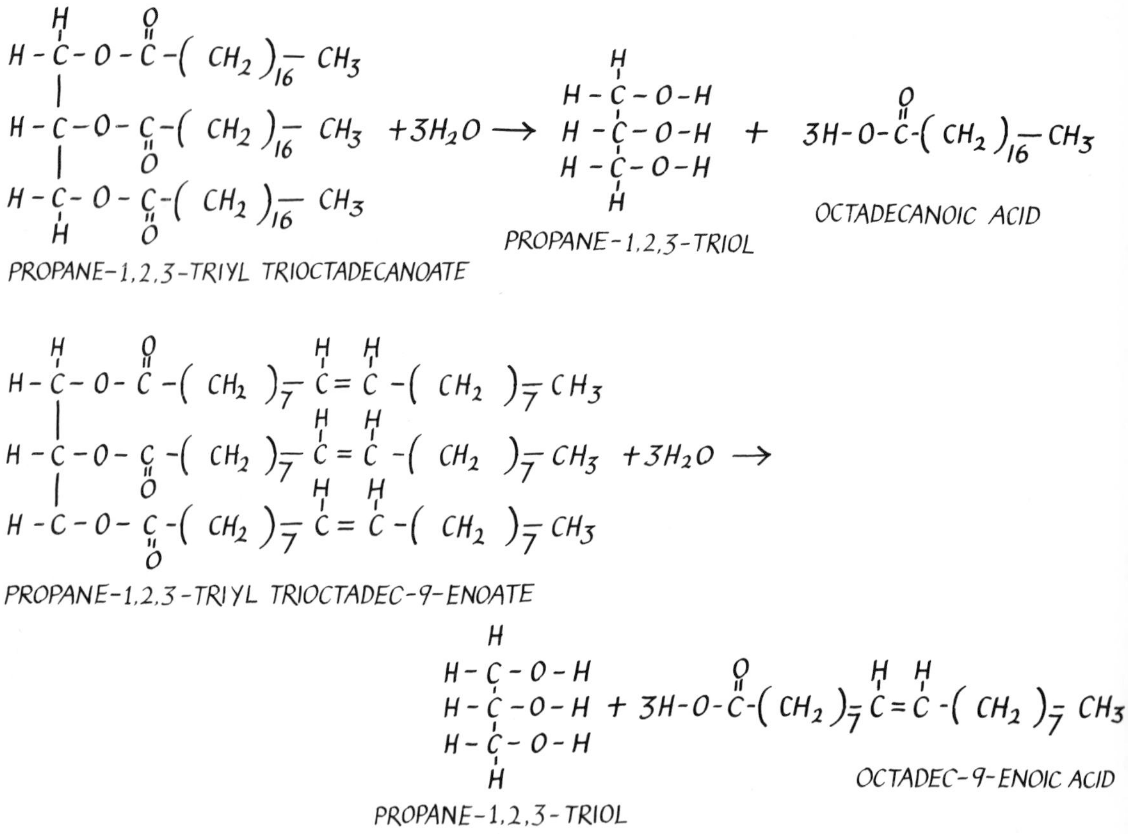

$$
\begin{array}{l}
H-\overset{\overset{H}{|}}{C}-O-\overset{\overset{O}{\|}}{C}-(CH_2)_{\overline{16}}\,CH_3 \\[2mm]
H-\overset{\overset{H}{|}}{C}-O-\underset{\underset{O}{\|}}{C}-(CH_2)_{\overline{16}}\,CH_3 \quad +3H_2O \longrightarrow \\[2mm]
H-\underset{\underset{H}{|}}{\overset{\overset{H}{|}}{C}}-O-\underset{\underset{O}{\|}}{C}-(CH_2)_{\overline{16}}\,CH_3
\end{array}
$$

$$
\begin{array}{l}
H-\overset{\overset{H}{|}}{C}-O-H \\[2mm]
H-\overset{\overset{H}{|}}{C}-O-H \quad + \quad 3H-O-\overset{\overset{O}{\|}}{C}-(CH_2)_{\overline{16}}\,CH_3 \\[2mm]
H-\underset{\underset{H}{|}}{\overset{\overset{H}{|}}{C}}-O-H
\end{array}
$$

PROPANE–1,2,3–TRIYL TRIOCTADECANOATE

PROPANE–1,2,3–TRIOL   OCTADECANOIC ACID

$$
\begin{array}{l}
H-\overset{\overset{H}{|}}{C}-O-\overset{\overset{O}{\|}}{C}-(CH_2)_{\overline{7}}\,\overset{\overset{H}{|}}{C}=\overset{\overset{H}{|}}{C}-(CH_2)_{\overline{7}}\,CH_3 \\[2mm]
H-\overset{\overset{H}{|}}{C}-O-\underset{\underset{O}{\|}}{C}-(CH_2)_{\overline{7}}\,\overset{\overset{H}{|}}{C}=\overset{\overset{H}{|}}{C}-(CH_2)_{\overline{7}}\,CH_3 \quad +3H_2O \longrightarrow \\[2mm]
H-\overset{\overset{H}{|}}{C}-O-\underset{\underset{O}{\|}}{C}-(CH_2)_{\overline{7}}\,\overset{\overset{H}{|}}{C}=\overset{\overset{H}{|}}{C}-(CH_2)_{\overline{7}}\,CH_3
\end{array}
$$

PROPANE–1,2,3–TRIYL TRIOCTADEC–9–ENOATE

$$
\begin{array}{l}
H-\overset{\overset{H}{|}}{C}-O-H \\[2mm]
H-\overset{\overset{H}{|}}{C}-O-H \quad + \quad 3H-O-\overset{\overset{O}{\|}}{C}-(CH_2)_{\overline{7}}\,\overset{\overset{H}{|}}{C}=\overset{\overset{H}{|}}{C}-(CH_2)_{\overline{7}}\,CH_3 \\[2mm]
H-\underset{\underset{H}{|}}{\overset{\overset{H}{|}}{C}}-O-H
\end{array}
$$

PROPANE–1,2,3–TRIOL   OCTADEC–9–ENOIC ACID

Figure 35.7 The hydrolysis of a fat and an oil

compounds and what happens to them when they are hydrolysed (that is when they react with water). Look carefully at this figure and see if you can see the difference between the two substances and the products of hydrolysis.

You will see that propane-1,2,3-triol has three alkanol ($-$OH) groups, so one molecule of oil or fat gives three molecules of acid. Notice that the acids are all similar to ethanoic acid, that is they contain the $-$COOH group (acids with this group are sometimes called 'fatty' acids).

Have you seen that the only difference between the two acids is that one acid has two more hydrogen atoms than the other? Because

388

of this, two of the carbon atoms in octadec-9-enoic acid are bonded together by a '*double bond*' (you have already met this type of bond in ethene). (Hydrocarbons with a 'double bond', like ethene, are sometimes called olefins, which means oil forming.)

If the 'missing' hydrogen atoms are added to an oil, it becomes a fat. This process, which is known as **hardening**, is used to convert animal and vegetable oils into margarine. Hydrogen gas is bubbled into the oil in the presence of a specially prepared nickel catalyst. When the oil has hardened sufficiently, the nickel is removed and the fat purified. Colouring, preservative and vitamins are then added before it is packed for sale. Your teacher may demonstrate this hardening process to you.

35.7. Tests for foodstuffs

Now that we have examined some of the important points about large organic molecules, and have seen that many of them are important as foodstuffs, we will see how you can test for certain basic foods.

Investigation 35i. Testing for reducing sugar (for example glucose)

Method 1: Benedict's test

Into a test-tube place about 2 cm³ of the blue liquid called Benedict's solution. Add as much glucose as you are able to pick up on the tip of a spatula. Shake, so as to mix the contents of the tube. Now heat the tube gently in a bunsen flame. What happens to the blue colour of the solution? Does this change happen if Benedict's solution is heated alone?

★ *Do not boil these solutions.*

Corrosive

Harmful

Method 2: Fehling's test

This test is carried out in the same manner as the test described using Benedict's solution, but the liquid used is Fehling's solution. Take a test-tube and place in it about 1 cm³ of Fehling's solution A (or 1) and add to it an equal volume of Fehling's solution B (or 2). It is important that these two solutions are not mixed until just before they are to be used. Add glucose as before and heat the solution. What change do you observe?

Investigation 35j. Testing for the presence of starch

Place a little starch on a watch-glass. Add a few drops of a solution of iodine in potassium iodide solution. How does the colour of the iodine solution change? Try this test with glucose.

Does the iodine change colour in the same way when it is added to glucose?

Investigation 35k. Testing for the presence of protein

Method 1: Millon's test

Into a test-tube place 2 cm³ of the clear liquid called Millon's reagent. With a spatula, add a little flour and then shake and heat the tube gently in a low bunsen flame. How does the reagent change in colour? Repeat the test, using glucose and starch instead of flour.

Method 2: the biuret test

Harmful   Corrosive

Place a little flour in a test-tube and add 1 cm³ of 1 per cent copper(II) sulphate(VI) solution, followed by 1 cm³ of 5 per cent sodium hydroxide solution. Do not heat. Notice any colour change in the reagents. Repeat the test, using glucose and starch in place of flour. (Note: since only part of the flour is protein, the final colour of the solution will depend on the concentration of protein in the food being tested.)

Investigation 35l. Testing for a fat

Method 1: the grease-spot test

Place a little margarine on a watch-glass. *Make sure that all bunsens are out and that there are no other naked flames near your bench.* Add a few drops of ethoxyethane to the margarine and mix the two with a spatula. Pour the liquid from the watch-glass on to the centre of a clean filter paper. Allow the paper to dry in the air. *Do not put the filter paper near a flame because ethoxyethane is highly inflammable.* Hold the dry paper up to the light and compare its appearance with that of a clean unused filter paper. How has the filter paper been affected by the fat?

Method 2: the emulsion test

Using a spatula, place a little fat in the bottom of a test-tube. Add 2 cm³ of ethanol. Shake well and then pour the resulting solution into a clean tube. To the ethanol – fat solution add 2 cm³ of cold water. How do you describe the appearance of the final liquid?

Notice that the test for sugars will only work with reducing sugars (simple sugars of the glucose type); it will not give a positive result with sucrose. To test a sugar of the sucrose type, it must

TABLE 35.2. SUMMARY OF RESULTS OF TESTS ON FOODSTUFFS

Test for	Reagent	Change if Substance is Present
Reducing sugar	Benedict's solution	Colour change: blue to green/red
Reducing sugar	Fehling's solution	Colour change: blue to orange/red
Starch	Iodine solution	Colour change: orange to blue/black
Protein	Millon's reagent	Colour change: colourless to deep red
Protein	Copper(II) sulphate(VI) and sodium hydroxide	Colour change: blue to pink/violet
Fat	Ethoxyethane	Translucent (grease) ring on paper
Fat	Ethanol and water	Cloudy emulsion produced

first be hydrolysed. This may be done by adding a few drops of dilute hydrochloric acid to a sucrose solution in a test-tube and by placing the test-tube in a beaker of warm water for about five minutes. After this time, drop in a small piece of litmus paper and add dilute sodium hydroxide solution, a drop at a time, until the litmus paper just turns blue. Now you may test as for a reducing sugar (Investigation 35i).

Harmful

Test your understanding

1. What is the purpose of the ptyalin in saliva?
2. What is the name of a group of compounds to which both starch and sugar belong?
3. What is the difference between starch and cellulose?
4. A certain form of sugar is supposed to be a quick source of energy. What is this form of sugar? Why do you think that its energy is more readily available than the energy in sucrose or starch?
5. Which chemical elements are always found in protein? (Four are required.)
6. Make a list of foods that you think would be good sources of protein.
7. From what simpler substances are proteins made?
8. Why do proteins have a definite structure?
9. What are the two components obtained when a fat is hydrolysed?
10. In what way do carbohydrates and fats differ in their liberation of energy?
11. How is margarine made?
12. How would you test for a starchy substance?
13. How would you test a liquid for the presence of cane sugar (sucrose)?

Chapter 36

Plastics

36.1. What are 'plastics'?

'It's made of plastic.' This is a statement that is often made, but there is no such substance as 'plastic'. On the other hand, there are many hundreds of substances which may be classed as **plastics**.

A plastic is a substance which, at some time during the course of its manufacture, could flow and be moulded into shape – that is it was plastic.

It is important to realize that although today there are many artificially made plastics, there are also many natural plastics.

36.2. Some natural plastics

Perhaps the best known natural plastic is **rubber**. Rubber is obtained from the bark of the rubber tree, **Hevea brasiliensis**. A thick liquid, the **latex**, trickles from the bark.

When the latex is heated, it is changed into rubber, but, in this form, it is of little use. To make it really useful, it is necessary to 'mill' it and to treat it with sulphur. Rubber which has been milled, that is, pulled about and torn up, becomes much more pliable, and may be moulded into shape. The addition of sulphur, followed by heating, produces a harder product. This sulphur treatment of rubber is known as **vulcanizing.**

Rubber that has been vulcanized becomes elastic and springy. The hardness of the product can be controlled by altering the proportion of sulphur used in the process.

Another natural plastic is **shellac**. This is formed as a protective coating by the larvae of the **lac insect** in India. Shellac is used as a varnish, for which purpose it is dissolved in ethanol. When the solution is painted on to a surface, the ethanol evaporates, leaving a hard protective film of the shellac. Shellac itself may be made to flow by heating, and was at one time used in the manufacture of gramophone records.

Sealing-wax is another substance formed from naturally occurring substances.

Investigation 36a. The use of sealing-wax

Take a stick of sealing-wax and examine it closely. Try to bend it. Now, warm it, by holding it near a flame. What happens? Let it cool again.

Make a small 'blob' of the wax on a tin tray, by heating the stick and allowing a drop or two of liquid to fall off the end of the stick on to the tray. Before the wax has had time to harden properly, press a signet ring, or some other piece of metal with a design cut into it, on to the blob of wax. Press firmly. When the wax is cool, examine it. You should find that the design from the metal has been transferred to the wax. Why is a *metal* usually used for making such impressions in wax? The impression produced in the wax is, of course, a 'negative' or reverse of the design in the metal.

36.3. The classification of plastics

You will find that it is possible to re-soften and re-harden sealing-wax, just as often as you heat it and cool it again. A plastic substance that behaves in this way is called a **thermoplastic**. Many of our artificial plastics are of this type.

However, not all plastics behave like this. There are many which, once they have been plastic under the action of heat, set permanently on cooling into a solid form which cannot be re-melted. Heating merely brings about the complete chemical decomposition of the substance. Such plastics are known as **thermosetting plastics**.

Investigation 36b. Examining some plastics

Take an iron sand-tray and put it on a tripod. Arrange a bunsen, with a small flame, so that the tray is heated near the edge. Now, away from that part of the edge, place on the tray small pieces of different plastic materials, for example, some polythene sheeting, a piece of the transparent casing of an old ball-point pen, a piece of Perspex, a piece of Bakelite, some nylon, etc.

Watch what happens to these materials as the heat reaches them by conduction through the tray. Which are thermoplastic? Which are thermosetting?

★ Warning – take great care not to inhale any of the fumes that may be formed from the heated plastics.

Toxic

36.4. The nature of plastics

When the gas ethene (C_2H_4) is heated to about 575 K (about 300 °C) and is subjected to a *very* high pressure, a wax-like solid is formed. This product is known as poly(ethene). It is a plastic of the thermoplastic type.

In the ethene molecule, the two carbon atoms are linked by

393

what is known as a 'double bond'. Figure 36.1 shows the arrangement of the atoms in an ethene molecule.

The double bond is easily 'opened-up'; this results in each of the carbon atoms having a 'spare' valency by which it can join on to other atoms. These other atoms may well be the carbon atoms of other 'opened-up' ethene molecules, and a whole long chain of such units can be formed (see Figure 36.2).

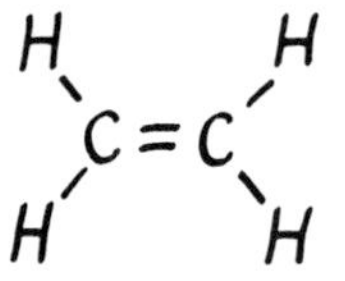

Figure 36.1 The ethene molecule

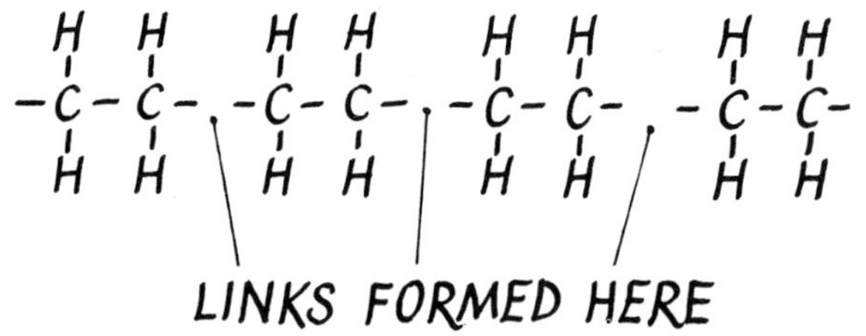

$$\begin{array}{c} H \\ \diagdown \\ \end{array} C = C \begin{array}{c} H \\ \diagup \\ \end{array}$$

Figure 36.2 How poly-(ethene) is formed

LINKS FORMED HERE

This process of the joining together of many identical units is known as **polymerization**. The original material, with the small separate molecules, is known as the **monomer**, while the final product, with the long chain, is known as the **polymer**. The word 'mono' means one; the word 'poly' means many. A chain may consist of many hundreds of monomer units.

The actual length of a particular chain will depend on how many units have joined together before a 'blocking unit' joins on to each end of the chain. In the case of poly(ethene), or **polythene** as it is more usually known, the blocking unit could be a hydrogen atom or a methyl group ($-CH_3$), formed by the breaking up of an ethene molecule under the conditions of the reaction. Figure 36.3 illustrates the idea of blocking the chain end.

Figure 36.3 Blocking the end of a chain

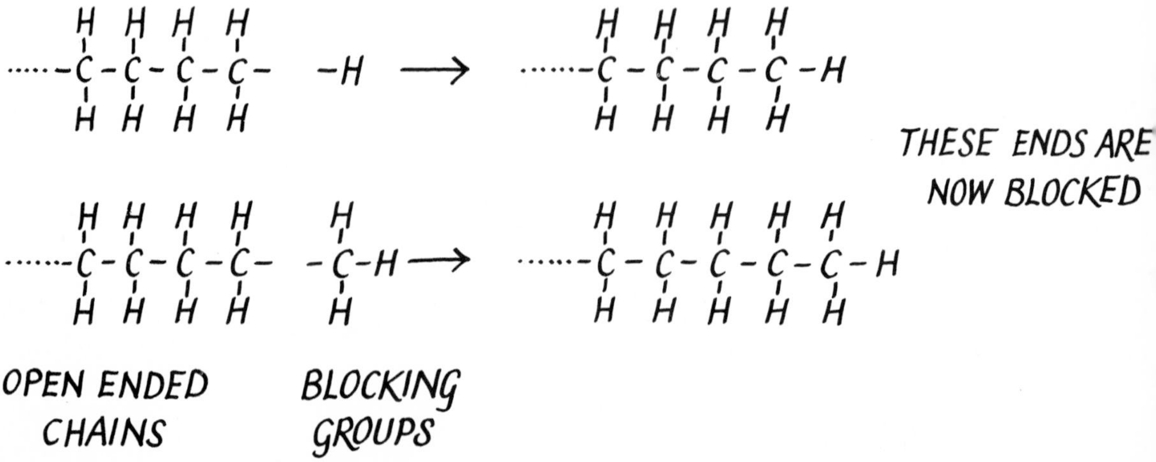

THESE ENDS ARE NOW BLOCKED

OPEN ENDED CHAINS   BLOCKING GROUPS

In general, all polymerization processes involve three clearly defined stages:

  a. **Initiation**. Here, the chain growth is started – often by means of another substance, which acts as a catalyst.
  b. **Propagation**. During this stage, the chain grows in length.
  c. **Termination**. Here, the chain growth is stopped by the addition of blocking units to the chain ends.

36.5. The cross-linking of polymer chains

In the polymer, the chains of atoms may be separate, in which case the chains are relatively free to move and to slide over each other, or the chains may be joined to each other by **cross-linking**, in which case the relative movement of the chains is greatly restricted.

Cross-linking can take place in many ways. For example, in the case of rubber, the sulphur which is added in the vulcanization process forms the links. In the case of poly(ethene), a link can be formed when two chains each lose a hydrogen atom and the two carbon atoms thus left with an unused valency join on to each other (see Figure 36.4).

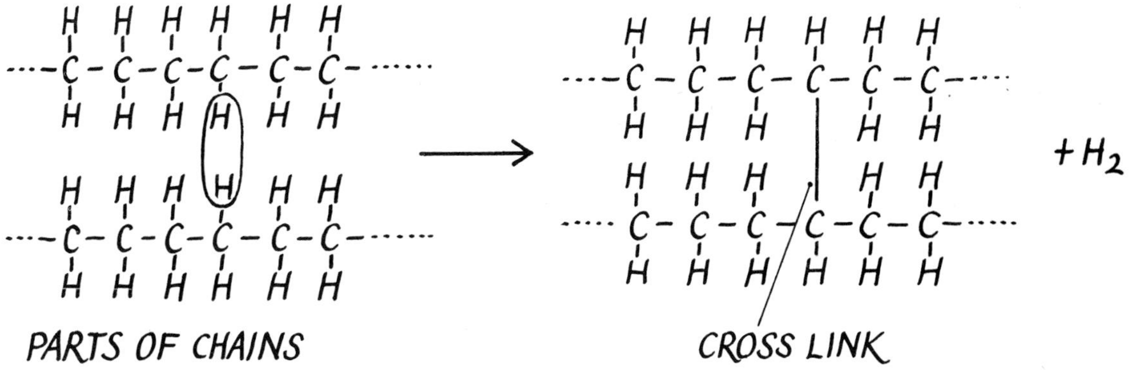

PARTS OF CHAINS   CROSS LINK

Figure 36.4   Cross-linking

Polymers in which there is little cross-linking are usually thermoplastic. With greater cross-linking, the product is often thermosetting and is harder, less pliable, and sometimes more brittle.

36.6. Polymerization and co-polymerization

In general, the process of polymerization can be represented as shown in Figure 36.5.

$$nA \longrightarrow (A)n$$

$$\text{MONOMER} \qquad \text{POLYMER}$$

$$\text{OR } A+A+A+A \cdots \longrightarrow A-A-A-A \cdots$$

Figure 36.5 Polymerization

Many polymers are made, however, by the joining together of *two different units* into chains. For example, when chloroethene, commonly known as vinyl chloride (see Section 36.9), and ethenyl ethanoate, commonly known as vinyl acetate, react together, the resulting polymer has both units in its chain. This may be represented as shown in Figure 36.6, and the process is known as **co-polymerization**. In some co-polymers, the units alternate down the chain, while in others, the order of the units in the chain is random.

Figure 36.6 Co-polymerization

$$nA + nB \longrightarrow (AB)_n$$

$$\text{OR} \quad A+A+A \cdots B+B+B \cdots \longrightarrow A-B-A-B-A-B \cdots$$

$$\text{OR} \quad A+A+A \cdots B+B+B \cdots \longrightarrow A-A-B-A-B-B \cdots$$

36.7. Condensation polymerization

In this form of polymerization, two different monomers are used, and a product rather like a co-polymer in structure is formed. In the process of the monomer units joining together, water, or some other substance with a small molecule, is eliminated. This may be represented as in Figure 36.7.

Specific examples of condensation polymers are referred to in Section 36.11.

$$nX + nY \rightarrow (XY)_n + nZ$$

$$X + X + X \cdots + Y + Y + Y \cdots \longrightarrow \overset{\uparrow Z}{X} - \overset{}{Y} - \overset{\uparrow Z}{X} - \overset{}{Y} - \overset{\uparrow Z}{X} - Y \cdots$$

Figure 36.7 Condensation polymerization

36.8. The advantages of synthetic plastics

Many synthetic plastic materials possess properties which are not often found in naturally occurring substances. Poly(ethene), for example, is light, strong, flexible, does not corrode, and is not attacked by animals, insects, fungi or bacteria. It is waterproof and is an excellent heat insulator, as well as being an excellent electrical insulator. Many other plastics possess similar valuable properties.

One of the most useful properties of plastics is the ease with which they may be shaped into the required form while they are in the liquid state. One common way of doing this is by **injection moulding**. The powdered monomer is heated until it becomes a liquid and is then forced under pressure into a cooled mould.

The change to the polymerized form takes place while the heat and pressure are being applied, and the final result is the complete article, which only requires finishing and removal of the 'stalk' or sprue. Alternatively, if a thermoplastic material is being used, it may be re-melted and put into the mould. Figure 36.8 shows the injection moulding process.

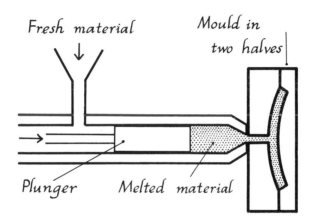

Figure 36.8 Injection moulding

In a very few cases, pure plastics are used to make the final products, but more often the properties are improved by adding plasticizers, fillers, dyes or pigments.

Plasticizers are substances which are added to help the product to form more easily, increase its flexibility and make it more resistant to moisture and chemical change.

Fillers are added to make the product cheaper, as well as to improve its physical properties. For this purpose, asbestos, cotton fibres, cork and wood powders are often used.

Dyes and **pigments** are used to give colour. The colour goes right through the finished article and is not just a surface cover.

36.9. Some important synthetic plastics

Celluloid

Although nearly all synthetic plastics are of relatively recent discovery, celluloid has been known for over a century. Cellulose, the substance which makes up the cell walls of plants, can be dissolved in a mixture of nitric and sulphuric acids to form cellulose nitrate. This is a highly inflammable material (sometimes called nitrocellulose or gun-cotton). By treating nitrocellulose with camphor and a little ethanol, and rolling the resulting dough-like mixture, sheets of material are formed. These are heated, and celluloid results. Despite its highly inflammable nature, celluloid is still made on a large scale because of its cheapness.

Cellulose acetate (correctly called cellulose ethanoate)

This product, which, unlike celluloid, is non-inflammable, is made by heating cotton fibres or wood pulp with a mixture of ethanoic acid and ethanoic anhydride, in the presence of a catalyst, such as a little concentrated sulphuric(VI) acid. The resulting product is known as **acetate rayon**.

Bakelite

This material, which is thermosetting, was first introduced in 1909 by a Belgian chemist, Dr. Baekeland, from whose name it takes its own. It is a condensation polymer (see Section 36.7), made from phenol and methanal. It is very hard, but rather brittle. It has a very high electrical resistance and is used as an insulator in electrical components.

Poly(ethene)

Originally, this was made by heating ethene gas under enormous pressures. Today, it is made much more easily (and more cheaply) by using catalysts known, after their discoverers, as **Ziegler-Natta catalysts**. Using these catalysts, the product is formed at much lower pressures. It is much more rigid, it is opaque, and it has longer chains than the original form. Poly(ethene) may also be made more resistant to the effects of temperature by treating it with the radiation from radioactive substances or from an atomic pile (see Sections 37.4 and 38.6). This **irradiation** causes more cross-links to be formed.

Poly(ethene) has many uses, and you will have come across many examples in your home. Buckets, washing-bowls, squeeze-bottles, toys and wrapping films are but a few of its uses. It is even used as a synthetic fibre (see Section 36.10), being sold in this country under the name of **Courlene**. Another use is as an electrical insulator, and it is also coming into general use for water-piping where only cold water is carried. It softens at too low a temperature for use with the hot water in a domestic hot-water system.

Poly(propene), commonly called polypropylene

It is made by the polymerization of propene gas. Propene is very similar to ethene and has the formula C_3H_6. Like ethene, it has

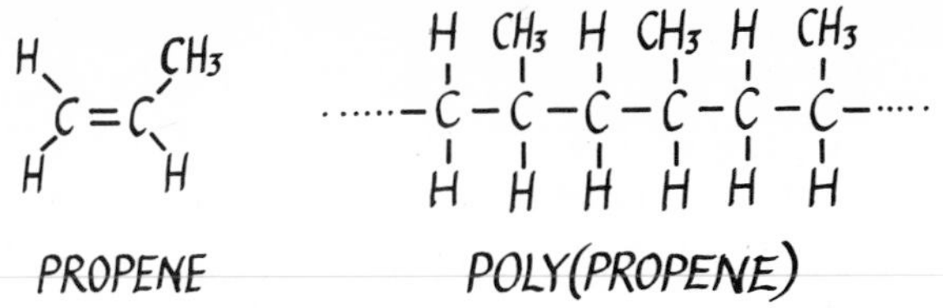

PROPENE   POLY(PROPENE)

Figure 36.9 The poly-
merization of propene

a double link between two of the carbon atoms in its molecule.
Figure 36.9 illustrates the polymerization of propene.

The discovery of the Ziegler-Natta catalysts made the commercial production of polypropylene possible. This polymer is able
to withstand a higher temperature than polythene without softening. Boiling water will not deform it. There is a strong possibility
that it may find a use in domestic hot-water systems.

Poly(chloroethene) P.V.C.

There are many different plastics based on compounds containing the group of atoms $H_2C = CH-$. This, you will see, is
ethene without one of its four hydrogen atoms. The group is called
the **ethenyl group.** They will polymerize in exactly the same way
as ethene. The polymerization of chloroethene is indicated in Figure
36.10.

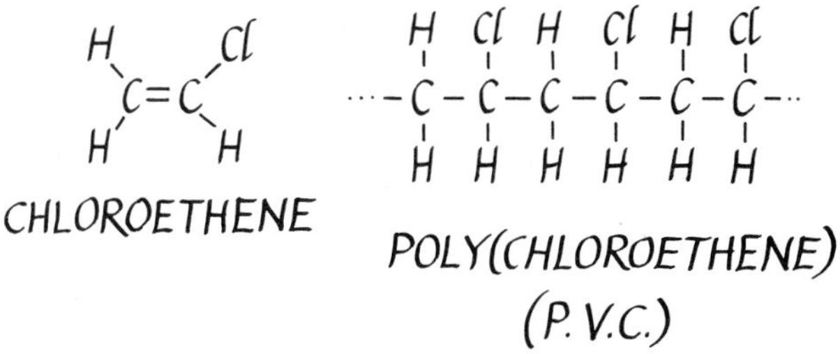

CHLOROETHENE   POLY(CHLOROETHENE)
(P.V.C.)

Figure 36.10 Ethenyl
chloride and P.V.C.

P.V.C. is particularly useful as a substitute for leather and
waterproof fabrics, and is used for upholstery, suitcases, raincoats,
water-piping and guttering. It is also used for flooring (where it
replaces linoleum), and is the material from which gramophone
discs are pressed. It is the insulating material on many insulated
wires. It is hard, but flexible, and has excellent wear resistance.

Poly(phenylethene), commonly called polystyrene

This material will probably be familiar to you as the substance
used for the 'kits' of parts from which plastic models are made.
Polystyrene will dissolve in suitable solvents, and pieces may be
joined together so strongly that the joint is at least as strong as
the original pieces. Like chloroethene, styrene is closely related to
ethene. Styrene, and the formation of polystyrene, are illustrated
in Figure 36.11. In this figure, the hexagonal ring represents the
phenyl group of atoms, C_6H_5-.

Polystyrene may also be obtained as a foam. Here, the plastic is formed from the liquid state at the same time that gas bubbles

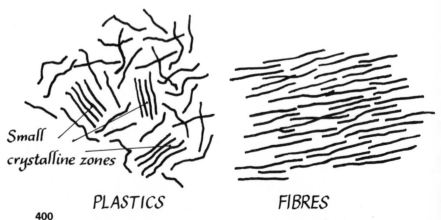

Figure 36.11 Styrene and polystyrene

STYRENE   POLYSTYRENE

(carbon dioxide) are formed, and the bubbles are trapped to give a solid foam. The resulting plastic is a particularly light material which is an excellent heat insulator. Polystyrene foam tiles are used to cover ceilings and walls.

Perspex

This is a trade-name for a very clear, tough material, which is very resistant to air and water. As a result, it finds application in windscreens and cockpit covers for aircraft, and in many optical arrangements. One disadvantage is that it is less resistant to scratching than glass.

36.10. Natural and synthetic fibres

Cotton, wool and silk are examples of naturally occurring fibres, and they have been used by man for many centuries. The search for *synthetic* fibres started about one hundred years ago, but it is only since the Second World War that synthetic fibres have been used on the large scale.

There is a close connection between fibres and plastics. As we saw in Section 36.9, celluloid is made from cellulose, which is also the chief part of cotton. Both fibres and plastics have chain-like molecules. The difference lies in the arrangement of those chains. In the plastics, the chains are in a more or less unordered arrangement, while in the fibres, the chains are lying more or less parallel to each other (see Figure 36.12).

Figure 36.12 Chain arrangements in the plastics and the fibres

Small crystalline zones

PLASTICS   FIBRES

Today, a whole range of synthetic fibres is available, possessing properties which the natural fibres do not have. Apart from silk, the natural fibres are rather short in total length.

36.11. Some synthetic fibres

An early attempt was made to make a synthetic fibre by dissolving cellulose nitrate (see Section 36.9) in a mixture of ethanol and ethoxyethane, and by squirting the solution through very fine holes into heated air so that the solvents evaporated and left threads of the material. The result was a silk-like thread, but it was highly inflammable. Today, non-inflammable fibres are made from cellulose ethanoate (see Section 36.9).

Rayon

This is not strictly a synthetic fibre. It is a general name for all artificial fibres which are made from celluloses and some other natural substances by treating them chemically and then re-forming them. As a result of the treatment, their properties are greatly improved. **Tricel** is the trade-name for a rayon made from cellulose via the ethanoate.

Nylon (a polyamide)

There are a number of different sorts of nylon. The most usual type is known as **Nylon 66**, which is a condensation polymer made from 1,6-diaminohexane and 1,6-hexanedioic acid. Each of these monomers has quite a long molecule, so the polymer has a very long molecule. Nylon is a thermoplastic material. Its formation is illustrated in Figure 36.13. Note that a **peptide link** can be formed at each end of both molecules.

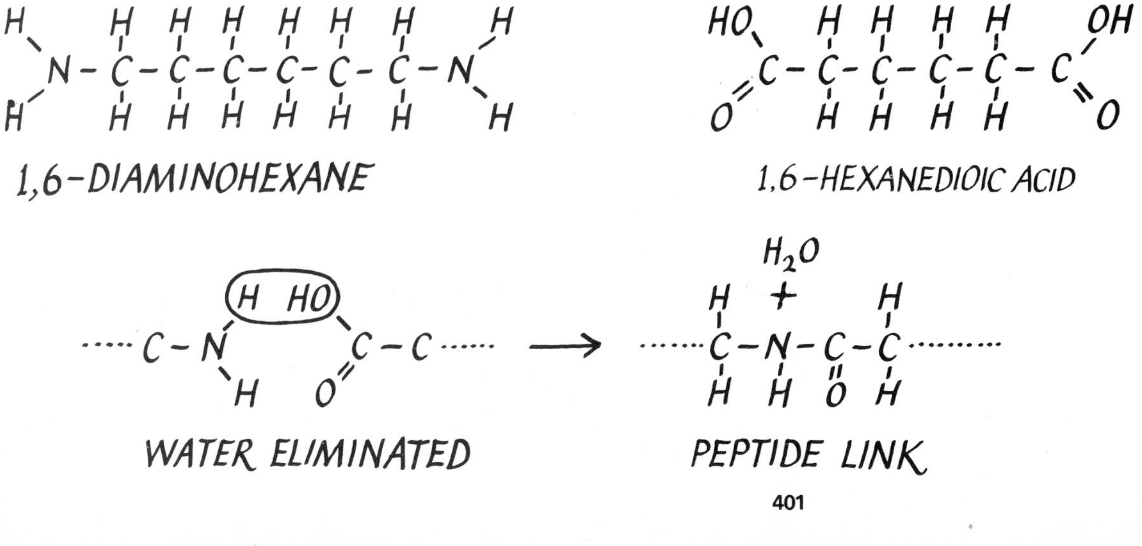

Figure 36.13 The formation of Nylon 66

1,6–DIAMINOHEXANE

1,6–HEXANEDIOIC ACID

WATER ELIMINATED

PEPTIDE LINK

Nylon 66 melts at about 530 K (about 260 °C). Filaments are formed by squirting the liquid polymer through very fine holes, known as **spinnerets**, and the solid is re-formed as it cools. The properties of these nylon filaments are greatly improved by pulling them out to about four times their original length. This 'lines-up' the polymer molecules and increases the strength of the filaments. The filaments may then be twisted into threads. **Monofilaments**, or single filaments of varying thickness, are also used, for example, in the manufacture of nylon stockings. The filaments are described as being '15 denier', '30 denier', etc. The denier is the mass in grams of 9 000 metres of filament!

Nylon has excellent water-resistance and elasticity. Apart from its use in clothing, it is used to make ropes and hawsers, tennis racquet strings and surgical sutures, that is the 'threads' which the surgeon uses for stitching up the incisions he makes. Thicker monofilaments are used as bristles for brushes of all types.

Nylon is also used (not as a fibre) for solid mouldings. It has a very low frictional resistance, and is made into gear wheels and bearings. For solid moulding, it is melted by heating and poured into heated moulds.

Investigation 36c. To make nylon

★ *Warning – this investigation must be done by the teacher. The substances used are poisonous and should not be inhaled or allowed to come into contact with the skin.*

Toxic

Corrosive

Prepare a solution of either 1,6-hexanedioyl dichloride or 1,10-decanedioyl dichloride by dissolving 1.5 cm³ in 100 cm³ of tetrachloromethane and put some of the solution into the bottom of a narrow beaker (50 cm³ if available). Prepare also a solution of 1,6-diaminohexane by dissolving 1.2 g in 100 cm³ of water containing 0.8 g of sodium hydroxide.

Pour some of this second solution *very carefully* on to the surface of the first solution in the beaker. At the interface between the two liquids (which do not mix), a vigorous reaction occurs, and nylon is formed, with the elimination of hydrogen chloride gas, which fumes in the moist air.

The solid polymer may be picked up from the interface with a pair of tongs and fed over a roller fixed above the beaker. A continuous tube of nylon may then be wound up on to the roller (see Figure 36.14).

Terylene (a polyester)

This, again, is a trade-name for a most useful synthetic fibre. It is rather like nylon in its general properties, but fabrics made from it may be given permanent creases, by heating well above any normal temperature of use, while holding the material in the required shape. At lower temperatures, the material is very crease-resistant. A mixture of 55 per cent terylene and 45 per cent wool combines the better characteristics of both materials.

402

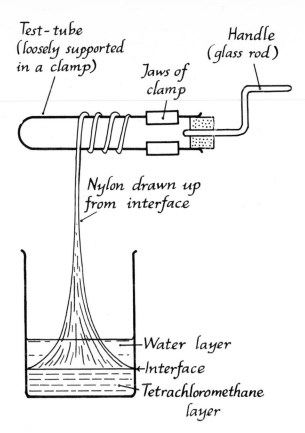

Figure 36.14 Making nylon in the laboratory

36.12. The silicone plastics

These remarkable materials, which are based not on carbon but on the element silicon, have been dealt with in detail in Sections 32.7 and 32.8 and Investigations 32g, 32h and 32i.

Test your understanding

1. Many different substances are shown as 'plastics'. What is the meaning of this term?
2. Name, and describe the sources of, THREE naturally occurring plastics.
3. How is natural rubber hardened? What is the hardening process called?
4. What is the difference between (a) a thermoplastic material, (b) a thermosetting material? Give ONE example of each type of substance.
5. What is the formula of ethene gas? Show, by means of diagrams, how ethene can polymerize.
6. What is the difference between an 'addition' polymer and a 'condensation' polymer?
7. Explain the following terms: (a) co-polymer, (b) cross-linking, (c) blocking group.
8. Describe some of the properties of poly(ethene) which make it such a useful material.

9. What are (a) 'fillers', (b) 'plasticizers'? Why are they used in the plastics industry?
10. What is 'celluloid'? How is it made? Why is its use banned in many countries?
11. One of the earliest polymers was Bakelite and it is still used today. From what substances is it made? What particular properties does it have that make it useful?
12. Poly(chloroethene), or P.V.C. as it is more often called, is used for many different purposes. Describe FOUR of them.
13. Outline the chief uses of (a) polystyrene, (b) polystyrene foam. How is the latter made?
14. Describe the molecular structure of a fibre and compare it with that of a plastic.
15. Describe some of the useful properties of a nylon fibre.
16. What special property of terylene fibres makes them of particular value to the clothing industry?
17. What are (a) Tricel, (b) Courlene, (c) rayon?
18. What is a 'peptide link'? How is it formed?

Chapter 37

Radioactivity

37.1. Stable and unstable nuclei

In Section 3.3, a simple picture of the structure of atoms was given. Atoms were said to be very small indeed, having a diameter of about 10^{-7} mm. Most of the mass of the atom is concentrated in the nucleus. The nucleus has a diameter of about 10^{-11} mm.

The nucleus is composed of protons and neutrons. These particles have approximately the same mass – the neutron is very slightly greater in mass – but, while the proton carries a positive charge of electricity, the neutron has no charge.

Surrounding the nucleus are the electrons. These are of much smaller mass (approximately 1/1 840 of a proton unit), but carry an electric charge, negative this time, exactly equal to the positive charge on the proton.

The particles in the nucleus are held together by extremely strong forces, the nature of which is not at present really understood, and there is no simple everyday comparison that may be used to explain them. They are about one hundred times stronger than the forces of electrical attraction and repulsion that we are used to. For want of a better term, they are often referred to as **short-range forces**. These forces only operate over very short distances, such as the distances inside atomic nuclei. Indeed, the particles in atomic nuclei are so close together that the density of the nucleus is about 10^{13} g cm^{-3}. A cubic centimetre of nuclear matter would weigh some 10 000 000 tonnes!

In order that a nucleus may be stable, there must be a balance between the numbers of protons and neutrons in it. In the case of the lighter nuclei, the numbers of protons and neutrons need to be roughly equal. As the mass of the nucleus increases, the relative number of neutrons becomes larger. In uranium, the 92 protons need about 138 neutrons for reasonable stability.

37.2. The magic numbers

Certain numbers, either of protons or of neutrons, give great stability to the nuclei in which they occur. This is even more true if the number of *both* sorts of particle in the nucleus is one of these special numbers. The special numbers are known as the **magic numbers**.

TABLE 37.1. THE MAGIC NUMBERS

Number	Element
2	Helium
8	Oxygen
20	Calcium
28	Nickel
50	Tin
82	Lead
126	—

The nucleus of helium, with two protons and two neutrons, is the most stable of all the nuclei. Other particularly stable nuclei are oxygen (with eight protons and eight neutrons), and calcium (with twenty protons and twenty neutrons).

Natural radioactive changes result in a lead nucleus as the final product. No element with an atomic number greater than that of lead (82) has a nucleus which is completely stable.

37.3. Natural radioactivity

The phenomenon of radioactivity was discovered in 1896 by Henri Becquerel. The discovery was due to a chance happening. By accident, Becquerel had left a sample of a **uranium** salt in a drawer which also contained some unused, wrapped photographic plates. Again, by accident, these plates were later developed. To Becquerel's surprise, the plates were marked exactly as if they had been exposed to light. Having a scientific training, Becquerel did not leave the matter there, but investigated it further, first repeating the experiment of putting the uranium salt near wrapped photographic plates. He found that they were again marked as if exposed.

Becquerel realized that something from the uranium salt had passed through the black wrapping paper and had affected the photographic emulsion.

It was soon discovered that the same sort of effect was produced by compounds of the element **thorium**.

In 1899, Rutherford (later Lord Rutherford), then a professor at Montreal University, showed that the **radiation**, as it was called, consisted of two parts, a less penetrating sort, which he called **alpha** (α), and a more penetrating sort, which he called **beta** (β). It was shown soon afterwards that there was a third sort of radiation, **gamma** (γ), which was still more penetrating.

37.4. The nature of radioactivity

In 1902, Rutherford and Soddy suggested that radioactivity is due to the nuclei of heavier atoms breaking up, leaving lighter atoms, and at the same time giving out electrically charged particles. This, in general, is our present-day picture of what happens.

Alpha radiation

This consists of relatively heavy particles which were shown by Rutherford to be the nuclei of helium atoms.

If a nucleus gives out an alpha particle, the remaining nucleus is lighter by four units of mass and also has an atomic number which is less by two units than that of the original nucleus. Thus a new element has been formed in the process. The change from one element to another is called **transmutation**. Examples of this sort of radioactivity are:

$$^{238}_{92}U \rightarrow {}^{234}_{90}Th + {}^{4}_{2}He$$

$$^{226}_{88}Ra \rightarrow {}^{222}_{86}Rn + {}^{4}_{2}He$$

The alpha particles are very easily stopped. A thin piece of paper will stop them completely, as will some 20 mm of air. Being helium nuclei, each alpha particle carries a double positive charge of electricity, and they are deflected by strong magnetic and electric fields.

Beta radiation

This radiation also consists of particles. The particles are electrons, and may be represented by the symbol $_{-1}^{0}e$. Their mass is negligible compared with that of the nucleus. Examples of beta radioactivity are:

$$^{234}_{90}Th \rightarrow {}^{234}_{91}Pa + {}_{-1}^{0}e$$

$$^{212}_{83}Bi \rightarrow {}^{212}_{84}Po + {}_{-1}^{0}e$$

It will be noticed that, in the case of beta radiation, the new element formed has one more proton, and hence one less neutron, than the parent element. The mass of the electron emitted is so small that it may be neglected in the mass balance.

Beta particles have a much greater penetrating power than alpha particles. They are ejected from nuclei with speeds approaching nine-tenths of the speed of light. They can easily penetrate paper and thin sheets of aluminium. They are stopped by thick sheets of aluminium and by thin sheets of lead. They are very easily deflected by magnetic and electric fields.

Gamma radiation

In this type of radiation, it is not particles which are given off, but electromagnetic waves of similar nature to light waves, but of much shorter wavelength. This radiation is very penetrating and can pass through thick lead sheets. In fact, gamma radiation is never completely stopped, although its intensity can be reduced more and more, according to the thickness and nature of the material in its path.

Gamma radiation is emitted when the particles in the nucleus of an atom re-arrange themselves, often at the time of emission of an alpha particle or a beta particle, to take up a more stable structure. When they do this, energy is lost, and this appears in the form of a **gamma ray**. Magnetic and electric fields have no effect on gamma radiation.

Figure 37.1 illustrates the effect of an electric field on the three types of radiation

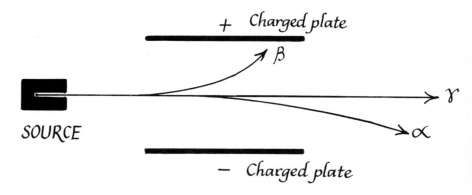

Figure 37.1 The effect of an electric field on the three types of radiation

37.5. How radioactivity is detected

Becquerel's discovery, that radioactivity will affect a photographic plate, gives us one method of detecting it. The use of photographic plates for this purpose is common practice today.

However, the radiations produce other effects which may be used to show their presence. One of these is that they cause **ionization** in the substances through which they pass – that is they remove electrons from atoms in their path. A number of detecting devices depend on this property.

   a. An **electroscope**, in its simplest form, consists of a plate of metal to which is attached a very thin piece of metal foil. The plate and leaf are fixed inside a metal container, but are very carefully insulated from it. The container has transparent sides so that the plate and leaf can be observed. The arrangement is shown in Figure 37.2.

     If an electric charge is placed on the plate and leaf, by rubbing a polythene rod on some cloth, for example, and then allowing the rod to touch the metal plate, the leaf will be repelled by the plate and will stand out from it (see Figure 37.3).

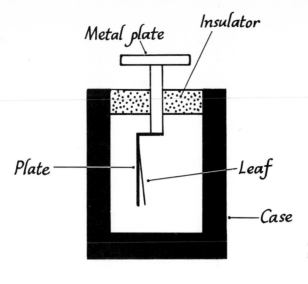

Figure 37.2　The electro-
scope

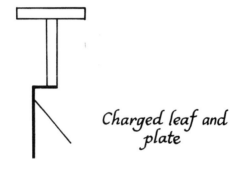

Charged leaf and
plate

Figure 37.3　A charged
electroscope

So long as the charge cannot leak away, the leaf will remain at an angle to the plate. If the charge leaks away, then the leaf will fall back towards the plate. The air in the case is normally a good insulator, but, if it becomes ionized, then it will conduct the charge away to the outside and the leaf will fall. Thus, if a radioactive substance is put in the case (or near it), its presence will be detected by the leaf falling.

b. **A Geiger–Müller tube** consists of a metal tube, down the centre of which, but insulated from it, is a wire. The end of the tube is closed by a thin 'window' of mica. The tube contains a mixture of gases and vapours at a pressure well below that of the atmosphere (see Figure 37.4).

A high voltage (400 V) is applied across the casing of the tube and the wire. Normally, the gas in the tube acts as an insulator, but if it becomes ionized, a current can flow momentarily between the wire and the tube. This 'pulse' of

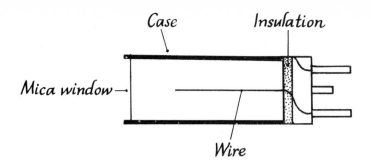

Figure 37.4 A Geiger–
Müller tube

current can be amplified and detected by a suitable electronic circuit.

Each particle (or ray) entering the tube may be separately detected, provided that they do not follow each other *too* closely. The total number may be shown on a **counter**, or the rate at which they are arriving may be given by a **ratemeter**.

c. A **cloud-chamber** is a very useful piece of apparatus which was invented by C. T. R. Wilson.

If a gas is saturated with vapour, and the pressure on it is suddenly lowered, it will be in a condition suitable for forming cloud. Cloud can be formed only if there are suitable 'centres' on which liquid droplets can form. Ions are very suitable for this purpose. If, at the moment when an ionizing radiation is passing through a saturated vapour, the pressure is suddenly lowered, a 'line' of cloud will fall along the track of the particle. It is possible to make the ionizing particle itself 'trigger off' the pressure reduction.

Another property possessed by radiation is that it can cause light to be given out when it strikes certain chemical substances, for example, zinc sulphide. This phenomenon is known as **scintillation**. Much of Rutherford's work involved counting by eye the scintillations produced in this way.

Today, we can use an electronic device called a **photomultiplier** to amplify the light produced sufficiently to make it record itself on a counter or ratemeter. The scintillation method is particularly useful for the detection of gamma rays.

37.6. The 'half-life' of a radioactive species

It is not possible to predict the time at which an unstable nucleus will break up; it might break up, or **decay**, in the next second, or it might not decay for many thousands of years. However, for any particular sort of nucleus, there is an *average time* which elapses before decay occurs. In the case of the more stable radioactive nuclei, this may be of the order of 10^{10} years; in the

410

case of very active nuclei, the order of time may be 10^{-6} seconds!

The average time needed for decay to occur is usually expressed as the **half-life**. This is the time needed for one-half of the nuclei originally present to break up. The time for this to happen is independent of the number of nuclei originally present. It takes the same time for 1 g of radium to leave 0.5 g, as it takes for 0.1 g to leave 0.05 g, or for 3 g to leave 1.5 g.

If the number of radioactive nuclei left is plotted against time, a typical **decay curve** is obtained. The decay curve for a radioactive substance with a half-life of one hour is shown in Figure 37.5, starting with 64×10^6 active nuclei.

Figure 37.5 A typical decay curve, illustrating half-life

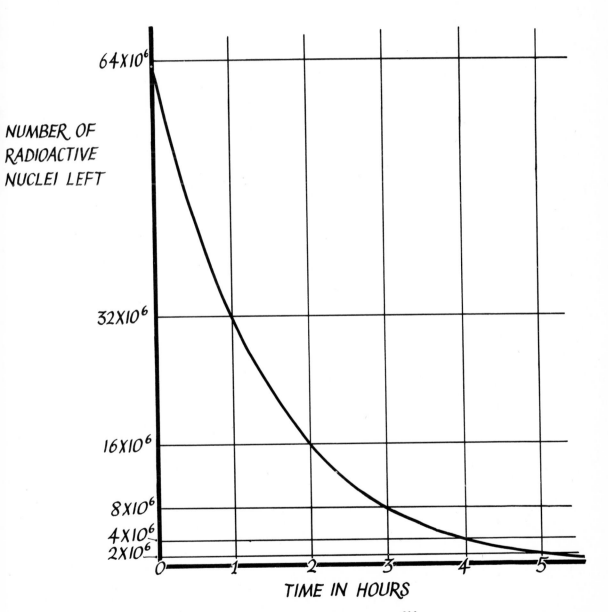

Radioactive

Corrosive

Toxic

Investigation 37a. The half-life of protoactinium 234

If you are fortunate enough to possess a counter or ratemeter and a G.M. tube or, better still, a G.M. liquid counting tube, it is a simple matter to measure the half-life of this radioactive isotope.

Dissolve 1 g of uranyl(VI) nitrate(V) in 3 cm³ of water and add 7 cm³ of concentrated hydrochloric acid. Transfer the solution to a small, cylindrical separating funnel (50 cm³) and add 15 cm³ of ethyl ethanoate. Shake well for about four minutes. The protoactinium, which is one of the decay products of the uranium (see Section 37.4), is soluble in the organic liquid and may be separated from the uranium and other decay products.

This is done by allowing the two layers to settle and by *immediately* running off the lower aqueous layer into a small beaker. The organic layer with the protoactinium is *at once* back-poured from the funnel into the liquid counting tube or, if you have not got one of these, into a specimen tube which has a flat bottom and is made of thin glass. The solution in the specimen tube can be counted by standing the tube on top of a thin piece of card which is placed above the window of an ordinary G.M. tube as shown in Figure 37.6.

As the half-life is so short (72 s), counting must be started rapidly after the liquids have been separated. The procedure is to count for 20 s, stop the counter for 10 s, then count for 20 s again, repeating the counts and pauses alternately for about eight minutes. A 'background' count, with the liquid counter or specimen tube filled with water, should be made. Results should be entered in a table as shown in Table 37.2.

TABLE 37.2. TABLE FOR HALF-LIFE INVESTIGATION READINGS

Times (s)	Total Count	Count Increase	Counts per minute (c.p.m.)	Corrected c.p.m.
0–20				
30–50				
60–80				
90–110				
120–140				
etc.				

Do *not* reset the counter after each twenty-seconds count. The counts per minute (c.p.m.) are obtained by multiplying the count

412

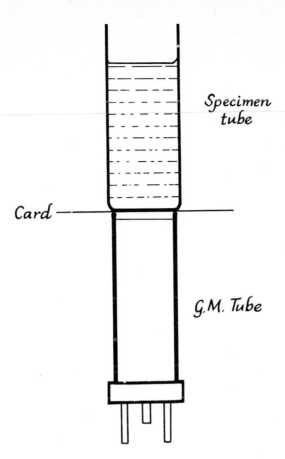

Figure 37.6 The arrangement for counting, using a specimen tube

increase by three. The background count is then subtracted in order to give the corrected counts per minute.

A graph of the corrected counts per minute against the time is plotted. The times used are the mid-times for each twenty-second count – that is, at 10 s for the 0–20 s period, at 40 s for the 30–50 s period, and so on.

Choose a suitable count-rate from your graph, 5 000 c.p.m. for example, and find how long it takes for the count-rate to fall to, say, 2 500 c.p.m. (one half-life), 1 250 c.p.m. (two half-lives) or 625 c.p.m. (three half-lives).

The organic liquid may be returned to the funnel, along with the aqueous layer. After standing for about ten minutes, another determination may be made. Indeed, the two layers may be kept in the funnel, ready for use, for many weeks.

37.7. Artificially produced radioactivity

It is possible to produce nuclei which are radioactive and which do not occur naturally. These artificial radioactive substances are often extremely useful. Alpha, beta and gamma emitters may be made, and it is also possible to produce an isotope which emits **positrons**, that is, positively charged particles, equal in mass to electrons.

Artificially radioactive nuclei are produced by bombarding stable nuclei with particles of high energy. These may be alpha particles or beta particles, moving at high speeds, or neutrons.

For example, radioactive phosphorus may be made from sulphur, by bombarding sulphur with neutrons:

$$^{32}_{16}S + ^{1}_{0}n \rightarrow ^{32}_{15}P + ^{1}_{1}p$$

The sulphur nucleus accepts the neutron and then immediately ejects a proton, leaving radioactive phosphorus. The phosphorus is a beta emitter, with a half-life of 14.2 days:

$$^{32}_{15}P \rightarrow ^{32}_{16}S + ^{0}_{-1}e$$

37.8. The use of radioisotopes

Radioisotopes, both natural and artificial, find many applications. A few examples of their uses are given below.

Measurement of thickness

In the production of thin sheets of material, for example, paper, plastics and metal foils, control of the thickness may be obtained by using radioisotopes. In one form of control, a radioactive source is placed on one side of the film as the film emerges from the rollers producing it, and a detector for the radiation is placed on the other side of the film.

The intensity of the radiation reaching the detector depends on the thickness of the film. If the radiation detected increases, then the film is getting thinner. The increased radiation may be used to control the pressure being applied to the rollers producing the film, and the pressure decreased, so that the film is produced thicker. Decreased radiation means that the film is getting thicker. Again, the change in detected radiation may be used to adjust the roller pressure. Figure 37.7 shows the arrangement.

Measurement of liquid depth

The depth of liquid in a tank may be found by using a technique similar to that used for controlling the thickness of a film. The source is placed on one side of the tank, and the detector on the other. They are arranged so that they may be moved up and down together. When the space between them is occupied by liquid, the radiation reaching the detector is far less than when air is the intervening material. By moving the source and detector up and

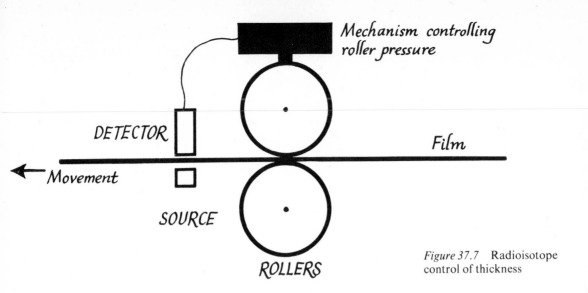

Figure 37.7 Radioisotope control of thickness

down, it is easy to locate the liquid surface. Figure 37.8 shows the arrangement and the result obtained.

The same technique may be used to ensure that packets and tins containing powders or liquids contain sufficient material. In this case, the source and detector are fixed at the height required for the surface. So long as the containers, which pass between the source and detector on a moving belt, contain sufficient depth of substance, little radiation will be detected. When a container with too little substance passes, the radiation detected will be greatly

Figure 37.8 Measuring liquid depth in a tank

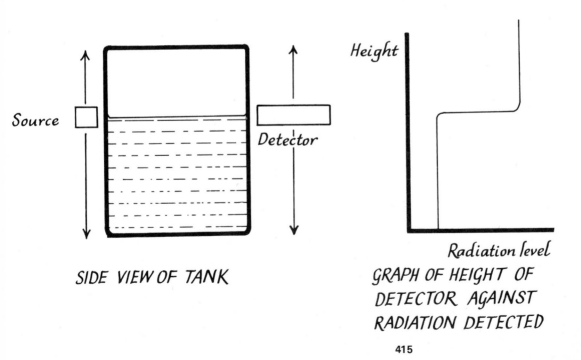

SIDE VIEW OF TANK

GRAPH OF HEIGHT OF DETECTOR AGAINST RADIATION DETECTED

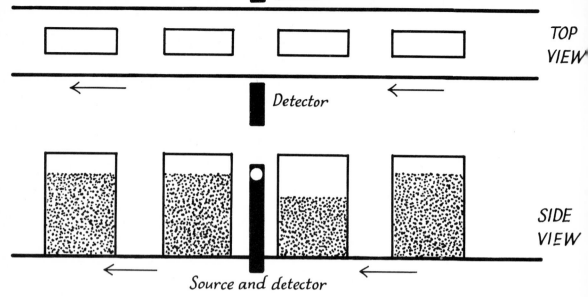

Figure 37.9 Checking the depth of material in a container

increased and an automatic apparatus is brought into action to remove that container from the production line (see Figure 37.9).

Altering the properties of plastics

The degree of cross-linking (see Section 36.5) of the chains of molecules in some plastics can be increased by exposing the material to radioactive radiation. This irradiation can knock atoms off the chains, leaving free valencies for the chains to link up. Irradiated polythene, for example, will withstand higher temperatures without becoming distorted than the untreated material.

Leak detection

Leaks in underground pipes can be detected by adding a radioactive substance to the fluid flowing through the pipes and by using detectors close to the positions where leaks are suspected.

Radiography

Flaws, and small cracks, in metal castings and welded joints, may be detected by using gamma radiation. The object to be inspected is placed in front of a sheet of photographic paper and the source is placed in front of the object. If there is a crack or flaw in the object, the gamma rays will pass more easily at the point. The crack or flaw will show up on the plate, when it is developed, as a darker line.

37.9. The dangers of radioactivity

Living cells are very sensitive to radiations from radioactive substances. The radiation, as it passes through the cells, may cause chemical changes to take place which alter the properties of the cells. This may result in the death of the cells, or may cause them to reproduce in an unusual way. This latter effect may result in the production of uncontrolled or **cancerous** growths. It is particularly important that vital organs should not receive large doses of radioactivity.

Persons handling radioactive substances, or using apparatus which may produce X-rays, must be **shielded** from the radiations. Both lead and thick concrete are used for this purpose. The substances and apparatus must be handled by remote control.

37.10. The use of radioisotopes in medicine

The fact that radiations can kill cells is of use in the treatment of cancer. Radiation (gamma rays) produced by a cobalt 60 source is carefully directed at the growth for a suitable time. In order to prevent healthy tissues from being effected, it is common practice to move the source round the patient in a circle, so that the growth being treated lies exactly at the centre of the circle and so receives continuous irradiation, while the surrounding tissues receive only a fraction of that dose.

Radioisotopes may also be used in the diagnosis of diseases. For example, radioactive iodine may be used to determine the size of the thyroid gland in the neck. This gland has the property of concentrating iodine in itself. If radioactive iodine is injected, it is possible to map out the area under the skin where it becomes concentrated, and so determine the size of the gland.

37.11. Experiments with radioactive substances

The conditions under which radioactive substances, and electrical sources producing X-rays, may be used in schools and Sixth Form colleges are laid down by the Department of Education and Science in their Administrative Memorandum 2/76, and in the Department of Employment Code of Practice for the protection of persons exposed to ionizing radiation in research and teaching (H.M.S.O.).

A number of useful experiments may be carried out using only the salts of potassium, thorium and uranium; the Association for Science Education publishes details of a number of these in their Modern Physical Science Report series.

The apparatus required consists of a counter (scaler) or a rate-meter. The former is of more general value. A Geiger-Müller tube and holder are also essential. It is useful, as well, to have a liquid G.M. counting tube.

Test your understanding

1. Some atomic nuclei are stable; others are unstable. What factor(s) control the stability of the nucleus?
2. What are 'magic numbers'?
3. Who first discovered the phenomenon of radioactivity? How did the discovery come about?
4. What are (a) alpha particles, (b) beta particles, (c) gamma rays?
5. Describe the construction of, and method of using, an electroscope.
6. How does a Geiger-Müller tube work?
7. Fill in the blanks in the following nuclear equations:

    (a) $^{234}_{91}Pa \rightarrow \, ^{234}_{92}U + ?$

    (b) $^{226}_{88}Ra \rightarrow \, ^{222}_{86}Rn + ?$

    (c) $^{222}_{86}Rn \rightarrow \, ^{?}_{?}Po + \, ^{4}_{2}He$

    (d) $^{206}_{81}Tl \rightarrow \, ^{?}_{?}Pb + \, ^{0}_{-1}e$

8. Explain the meaning of the term 'half-life' with regard to a radioactive isotope.
9. A count is made of the radiation coming from a radioactive isotope. It is found to be 7 680 counts per minute. Ten minutes later the count has fallen to 3 840 counts per minute. What will the count rate be after another thirty minutes?
10. What is a 'positron'?
11. Describe THREE practical applications of radioactive isotopes.
12. Why are radioactive isotopes dangerous to handle?
13. Describe some of the uses of radioisotopes in medicine.

Chapter 38

Atomic Energy

38.1. Nuclear fission

When atomic nuclei are hit by particles such as protons, neutrons or alpha particles, a new sort of nucleus, often radio-active, is usually formed (see Section 37.7).

However, it was discovered in 1938 that when the nucleus of the uranium isotope $^{235}_{92}U$ is struck by a neutron, an entirely different sort of change can take place. The neutron is captured by the nucleus, forming $^{236}_{92}U$, which then breaks up into pieces, liberating at the same time, a great deal of energy.

This type of change is known as **nuclear fission**. Other nuclei have since been found which undergo the same sort of change when struck by neutrons. Figure 38.1 illustrates the idea in diagram form.

The energy comes from the change of mass that accompanies the reaction. The total mass of the particles forming the products is less than the total mass of the original nucleus and the neutron. It was Albert Einstein who, in 1907, predicted that mass could be converted into energy and gave the now well-known 'Einstein Equation':

$$E = mc^2$$

In this relation, E is the energy liberated, m is the mass lost, and c is the velocity of light. As the velocity of light is so great

Figure 38.1 The fission of a $^{235}_{92}U$ nucleus

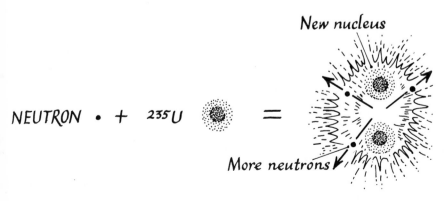

NEUTRON $\bullet$ + $235U$ = New nucleus / More neutrons

$(3 \times 10^8 \text{ m s}^{-1})$, and the velocity is squared in the equation, it is clear that a great deal of energy can be obtained from a small mass loss. If one gram of matter could be turned completely into energy, 9×10^{13} J could be obtained.

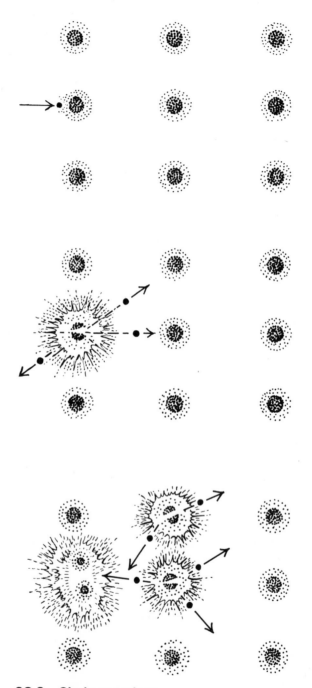

Figure 38.2 A chain reaction

38.2. Chain reactions

When the $^{235}_{92}$U nucleus is broken up in this way, the products include several neutrons, each of which, given the right conditions, is capable of breaking up another $^{235}_{92}$U nucleus.

Suppose, for example, that the fission results in three neutrons being formed, and that each of these strikes another $^{235}_{92}$U nucleus, again liberating three more neutrons, there will now be nine neutrons available. If in turn these break up more nuclei, there will be twenty-seven neutrons freed. A process which goes on in this way is known as a **chain reaction.** The process is illustrated in Figure 38.2.

In a very short time indeed (a small fraction of a second), as more and more neutrons are liberated, all the energy in a suitable piece of uranium 235 may be released. The energy available is very great, and if it is all liberated at once, an explosion of considerable magnitude occurs – we have an **atomic fission bomb.**

38.3. The conditions needed for a chain reaction in $^{235}_{92}$U

Fortunately, a chain reaction cannot take place in an ordinary piece of uranium. This is because:
  a. ordinary uranium contains less than 1 per cent of the $^{235}_{92}$U isotope. The remainder is $^{238}_{92}$U, and this nucleus is not fissile.
  b. unless the piece of uranium is large enough, most of the neutrons formed in the early stages will escape from the piece before they meet another nucleus. If this happens, no chain can be established.
  c. if the liberated neutrons are travelling fast, they are unlikely to cause fission. Slowly moving neutrons are more effective.

In order, therefore, to get liberation of energy by fission from uranium, it is necessary to concentrate the uranium 235 and to fashion a piece or pieces of shape and size such that the neutrons liberated do not escape before they can meet other nuclei. In the construction of an atomic fission bomb, the speed of the neutrons is relatively unimportant. In the case of an **atomic reactor** (see Section 38.6), they have to be slowed down.

The concentration of the uranium 235 from natural uranium is a complicated and expensive matter, but it is possible.

The minimum sized piece of the material needed before a chain reaction can start depends on the proportion of uranium 235 in the concentrated uranium and on the geometrical shape of the piece. The term **critical mass** is used for the smallest piece which will sustain a chain reaction.

38.4. The original fission bomb

The first atomic bomb was constructed by concentrating the uranium 235 and making from the concentrated material separate pieces, each of which was less than the critical mass. The pieces were mounted in an outer container in such a way that they could be brought together instantaneously. This was achieved by firing one or more pieces into a 'target' consisting of another piece. The combined material was greater than the critical mass, the chain reaction occurred and the explosion followed. A possible arrangement for a fission bomb is shown in Figure 38.3.

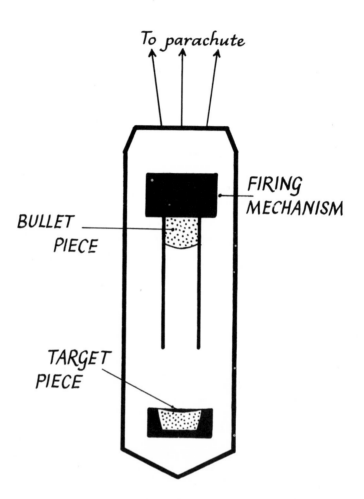

Figure 38.3 A possible arrangement for a fission bomb

38.5. The peaceful use of atomic energy

The destructive power of atomic energy was demonstrated only too catastrophically when the first fission bomb was dropped on Hiroshima in 1945.

However, there is another side to the picture. Instead of all the available energy being liberated in a fraction of a second in a devastating manner, it may be liberated in a controlled way, so that it can be used beneficially. The plant in which the controlled liberation of energy is achieved is known as an atomic **reactor**.

38.6. A simple atomic reactor

The very first atomic reactor, or atomic **pile** as it was called, was erected in 1942 in a squash court at Chicago University, and a controlled chain reaction was first obtained on 2 December in that year. The leader of the team constructing this reactor was Enrico Fermi who, in 1938, was awarded the Nobel Prize for Physics for his work in the field of atomic physics.

This first pile consisted of some 290 tonnes of very highly purified graphite. Embedded in the graphite, in layers, was some 40 tonnes of enriched uranium and uranium oxide. At the bottom of the pile was a source of neutrons to start the chain reaction. For controlling the rate of the reaction, rods of cadmium metal were placed in the pile.

The graphite was needed as a **moderator**, to slow down the neutrons so that they would be more effective in bringing about fission. Cadmium is an excellent absorber of neutrons.

When the pile was complete, the cadmium rods were slowly withdrawn until enough slow neutrons were present to start the chain reaction. As the reaction started to get faster, the rods were pushed into the pile again. By continual adjustment of the position of the cadmium **control rods**, the reaction was kept going at a steady rate.

For safety, additional cadmium rods were held suspended above holes in the pile so that, if the temperature rise showed that things were getting out of hand, they would fall into the pile and shut it down. A simplified diagram of a graphite-moderated reactor is given in Figure 38.4.

Since the first pile was constructed, the design of atomic reactors has advanced considerably, but the use of a cooling fluid to take the heat from the reactor core, followed by the use of the heated fluid to raise steam in a **heat exchanger**, is still used. The steam may then be used to drive electric generators.

38.7. Nuclear fusion

Energy is also liberated when the nuclei of very light atoms, such as hydrogen atoms, join together. The energy released is proportionately far greater than that produced in atomic fission reactions.

Heavy hydrogen nuclei consist of a proton and a neutron ($^2_1$H). This form of hydrogen is also known as **deuterium**, and is given a special symbol, the letter D. It is far easier to separate deuterium from ordinary hydrogen than it is to separate the isotopes of uranium.

A typical fusion reaction between two heavy hydrogen nuclei

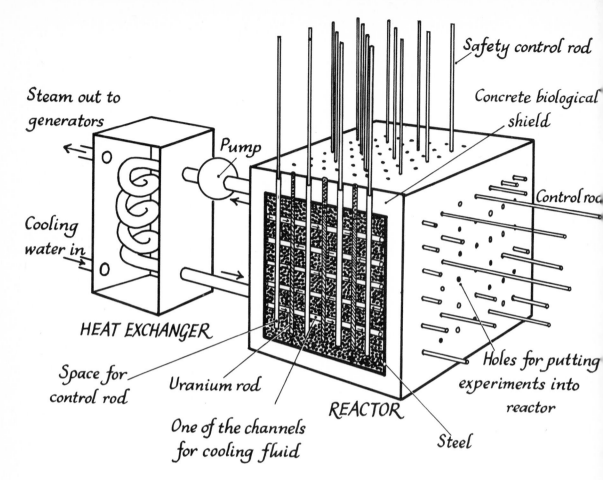

Steam out to generators

Cooling water in

HEAT EXCHANGER

Pump

Safety control rod

Concrete biological shield

Control rod

Holes for putting experiments into reactor

Space for control rod

Uranium rod

One of the channels for cooling fluid

REACTOR

Steel

Figure 38.4 Simplified diagram of an atomic reactor

brings about the formation of a light isotope of helium, and liberation of a neutron and a large quantity of energy is freed, the total mass of the products being less than the total mass of the reactants:

$$_1^2H + {}_1^2H \rightarrow {}_2^3He + {}_0^1n + energy$$

In order for the reaction to go at all, the heavy hydrogen nuclei have to be given a large quantity of activation energy. A temperature of several million degrees is needed!

Temperatures of this order are found in the stars. It is due to reactions like the one mentioned above that our own sun can provide the huge output of energy that it does.

On earth, the attainment of such high temperatures is very difficult. So far, the only conditions under which a fusion reaction has been carried out are those found in an atomic fission bomb explosion.

If a fission bomb is surrounded by heavy hydrogen, then, when the fission reaction occurs, the energy liberated will trigger-off the fusion reaction and the energy of this will be added to the fission energy. This is the so-called **hydrogen bomb**.

Whereas a fission bomb has an energy equivalent of, say, 20 000 tonnes of T.N.T., a fusion bomb may have an energy equivalent of, perhaps, 100 000 000 tonnes of T.N.T.

The search for the conditions under which a **controlled** fusion reaction can be obtained are continuing. If they are found, mankind will have at his disposal an almost limitless supply of energy.

38.8. New elements

When a neutron strikes the nucleus of a $^{235}_{92}U$ atom, then, as we have seen, fission may occur. If, however, a neutron strikes the nucleus of a $^{238}_{92}U$ atom (the more plentiful isotope), fission does not occur, but a new nucleus is formed which is radioactive:

$$^{238}_{92}U + ^{1}_{0}n \rightarrow ^{239}_{92}U$$

Then:

$$^{239}_{92}U \rightarrow ^{239}_{93}Np + ^{0}_{-1}e$$

A new element, **neptunium**, with an atomic number of 93 has been made. This element is not found in nature, but it may be made by bombarding ordinary uranium with neutrons in an atomic reactor.

The neptunium itself is radioactive. It has a half-life of 2.3 days and is a beta emitter:

$$^{239}_{93}Np \rightarrow ^{239}_{94}Pu + ^{0}_{-1}e$$

The resulting element, with an atomic number of 94, is called **plutonium**. Its nucleus is relatively stable, and it has a half-life of 24 000 years. The really important point, however, is that plutonium nuclei can undergo fission, in the same way as the $^{235}_{92}U$ nuclei.

Thus, by including some uranium 238 in a reactor, it is possible to obtain more nuclear 'fuel' from the working of the reactor itself. This type of reactor is known as a **breeder reactor**.

Man-made elements, such as neptunium and plutonium, have atomic numbers greater than that of uranium, the naturally occurring element with the highest atomic number. They are known as **trans-uranic elements**. To date, elements with atomic numbers up to 105 have been isolated.

Test your understanding

1. Explain, using a labelled diagram, what is meant by 'fission' in connection with an atomic nucleus.
2. What is a 'nuclear chain reaction'? What conditions are necessary for such a reaction to occur?
3. Where does the energy liberated in a fission reaction come from?
4. Explain the meaning of the terms (a) critical mass, (b) moderator, (c) control rod, in connection with a nuclear reactor.
5. What is 'deuterium'?
6. What is meant by the term 'nuclear fusion'? Where does the energy liberated in nuclear fusion come from?
7. What happens when a neutron is absorbed by a $^{238}_{92}U$ nucleus? What is the importance of this in the production of nuclear energy?
8. What is a 'trans-uranic' element? Make a list of the known transuranic elements, giving their names and their atomic numbers.
9. What is a 'hydrogen bomb'?
10. The 'Einstein equation', $E = mc^2$, was proposed by Albert Einstein in 1907. What do the letters E, m and c stand for?
11. What is the main problem that has to be overcome if a controlled nuclear fusion reactor is to be designed?

Chapter 39

Chemical Arithmetic

39.1. Symbols, formulae and mole masses

As we saw in Chapter 3, the symbol for an element stands not only for the element, but for a particular amount of the element, namely one mole or 6×10^{23} atoms. We also saw that the number 6×10^{23} is the number of $^{12}_{6}C$ atoms in exactly 12 g of that isotope and that for all the other elements the mole mass is the relative atomic mass in grams. For example, the relative atomic mass of sulphur is 32, and so the mole mass is 32 g. The relative atomic mass of iron is 56, and so the mole mass of iron is 56 g.

TABLE 39.1.   SOME APPROXIMATE MOLE MASSES

Element	Symbol	Mole Mass/g	Element	Symbol	Mole Mass/g
Hydrogen	H	1	Potassium	K	39
Carbon	C	12	Calcium	Ca	40
Nitrogen	N	14	Iron	Fe	56
Oxygen	O	16	Copper	Cu	64
Fluorine	F	19	Zinc	Zn	65
Sodium	Na	23	Bromine	Br	80
Magnesium	Mg	24	Silver	Ag	108
Aluminium	Al	27	Iodine	I	127
Silicon	Si	28	Barium	Ba	137
Phosphorus	P	31	Mercury	Hg	201
Sulphur	S	32	Lead	Pb	207
Chlorine	Cl	35.5	Uranium	U	238

The mole masses of compounds are found by adding up the mole masses of the atoms in the formula for the compound. Water, for example, has the formula H_2O, so each mole of water has in it two moles of hydrogen atoms and one mole of oxygen atoms. The mole mass of water is thus $(2 \times 1 \text{ g}) + (1 \times 16 \text{ g})$ which makes 18 g. The mole mass of copper(II) sulphate(VI) crystals, $CuSO_4.5H_2O$, is $64 \text{ g} + 32 \text{ g} + (4 \times 16 \text{ g}) + (5 \times 18 \text{ g})$, which comes to 250 g. Work out for yourself the mole masses of calcium carbonate ($CaCO_3$), lead(II) nitrate(V) ($Pb(NO_3)_2$), barium chloride ($BaCl_2.2H_2O$) and washing soda ($Na_2CO_3.10H_2O$).

39.2. Simple calculations using formulae

In Section 8.6, we saw how it is possible to calculate the empirical formula of a compound from its percentage composition. Once we know the empirical formula we are in a position to calculate the mass of one of the elements, or parts of the compound, in any given mass of the compound. For example, if we want to know the mass of sulphur in, say, 4.9 g of sulphuric (VI) acid (H_2SO_4), we first work out the mole mass of the acid, which comes to 98 g. The sulphur in the acid contributes 32 g to this 98 g. In other words, the sulphur is 32/98 of the whole amount. It follows that there is $32/98 \times 4.9$ g of sulphur in the 4.9 g of the sulphuric(VI) acid:

$$\text{Mole mass of sulphuric(VI) acid} = 98\,\text{g}$$
$$\text{Mass of sulphur in mole mass} = 32\,\text{g}$$
$$\therefore \text{Proportion of sulphur in mole mass} = \frac{32}{98}$$
$$\therefore \text{Mass of sulphur in 4.9 g of the acid} = \frac{32}{98} \times 4.9\,\text{g}$$
$$= 1.6\,\text{g}$$

Another example: What is the mass of water in 50 g of copper(II) sulphate(VI) crystals ($CuSO_4 . 5H_2O$)?

$$\text{Mole mass of copper(II) sulphate(VI) crystals} = 250\,\text{g}$$
$$\text{Mass of water in mole mass} = 5 \times 18\,\text{g}$$
$$= 90\,\text{g}$$
$$\therefore \text{Proportion of water in mole mass} = \frac{90}{250}$$
$$\therefore \text{Mass of water in 50 g of crystals} = \frac{90}{250} \times 50\,\text{g}$$
$$= 18\,\text{g}$$

Work out for yourself: (i) the mass of lead in 23.9 g of lead(IV) oxide (PbO_2); (ii) the mass of nitrogen in 4.0 g of ammonium nitrate(V) (NH_4NO_3); (iii) the mass of water in 12.2 g of barium chloride crystals ($BaCl_2 . 2H_2O$).

39.3. Calculations using equations

A chemical equation tells us the numbers of moles of substances of the reactants and of the products concerned in the reaction. Thus, if we know the equation for the reaction, we can calculate their masses. For example, suppose we want to know the mass of nitric(V) acid needed to neutralize 4 g of sodium hydroxide, and also want to know the mass of sodium nitrate(V) that will be found in the reaction:

$$HNO_3(aq) + NaOH(aq) \rightarrow NaNO_3(aq) + H_2O(l)$$

We first work out the mole masses of the substances. These are

$HNO_3 = 63$ g; $NaOH = 40$ g; $NaNO_3 = 85$ g. There is no need to work out the mole mass of water (18 g) as we do not want to calculate the mass of water formed, but, if we do work it out, then we have a check on our arithmetic as the total masses of the products must equal the total masses of the reactants; i.e. 85 g + 18 g should be equal to 63 g + 40 g, which is indeed the case.

So we can see that it takes 63 g of nitric(V) acid to neutralize 40 g of sodium hydroxide and that, as a result, 85 g of sodium nitrate(V) are produced. Now, 4 g of sodium hydroxide is 4/40 of a mole and so our answer requires 4/40 of each of the other masses as shown by the equation, i.e. $4/40 \times 63$ g of nitric(V) acid are needed and $4/40 \times 85$ g of sodium nitrate(V) are produced:

$$HNO_3(aq) + NaOH(aq) \rightarrow NaNO_3(aq) + H_2O(l)$$

Mole masses: 63 g   40 g   85 g   18 g

To neutralize 4 g of sodium hydroxide we need (and get):

$$\frac{4}{40} \times 63 \text{ g} = 6.3 \text{ g of nitric(V) acid}$$

$$\frac{4}{40} \times 85 \text{ g} = 8.5 \text{ g of sodium nitrate(V)}$$

When there is more than one mole of a substance shown in the equation, then, of course, allowance must be made: e.g. if we need to know the mass of ammonia that *might* be formed from 28 g of nitrogen gas, using the Haber process:

$$N_2(g) + 3H_2(g) \rightarrow 2NH_3(g)$$

28 g 3×2 g 2×17 g

28 g   6 g   34 g

39.4 The mole volumes of gases

Equal volumes of **all** gases, providing that they are at the same temperature and at the same pressure, contain equal numbers of molecules. This statement of fact is known as **Avogadro's law**, after Amadeo Avogadro, who first suggested that this is so. It follows that one mole of any gas at a particular temperature and at a particular pressure will have the same volume as one mole of any other gas under the same conditions. Experiment shows that this volume is very nearly 22.4 dm^3 at 273K and 101.3 kN m^{-2} (known as standard temperature and pressure or s.t.p.), or about 24 dm^3 at room conditions.

Thus 2 g of hydrogen (H_2) will have a volume of 22.4 dm^3 at s.t.p., as will 44 g of carbon dioxide (CO_2), 17 g of ammonia (NH_3), etc. Other masses of these gases will have proportional volumes; thus 11 g of carbon dioxide gas will have a volume of $11/44 \times 24$ dm^3 under room conditions, i.e. 6 dm^3.

Work out for yourself the volumes of (i) 8 g of oxygen; (ii) 14 g of nitrogen gas; (iii) 3.55 g of chlorine gas and (iv) 2.2 g of propane gas (C_3H_8), first under room conditions, and then at s.t.p.

39.5 Equations and mole volumes of gases

Avogadro's law can be used with equations to find the volumes of gases taking part in, and being formed in, a reaction. Suppose we need to know the volume of carbon dioxide gas given off (under room conditions) when 5.0 g of calcium carbonate are dissolved in excess of hydrochloric acid:

$$CaCO_3(s) + 2HCl(aq) \rightarrow CaCl_2(aq) + CO_2(g) + H_2O(l)$$
$$100\,g \qquad 2 \times 36.5\,g \qquad 111\,g \qquad 44\,g \qquad 18\,g$$

As 44 g of carbon dioxide is one mole, we can substitute 24 dm³ (under room conditions) and we see that:

100 g of calcium carbonate produce 24 dm³ of carbon dioxide

$\therefore$ 5 g of calcium carbonate will produce $\dfrac{5}{100} \times 24$ dm³

$\underline{= 1.2\,dm^3}$ under room conditions

Again, if we want to know the volume of ammonia gas (measured at s.t.p.) needed to neutralize 9.8 g of sulphuric(VI) acid:

$$H_2SO_4(aq) + \quad 2NH_3(g) \quad \rightarrow (NH_4)_2SO_4(aq)$$
$$98\,g \qquad 2 \times 22.4\ dm^3$$
$$(at\ s.t.p.)$$

9.8 g of sulphuric(VI) acid react with:

$\dfrac{9.8}{98} \times 2 \times 22.4$ dm³ of ammonia at s.t.p.

$\underline{= 4.48\ dm^3}$ at s.t.p.

39.6. A further use of Avogadro's law

If the volumes of **all** the gases in a reaction are measured at the **same** temperature and at the **same** pressure then, if we need to know the **relative** volumes, we do not have to know the actual temperature or the actual pressure. For example, if we want to know the volume of oxygen gas needed to burn completely 100 cm³ of ethane then:

$C_2H_6(g)$	$+ 3\frac{1}{2}O_2(g)$	$\rightarrow 2CO_2(g)$	$+ 3H_2O(g\ or\ l)$
1 mole	$3\frac{1}{2}$ moles	2 moles	
1 volume	$3\frac{1}{2}$ volumes	2 volumes	from Avogadro's law
$\therefore$ 100 cm³	350 cm³	200 cm³	by proportion
			(providing all are at the *same* temperature and pressure)

The volume of carbon dioxide gas formed is also given in the same simple way.

With regard to the water, the physical state will depend on the actual temperature and pressure: it could be (g) or (l). If the temperature of the reaction is above 100 °C, then the water will be in the gaseous state and the volume of vapour will be 300 cm^3. If the temperature is below 100 °C, the water will be in the liquid (or solid) state and its volume will be negligible compared with the volumes of the gases.

39.7. Calculations on electrolysis

It was Michael Faraday who discovered the laws of electrolysis. The first of the two laws states that the mass of a substance liberated at one of the electrodes during an electrolysis depends on the quantity of electricity that passes through the voltameter.

The quantity of electricity depends on the strength of the current being used and on the time for which it is passed. In SI units, the current is measured in **amperes** and the time in **seconds**. The unit for the quantity of electricity is called the **coulomb**.

$$1 \text{ coulomb} = 1 \text{ ampere second}$$
$$\therefore \text{number of coulombs} = \text{number of amperes} \times \text{number of seconds}$$

Faraday's second law of electrolysis, expressed in modern terms, is that the mass of a substance liberated during an electrolysis depends on its mole mass and on the charge carried by its ion in the electrolyte. The following examples should make this clear:

$$\begin{array}{cccc} Ag^+(aq) + & e^- & \to & Ag(s) \\ 1 \text{ mole} & 1 \text{ mole} & & 1 \text{ mole} \end{array}$$

$$\begin{array}{cccc} Cu^{2+}(aq) + & 2e^- & \to & Cu(s) \\ 1 \text{ mole} & 2 \text{ moles} & & 1 \text{ mole} \end{array}$$ Where e^- represents one mole of electrons

$$\begin{array}{cccc} Al^{3+}(aq) + & 3e^- & \to & Al(s) \\ 1 \text{ mole} & 3 \text{ moles} & & 1 \text{ mole} \end{array}$$

One mole of electrons (6×10^{23}) is about 96 000 coulombs (C). It follows that the mass of any substance liberated by 96 000 C will be the mole mass of the substance divided by the charge carried by its ion in the electrolyte:

$$\text{Mass liberated by 96 000 C} = \frac{\text{mole mass}}{\text{charge on ion}}$$

$$\therefore \text{Mass liberated by 1 C} = \frac{\text{mole mass}}{\text{charge on ion} \times 96\,000}$$

$\therefore$ Mass of substance liberated during an electrolysis (in grams)

$$= \frac{\text{mole mass}}{\text{charge on ion} \times 96\,000} \times \text{coulombs passed}$$

$$= \frac{\text{mole mass}}{\text{charge on ion} \times 96\,000} \times \text{amperes} \times \text{seconds g} \qquad (1)$$

Example: What mass of zinc is deposited on the cathode of a zinc voltameter by a current of 1.5 amperes passed for 3 hours? (Mole mass of zinc is 65 g; the zinc ion is Zn^{2+}.)

$$\text{Mass of zinc deposited} = \frac{65}{2 \times 96\,000} \times 1.5 \times 3 \times 60 \times 60 \text{ g}$$

$$= 5.484 \text{ g}$$

The quantity $\dfrac{\text{mole mass}}{\text{charge on ion} \times 96\,000}$ (that is, the mass of a substance liberated by one coulomb) is known as the **electrochemical equivalent** (E.C.E. for short) of the substance, and equation (1) above can be summarized as:

$$\text{Mass liberated (in grams)} = \text{E.C.E.} \times \text{amperes} \times \text{seconds}$$

39.8. Calculations in volumetric analysis

These are based on the idea of the **molarity** of a solution. When one mole of a substance is dissolved in a suitable solvent (usually water) and the *total volume* is made up to 1 dm³, then the solution is said to be 1 molar (1 M). If, say, 0.5 mole of substance is used in 1 dm³ of solution, then the solution is said to have a molarity of 0.5, i.e. it is 0.5 M. The molarity of a solution is thus seen to be the number of moles of dissolved substance in each dm³ of the solution.

The **concentration** of a solution used in volumetric analysis is the mass in grams of substance dissolved in each dm³ of the solution. It follows that:

$$\text{Molarity of a solution} = \frac{\text{concentration of dissolved substance}}{\text{mole mass of dissolved substance}} \qquad (2)$$

Example 1   What is the molarity of a solution made by dissolving 2.0 g of sodium hydroxide in 100 cm³ of solution?

Mole mass of sodium hydroxide (NaOH) is 40 g
As 2.0 g are dissolved in 100 cm³, the concentration of the solution is:

$$\frac{1\,000}{100} \times 2 \text{ g dm}^{-3}$$

$$= 20 \text{ g dm}^{-3}$$

$$\text{Molarity of the solution is } \frac{20}{40}, \quad \text{i.e. } \underline{0.5\,M}$$

Example 2   What mass of potassium hydroxide is needed to neutralize 250 cm³ of a solution of sulphuric(VI) acid containing 4.9 g of acid in each dm³ of solution?

Mole mass of sulphuric(VI) acid is 98 g

$\therefore$ 4.9 g of the acid is $\dfrac{4.9}{98}$ mole

$\therefore$ in 250 cm³ of the solution there is $\dfrac{4.9}{98} \times \dfrac{250}{1\,000}$ moles

$$= 0.0125 \text{ mole}$$

The equation for the reaction is:

$$H_2SO_4(aq) + 2KOH(aq) \rightarrow K_2SO_4(aq) + 2H_2O(l)$$

so one mole of sulphuric(VI) acid needs two moles of potassium hydroxide for neutralization. It follows that 0.0125 mole of the acid needs 2×0.0125 mole of potassium hydroxide and, as the mole mass of potassium hydroxide is 56 g, the mass needed to neutralize the acid is:

$$2 \times 0.0125 \times 56 \text{ g}$$
$$= 1.40 \text{ g}$$

Example 3 20 cm³ of a solution of nitric(V) acid are neutralized by 25 cm³ of a solution of barium hydroxide ($Ba(OH)_2$). If the latter solution has a concentration of 17.1 g dm⁻³, what is (i) the molarity and (ii) the concentration of the nitric(V) acid solution?

Mole mass of barium hydroxide is 171 g
$\therefore$ Molarity of the barium hydroxide solution

$$= \dfrac{17.1}{171}$$
$$= 0.1 \text{ M}$$

25 cm³ of the barium hydroxide solution contain

$$\dfrac{25}{1\,000} \times 0.1 \text{ mole of } Ba(OH)_2$$
$$= 0.0025 \text{ mole}$$

The equation for the reaction is:

$$Ba(OH)_2(aq) + 2HNO_3(aq) \rightarrow Ba(NO_3)_2(aq) + 2H_2O(l)$$

so 0.0025 mole of the barium hydroxide reacts with 2×0.0025 mole of acid.

$\therefore$ 20 cm³ of nitric(V) acid contain 2×0.0025 mole
$\therefore$ 1 000 cm³ of the acid contain $2 \times 0.0025 \times \dfrac{1\,000}{20}$ mole
$\therefore$ Molarity $= 0.25 \text{ M}$

As the mole mass of nitric(V) acid is 63 g, the concentration of the acid solution (see equation (2) above) is:

$$0.25 \times 63 \text{ g dm}^{-3}$$
$$= 15.75 \text{ g dm}^{-3}$$

Test your understanding

USE THE RELATIVE ATOMIC MASSES GIVEN AT THE START
OF THIS CHAPTER.

1. Calculate (i) the mass of hydrogen in 2.8 g of ethene (C_2H_4); (ii) the mass of oxygen in 17 g of sodium nitrate(V) ($NaNO_3$); (iii) the mass of copper in 6.25 g of copper(II) sulphate(VI)-5-water ($CuSO_4.5H_2O$); (iv) the mass of nitrogen in 33.1 kg of lead(II) nitrate(V) ($Pb(NO_3)_2$); (v) the mass of water in 143 kg of washing soda ($Na_2CO_3 . 10H_2O$).

2. What is the percentage of (i) sulphur in sulphur dioxide gas (SO_2); (ii) sodium in sodium hydroxide ($NaOH$); (iii) carbon in ethane (C_2H_6); (iv) nitrogen in ammonium sulphate(VI) (($NH_4)_2SO_4$); (v) zinc in zinc carbonate ($ZnCO_3$)?

3. Calculate the mass of lead(II) sulphate(VI) formed when an excess of dilute sulphuric(VI) acid is added to a solution containing 2.07 g of lead(II) ions:

$$Pb^{2+}(aq) + H_2SO_4(aq) \rightarrow PbSO_4(s) + 2H^+(aq)$$

4. What mass of sodium nitrate(V) could be obtained from the solution made by adding the correct quantity of nitric(V) acid solution to a solution containing 4.0 g of sodium hydroxide? What mass of nitric(V) acid is required?

$$NaOH(aq) + HNO_3(aq) \rightarrow NaNO_3(aq) + H_2O(l)$$

5. What is the volume, under room conditions, of each of the following: (i) 2 g of hydrogen gas; (ii) 16 g of oxygen gas; (iii) 56 g of nitrogen gas; (iv) 4.4 g of propane gas (C_3H_8); (v) 3.3 g of carbon dioxide gas?

6. Calculate the volume of ammonia gas, measured at s.t.p., given off when 5.35 g of ammonium chloride are warmed with excess of potassium hydroxide solution:

$$NH_4Cl(s) + KOH(aq) \rightarrow KCl(aq) + H_2O(l) + NH_3(g)$$

7. What volume of carbon dioxide gas, measured under room conditions, is needed to precipitate 10 g of calcium carbonate from a lime-water solution?

$$Ca(OH)_2(aq) + CO_2(g) \rightarrow CaCO_3(s) + H_2O(l)$$

8. What volume of oxygen gas is needed to burn completely 30 cm³ of ethane? What will be the volume of the products? (All measurements are at the same temperature and pressure.)

$$C_2H_6(g) + 3\tfrac{1}{2}O_2(g) \rightarrow 2CO_2(g) + 3H_2O(l)$$

9. What volume of oxygen gas is needed to burn completely 50 cm³ of hydrogen sulphide gas? What will be the volume of the products? (All measurements are made at room temperature and pressure.)

$$H_2S(g) + 1\tfrac{1}{2}O_2(g) \rightarrow H_2O(l) + SO_2(g)$$

10. If the temperature in question 9 had been 105 °C what would the volume of the products have been?

11. What mass of silver would be deposited on the cathode of a silver voltameter by a current of 3 A running for 100 minutes?
12. What current is needed to deposit 3.2 g of copper on the cathode of a copper voltameter in a period of 2 hours?
13. Calculate the electrochemical equivalent of (i) nickel, (ii) aluminium, (iii) sodium. The ions are Ni^{2+}, Al^{3+}, and Na^+.
14. What are the molarities of the following: (i) a solution of hydrochloric acid containing 3.65 g in 1 dm³; (ii) a solution of sodium hydroxide containing 20 g in 500 cm³; (iii) a solution of lead(II) nitrate(V) containing 66.2 g in 250 cm³; (iv) a solution of sulphuric(VI) acid containing 98 g in 500 cm³?
15. What mass of solute is contained in each of the following solutions: (i) 1 dm³ of 1 M potassium hydroxide; (ii) 400 cm³ of 1 M sodium hydroxide; (iii) 2 dm³ of 0.5 M sodium chloride; (iv) 100 cm³ of 0.1 M nitric(V) acid; (v) 30 cm³ of 0.4 M ethanoic acid $(CH_3.COOH)$?
16. 20 cm³ of a solution of sodium hydroxide, containing 8 g in each dm³, are neutralized by 40 cm³ of a solution of hydrochloric acid.
    (i) What is the molarity of the sodium hydroxide solution?
    (ii) How many moles of sodium hydroxide are there in the 20 cm³?
    (iii) If the equation for the reaction is

$$NaOH(aq) + HCl(aq) \rightarrow NaCl(aq) + H_2O(l)$$

    how many moles of hydrogen chloride are there in the 40 cm³ of hydrochloric acid?
    (iv) What, then, is the molarity of the hydrochloric acid?
    (v) What, then, is the concentration of the hydrochloric acid?
17. What volume of 0.5 M sulphuric(VI) acid is needed to neutralize 30 cm³ of 2.0 M potassium hydroxide solution?

$$2KOH(aq) + H_2SO_4(aq) \rightarrow K_2SO_4(aq) + 2H_2O(l)$$

18. If 25 cm³ of a solution of nitric(V) acid are neutralized by 45 cm³ of a solution of sodium carbonate containing 10.6 g of Na_2CO_3 in each dm³, what is (i) the molarity, (ii) the concentration of the nitric(V) acid solution? The equation for the reaction is

$$2HNO_3(aq) + Na_2CO_3(aq) \rightarrow 2NaNO_3(aq) + CO_2(g) + H_2O(l)$$

Chapter 40

Periodicity

When you visit the public library to borrow a book you go to that part of the library in which the books you are interested in are kept. For example, if you want a book on chemistry you do not start inside the door of the library and walk round until you find a book about chemistry; instead you look for the shelves labelled chemistry. In the same way the Periodic Table (Table 3.2) collects together all the chemical elements so that those that are similar are close together.

40.1. The alkali metals

In Chapter 11, we discovered that potassium and sodium metals react with cold water to form hydrogen. The reaction is rapid and can be violent. The metals float on water and produce an alkaline solution without a precipitate. You also discovered that there were a few other metals that reacted with water, but only sodium and potassium react so vigorously.

Investigation 40a. The action of lithium on cold water

★ *Warning – it is important that clean basins are used. The lithium must not be bigger than stated. Safety spectacles must be worn.*

Flammable

Corrosive

In this investigation you are going to use a very reactive metal. Put on safety spectacles or eye-shields. Take an evaporating basin that is at least 50 mm in diameter and three-quarters fill it with water. Collect (on a watch glass) a small piece of lithium, a cube of not more than 5 mm side, from your teacher. Using tongs, lift the piece of lithium from the watch glass and drop it into the water. Observe what happens.

You will see that the lithium reacts in a similar way to the sodium and potassium that your teacher demonstrated to you. Which of the three metals do you think was most reactive?

These elements show properties that are very similar to each other with some gradual changes from one element to the next, for example melting-point. Elements that show this sort of similarity are often called families of elements. You will find information about another family of elements, the halogens, in Table 31.1.

TABLE 40.1. THE ALKALI METALS

Metal	Melting-point (K)	Density (g cm^{-3})	Valency
Lithium	454	0.53	+ 1
Sodium	371	0.97	+ 1
Potassium	336	0.86	+ 1
Rubidium	312	1.53	+ 1
Caesium	302	1.87	+ 1

40.2. The Periodic Table

As chemists discovered more and more elements during the nineteenth century, it became obvious that there were groups, or families, of similar elements, like those we have seen in the alkali metal and halogen families. Many chemists made helpful suggestions about ways in which the elements could be grouped together, until in 1869 a Russian chemist called Dimitri Mendeleev drew up a chart very similar in form to the Periodic Table that we use today.

Mendeleev assembled the elements in order of increasing relative atomic mass and placed similar elements under one another. As he did this he discovered that there appeared to be some missing elements; for example, at the time when he drew up his chart the element germanium had not been discovered. When he thought that there was an element missing he left a gap in the chart. From the properties of the other elements in the chart he was able to predict the properties of the missing elements – the accuracy of his predictions was very good.

In the modern form of the Periodic Table, the vertical columns are called **groups** and the horizontal rows are called **periods**. You will see that groups contain elements that have similar properties.

40.3. Using the Periodic Table

We have discovered that the elements may be divided into metals and non-metals and that the metals form basic oxides and the non-metals form acidic oxides. Figure 40.1 shows how the Periodic Table may be divided up to show metal and non-metal elements. Figure 40.1 also shows the names given to elements in various parts of the Table. The group number is often used to refer to a particular set of elements – for example, the group 4 elements are carbon, silicon, germanium, tin and lead – but it also tells us how many outer electrons there are in an atom of the element. You will remember from Chapter 5 that we need to know how many electrons there are in the highest energy level if we are to work out how many atoms join together to form compounds.

Elements in groups 1 and 2 always form ionic compounds.

PERIODIC TABLE OF THE ELEMENTS

Group number 1 2 3 4 5 6 7

1	2											3	4	5	6	7	
1 Hydrogen H	2 Helium He																
3 Lithium Li	4 Beryllium Be											5 Boron B	6 Carbon C	7 Nitrogen N	8 Oxygen O	9 Fluorine F	10 Neon Ne
11 Sodium Na	12 Magnesium Mg											13 Aluminium Al	14 Silicon Si	15 Phosphorus P	16 Sulphur S	17 Chlorine Cl	18 Argon Ar
19 Potassium K	20 Calcium Ca	21 Scandium Sc	22 Titanium Ti	23 Vanadium V	24 Chromium Cr	25 Manganese Mn	26 Iron Fe	27 Cobalt Co	28 Nickel Ni	29 Copper Cu	30 Zinc Zn	31 Gallium Ga	32 Germanium Ge	33 Arsenic As	34 Selenium Se	35 Bromine Br	36 Krypton Kr
37 Rubidium Rb	38 Strontium Sr	39 Yttrium Y	40 Zirconium Zr	41 Niobium Nb	42 Molybdenum Mo	43 Technetium Tc	44 Ruthenium Ru	45 Rhodium Rh	46 Palladium Pd	47 Silver Ag	48 Cadmium Cd	49 Indium In	50 Tin Sn	51 Antimony Sb	52 Tellurium Te	53 Iodine I	54 Xenon Xe
55 Caesium Cs	56 Barium Ba	57 Lanthanum La	72 Hafnium Hf	73 Tantalum Ta	74 Tungsten W	75 Rhenium Re	76 Osmium Os	77 Iridium Ir	78 Platinum Pt	79 Gold Au	80 Mercury Hg	81 Thallium Tl	82 Lead Pb	83 Bismuth Bi	84 Polonium Po	85 Astatine At	86 Radon Rn

TRANSITION ELEMENTS

NOBLE GASES

HALOGENS

ALKALI METALS

ALKALINE EARTH METALS

Figure 40.1 Using the Periodic Table

438

Since all elements in the same group have similar properties it is only necessary to study one element in a group to have a reasonable idea of the properties of all the elements in the group. When Mendeleev drew up the Periodic Table, the element germanium had not been discovered, but Mendeleev was so certain that such an element ought to exist that he predicted the properties of the element, which he called eka-silicon (like silicon). Table 40.2 lists some of the properties predicted by Mendeleev, together with the properties of the element germanium.

TABLE 40.2. PROPERTIES OF GERMANIUM

Properties predicted by Mendeleev for eka-silicon	Properties of the element germanium

Relative atomic mass 72
Density 5.5 g cm^{-3}
Will form a chloride $EsCl_4$,
  which will be a liquid, boiling
  below 373 K with a density of
  1.9 g cm^{-3}
	Relative atomic mass 72.6
	Density 5.47 g cm^{-3}
	Forms a chloride $GeCl_4$,
	  which is a liquid, boiling at
	  359.5 K with a density of
	  1.887 g cm^{-3}

Test your understanding

1. Make a list of your reasons for placing the alkali metals in the same group of the Period Table. You should include additional information to that given in Table 40.1.
2. Select any other elements that you have studied, except for the alkali metals and the halogens, that are in the same group in the Periodic Table. Give your reasons for placing them in the same group.
3. From your knowledge of the chemistry of sulphur, suggest the properties of selenium.

Analytical Contents List

1 Matter

1.1 Particulate state of matter   1.2 Movement of particles
1.3 States of matter   1.4 Boiling   1.5 Solids to gases
1.6 The use of physical properties

2 Mixtures and Solutions

2.1 Solutions   2.2 Solutions of solids in liquids   2.3 Crystallization
2.4 Separation of solids from liquids   2.5 Division of solutes between solvents
2.6 Some solvents   2.7 Osmosis

3 Atoms

3.1 Scientific models   3.2 The idea of atoms   3.3 An atomic model
3.4 Some evidence for atomic structure   3.5 The use of symbols
3.6 Atomic mass and atomic weight   3.7 The mole
3.8 Stable and unstable nuclei   3.9 The arrangement of the electrons in atoms
3.10 The shapes of electron clouds
3.11 The occurrence of the elements on earth
3.12 Metals and non-metals

4 Energy

4.1 The meaning of the term 'energy'
4.2 The law of conservation of energy   4.3 Chemical energy
4.4 Energy is needed to start chemical changes
4.5 Endothermic chemical changes
4.6 Exothermic and endothermic: an industrial example
4.7 Spontaneous chemical changes

5 The Result of Atoms Reacting – Ions

5.1 Ion formation   5.2 The electrical properties of ionic compounds
5.3 Water and ionic compounds   5.4 Water in the air

6 The Result of Atoms Reacting – Molecules

6.1 A simple combination   6.2 The directional properties of an atom
6.3 Formation of a molecule   6.4 The shape of a molecule
6.5 Distribution of charge   6.6 Properties of covalent compounds
6.7 Larger covalent compounds   6.8 Co-ordinate bonds
6.9 Bonds in metals

7 Compounds and Valency

7.1 Covalent and ionic compounds compared
7.2 Water as a solvent   7.3 Electrons gained or lost?
7.4 Complex ions   7.5 Valency

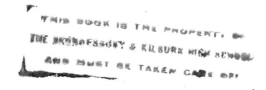

Index